Praise for *Almost Chimpanzee*

"[An] engrossing new book . . . like something out of Beckett, or maybe the Marx Brothers . . . deeply skeptical."
—Jennifer Schuessler, *The New York Times Book Review*

"A meticulous exploration of how both small quirks and large kinks in biology and culture led to such different destinations . . . Wonderfully weird . . . A briskly told, clearheaded survey of research that looks at the innate differences between two closely linked species, never forgetting that one of those species—at least for now—stands as the most successful primate in the planet's history."
—Deborah Blum, *The Washington Post*

"*Almost Chimpanzee* is an extraordinary journey into a world of great interest but—until now—little understanding. An astute observer and engaging writer on complex issues at the intersection of science and society, Cohen summons his prodigious talents in this examination of chimpanzee research and conservation. . . . [He] demonstrates how much we can learn about chimpanzees—and ourselves—by exploring their unique qualities."
—Paul Farmer, M.D., Ph.D.,
Partners In Health, Harvard Medical School

"How human are chimpanzees? Jon Cohen, in his well-written and carefully argued report, provides an up-to-date examination of the question. The bottom line is that we are far from understanding chimp/human relationships, but *Almost Chimpanzee* is a fascinating look at how investigators are probing the unknowns and searching for definitive answers." —David Baltimore, Nobel Laureate and
professor, California Institute of Technology

"It's often been said that we can look into a chimpanzee's eyes and see ourselves. Well . . . almost. And it's that very big almost that Jon Cohen so tenaciously explores in this extraordinary scientific odyssey. . . . To get at the truth, Cohen clomps through malarial jungles, travels in an RV with a baby orangutan, even handles fresh chimp sperm. The result is world-class science writing that is also a rollicking adventure story—one that takes us to the ends of the earth and to the margins of our species."

—Hampton Sides, editor at large, *Outside* magazine, and author of *Hellhound on His Trail* and *Ghost Soldiers*

"A dazzling look at a field in which no two scientists seem to agree on what makes us either human, animal, or both. Jon Cohen has a gift for bringing this issue to life: He gives our species its due without losing respect for our fellow evolutionary travelers, the apes."

—Frans de Waal, author of *The Age of Empathy: Nature's Lessons for a Kinder Society*

"Precious animals are rapidly disappearing, victimized by hunters, collectors, and habitat encroachment. Among them, chimpanzees have earned a special place in humanity's imagination because we look in their eyes and see ourselves. Jon Cohen has done a magnificent, masterful job of showing us why chimps are not like humans, yet—or because of the differences—they must be saved, in the wild. Controversial? You bet. But it's high time humanity takes responsibility for both its sins: killing species with the 'kindness' of making them us and through outright ruthless slaughter."

—Laurie Garrett, Pulitzer Prize winner and author of *The Coming Plague: Newly Emerging Diseases in a World Out of Balance*

"How are we different from chimps? That's the question that Cohen (*Shots in the Dark*) sets out to answer in his absorbing account of current chimpanzee research. . . . Readable and replete with surprising theories for the origins of human traits from 'concealed ovulation' to endurance running."

—*Publishers Weekly*

ALMOST
CHIMPANZEE

REDRAWING THE LINES THAT
SEPARATE US FROM THEM

JON COHEN

St. Martin's Griffin

New York

www.stmartins.com

Design by Meryl Sussman Levavi
Charts by Bob Roman/Panverde Design & Technology
Title page and parts one and three photographs by Malcolm Linton
Part two photograph by Tetsuro Matsuzawa

The Library of Congress has cataloged the Henry Holt edition as follows:

Cohen, Jon, 1958–
   Almost chimpanzee : searching for what makes us human, in rainforests, labs, sanctuaries, and zoos / Jon Cohen.
      p. ; cm.
   Includes bibliographical references and index.
   ISBN 978-0-8050-8307-1
   1. Human evolution.   2. Chimpanzees.   I. Title.
   [DNLM: 1. Evolution—Popular Works.   2. Hominidae—Popular Works.
GN 281 C678a 2010]
   GN281.C56 2010
   599.93'8—dc22   2010000813

ISBN 978-0-312-61176-7 (trade paperback)

Originally published in hardcover format by Times Books, an imprint of Henry Holt and Company

First St. Martin's Griffin Edition: December 2011

P1

To Esther, Avshalom, and Ronnie,
my first family

But there is now a growing interest in biological problems which the primates may help to solve. Consequently, in the future, we may come to regard them rather as valuable objects of scientific study than as pets, curiosities, or inventions of the devil.

—ROBERT M. YERKES,
Almost Human, 1925

CONTENTS

ALMOST CHIMPANZEE

INTRODUCTION:
IN THEIR HABITAT

WHEN JANE GOODALL FIRST HEADED TO THE GOMBE STREAM Chimpanzee Reserve at the behest of the archaeologist Louis Leakey, it was July 1960 and the country was called Tanganyika. It took four months of watching chimpanzees through binoculars before Goodall finally managed to win their trust enough that a male with a gray beard let her observe him at close range.

National Geographic three years later introduced the world to Goodall and the chimps she named David Greybeard, Mrs. Maggs, Count Dracula, Huxley, and Goliath. She was not the first person to study chimpanzees in the wild. But with her patient constitution and the help of bananas, she *was* the first person from whom wild chimpanzees did not run, allowing her to document in detail their daily lives, social structure, tool use, hunting, and emotions.[1] Like an explorer who makes first contact with a remote tribe, Goodall penetrated a community, complete with its own culture, that until then had been known only to its own members.

Nearly half a century has passed since Goodall made her initial visit to what is now the Gombe Stream National Park in Tanzania, and the catalog of sites conducting long-term studies of wild chimpanzees by 2010 included Mahale Mountains National Park in Tanzania, Kibale National Park and Budongo Forest Reserve in Uganda, the Taï National Park in Côte d'Ivoire, Bossou in Guinea, Fongoli in Senegal, and the Goualougo Triangle in the Republic of Congo's Nouabalé-Ndoki National Park. The many other researchers who "habituated" wild chimpanzee communities also needed much patience, especially since the field now looks askance at making chimps feel comfortable with the presence of humans by provisioning bananas and other foods, which Goodall and others later recognized disrupted the animals' natural behavior. So while it is not easy to gain the trust of wild chimpanzees and to observe them at close range for extended periods—habituation took five years in Taï—clearly, humans have figured out how to do it.[2]

In 2006, I set out to observe the world of chimpanzee research at close range, and that, too, required that I slowly build the trust of individuals in a foreign community. It was hardly the first contact between a journalist and primate researchers—which in part explained why my many phone calls and e-mails for a time went unanswered. I also did not want to simply interview people for a few minutes on the phone. I wanted to see the scientists doing what they do in their natural environments, from the rainforests of Africa to the laboratories, zoos, and sanctuaries where captive chimpanzees live in many countries. That added another obstacle: great apes are endangered, which means that few exist, and even fewer humans devote their lives to studying them. Apes also live in places where entry is tightly restricted, and it can be just as difficult to receive permission to visit a protected national forest to see a wild chimp community as it is to be invited to a biomedical research facility to see a captive one. So I had to be patient with my requests.

From the outset, I understood that many in the community had received too much media interest for their own tastes. They did not exactly run and hide away in the trees when they saw a journalist coming, but they knew both the benefits and the perils of publicity, and they preciously guarded their time—and access to the animals they studied. It made good sense. Goodall is arguably the world's most famous living scientist, and many of those who have followed in her footsteps—including the ethologist Frans de Waal, the evolutionary biologist Richard Wrangham, and the psychologist Roger Fouts—have written popular books and magazine articles themselves and been featured prominently in an endless stream of documentaries. Dian Fossey, who studied gorillas in Rwanda also at the behest of Leakey, became the subject of a Hollywood blockbuster, *Gorillas in the Mist*. Biruté Galdikas, the "orangutan lady" and the third of "Leakey's Angels," has garnered two appearances on the cover of *National Geographic*, written four books, and had several books written about her. In his native Switzerland and elsewhere in Europe, the chimp researcher Christophe Boesch has a high profile, and the same is true of the primatologists Tetsuro Matsuzawa, Toshisada Nishida, and Kinji Imanishi in their native Japan.

In time, most every request I made was granted, and people graciously invited me to watch them hike through rainforests, conduct biomedical and cognitive experiments with live animals, examine bones and brains from dead ones, and even collect and then study chimp sperm under a microscope. My travels took me to Uganda, Japan, Germany, England, Russia, and all over the United States. Only one place I wanted to visit ultimately froze me out: *National Geographic* actually blocked a trip I had planned to the Republic of Congo to observe researchers studying chimpanzees and gorillas in the remote Goualougo Triangle; the husband-and-wife team of scientists working there had welcomed me, but then were forced to rescind their invitation because the magazine funded them and wanted to monopolize their research for a story. I mention this not to complain, but to illustrate the peculiar inner workings of the community, as well as to explain the great satisfaction I felt each time I managed to finally visit researchers at their work sites. They had, in a sense, become habituated to my presence. At the very least, they tolerated my watching them do their business, and more often than not, I was greeted with enthusiasm and a genuine eagerness to help me understand and communicate what I was seeing.

—◦◦◦—

UNLIKE many authors of books about chimpanzees, I am not a researcher or even connected to one. That means I do not emphasize my own original findings, I have no agenda, and I, too, struggle to cut through the scientific jargon to distill the significance of specific studies. My aim from the start was simply to explore anew a question that tickles at the human mind, and which with the publication of the chimpanzee genome in 2005 was pushing answers into novel directions: What are the dividing lines between humans and chimpanzees, between us and them?

In 1925, the psychologist Robert Yerkes teased open many minds with his book *Almost Human*, which argued that humans and chimpanzees had so many similarities that much could be learned if we studied them more carefully. *Almost Human* had another agenda that from today's vantage seems unnecessary and downright absurd: to convince people not to revile chimpanzees. "There is intense and well-nigh universal curiosity about these animals, but it is often coupled with strong dislike or repulsion," he wrote. "Perhaps as our ignorance disappears we shall lose also the prejudice and unreasonable dislike which makes many feel that genetic relationship with the monkeys or apes is belittling."[3] The fact that chimpanzees were seen by some at the time as "inventions of the devil," as Yerkes noted, could be traced to creationism: *Almost Human* appeared the same year as *Scopes v. the State of Tennessee*, which famously put the teaching of Darwinian evolution on trial. Yerkes was doing his bit to combat the creationists of his day, and the mounting evidence for the chimp's humanlike appearance, behavior, and biology made a strong case for the Darwinians, one that continues to resonate.

Goodall, fifty years later, pushed this "almost human" viewpoint to forward an entirely separate agenda. As a leading advocate for protecting the habitats of wild chimps and a foe of researchers who housed chimps in small cages and conducted invasive biomedical experimentation, she believed that a critical mass of humans would most likely come to her cause if they imagined their own hands reaching for the curl of a chimpanzee's finger. Goodall was pursuing noble and worthwhile goals, and indeed she, along with Yerkes and other pioneering chimpanzee researchers, deserves much credit for making people more aware of the intelligence, social needs, and emotional depth of our closest cousins. But I think the need to emphasize our similarities has abated.

With the flood of genetic information now available from many species, the argument for Darwinian evolution no longer requires the chimpanzee-human connection as its linchpin. Advocates also have made much headway in persuading humans to treat chimpanzees more humanely, with invasive research steadily becoming less common, housing for captive chimps improving, and more people recognizing the plight that wild chimpanzees face because of our disregard for their well-being. And there is something fundamentally backward about the "almost human" rubric for chimps. From everything I can tell, no chimpanzee looks at a human and wonders, Is that where I came from? Nor do chimps ponder the possibility that we represent where they are heading. Yet humans from every culture look at chimpanzees and see hints of their more primitive selves.

"Almost human" inherently pushes people to look for similarities. Yet, while we have a lot of chimpanzee in us, we cannot hope to see that clearly unless we can identify the specific features and forces that separate us from them. We have bigger and more complex brains, full-fledged language and writing, sophisticated tools, the control of fire, cultures that become increasingly complex, permanent structures in which to live and work, and the ability to walk upright and travel far and wide. It also is important to recognize that we are not "better" in every regard. No human stands a chance against an adult chimpanzee in a fight or a tree-climbing contest; many diseases that devastate us spare them; and one study even suggests that they have much better short-term memory.

I am not arguing that we should treat chimpanzees with any less respect, that the case for evolution is any less compelling, or that we should conduct studies of humans to better understand chimpanzees. "Almost human" and "almost chimpanzee" represent two sides of the same coin. But people have many misunderstandings about our relationship to chimpanzees, and I am convinced that we have focused too much attention on the heads rather than the tails.

⟼⟀⟻

H. A. Rey's children's classic, *Curious George*, opens with an illustration of George swinging on a vine and eating a banana. "This is George," the text reads. "He lived in Africa. He was a good little monkey and always very curious."

George, as the drawing clearly shows, is a chimpanzee, not a monkey.

This fundamental mistake riles people who study chimpanzees, and with good reason. In evolutionary time, monkeys and apes diverged from each other about 25 million years ago. Humans and chimpanzees split from a common ancestor at most 7 million years ago. So to understand the difference between humans and chimpanzees, as a starting point, it is critical to recognize that monkeys are not apes.

How do we know George is not a monkey? He does not have a

tail. All monkeys, save for the misnamed Barbary ape and the Sulawesi black ape (which look nothing like George), have tails. In contrast, no ape—a family that includes chimpanzees, humans, gorillas, bonobos, orangutans, gibbons, and siamangs—has a tail.

The differences between apes and monkeys reach far beyond the tail. But the fact that many people don't know that simple distinction—and that a popular children's book has corrupted young minds since 1941—underscores the confusion about where to draw the dividing lines that separate humans from other species.

As I tour the world of chimpanzees, I do not address every dividing line between us and them, nor do I become chummy with the likes of David Greybeard, the extraordinarily communicative Washoe and Kanzi, or any other knuckle-walking ape. But I introduce many chimpanzees, as well as other apes, including bonobos and orangutans, focusing on the differences that matter. I begin at the most microscopic of differences, looking into the blood, then slowly zooming out for the successively broader, and more encompassing, vantages of the brain and finally the body. I mesh little-known historical tales with the most cutting-edge science to explain the origins of chimpanzee research and reveal where the field is heading. And while few primatologists study both captive and wild chimpanzees, I blend findings from both areas, which often complement or challenge each other in surprising ways.

George, the chimpanzee, is defined by his curiosity. Humans are an extraordinarily curious species, too. But the range of our curiosities and how we satisfy them are distinct, and since the dawn of our consciousness, since the first myth of our origins was passed on, we have explored that particular difference and pondered what, exactly, it means to be human. Chimpanzees are a unique—and rapidly closing—window into our answer.

ONE

BLOOD

The persistence of the chemical-blood relationship
between the various groups of animals serves to carry
us back into geological times, and I believe we have
but begun the work along these lines, and that it will
lead to valuable results in the study of various problems of evolution.

—George Nuttall,
Blood Immunity and Blood Relationship, 1904

1

THE FAMILY TREE

A DOMINANT "ALPHA" MALE TYPICALLY RULES A CHIMPANZEE community, and at the Detroit Zoo, Chuck and Joe-Joe were the only two contenders. The two had been at the zoo since they were infants, but when the new "Chimps of Harambee"—Swahili for "let's all pull together"—exhibit put "the boys" in closer contact with each other than ever before, Joe-Joe did not stand a chance. Chuck constantly harassed Joe-Joe, tearing off fingers and much of his hair in their fights. "Joe was very stressed because of battling with Chuck and the whole introduction process," said Susan McDonald, who oversaw the exhibit's development. "He was getting the short end of the stick."

Rick Swope, his wife, Cynthia, their three children, and his niece spent a half hour at the Chimps of Harambee on Sunday, July 29, 1990, watching Chuck repeatedly challenge Joe-Joe. Just as Swope, then a thirty-three-year-old truck driver, and his family were ready to leave for the day and head back to their home ninety minutes away in Cement City, Joe-Joe ran into the moat surrounding the exhibit. "Just out of the corner of my eye, I caught a glimpse of one of the chimps flying through the air," Swope later recalled. "And I turned and looked, and all I seen was this splash, and people were hollering, and then I seen him splashing around in the water." He had not heard the news that another chimp at the zoo, Tanya, had recently drowned, and he had no idea that chimpanzees could not swim.

Swope watched as Joe-Joe twice surfaced from beneath the water. "Then the third time he was going down, and as I watched, I could see him facing a crowd of people," said Swope. "He had both arms, hands sticking right out of the water. It was like in slow motion, and he just sunk right down. I could see him looking at people, and it just looked pitiful. And I'm just, 'Oh, geez, nobody's doing nothing.'"

Ignoring his family, Swope circled the exhibit until he could get closer to Joe-Joe, and then hopped a short fence made of lifeline wire and dashed down to the moat. "I just took off because I knew this thing was drowning, and when nobody went in, I felt like I had a bit of an obligation there."

At that point, Joe-Joe had sunk, and Swope did not know where he was. So he dove into the murky water and ran his hands along the clay liner at the bottom of the moat. "I'm diving down in that stuff, and I'm feeling around down there on the bottom, and I cannot find this thing," said Swope. "And I come back up, and people were, 'No, no, he's over there.' And they're pointing to the left. So I go down again, and I find him. And he was facedown, just laying right on the bottom, belly right on the bottom."

A woman visiting the zoo with her grandchildren videotaped the rescue.[1] Swope was an unusual-looking lifeguard. He was wearing a light blue T-shirt and darker blue shorts with a black belt. His bald dome of a head, fringed with dark hair, was set off by his black beard. A fit five foot ten and two hundred pounds, he was a chunk of muscle, not unlike his objective, the 135-pound Joe-Joe. He lifted Joe-Joe from the bottom, the chimp's back to his chest. When Swope surfaced, he did a sidestroke with his right arm, holding the lifeless animal under its armpit with his left hand. As he approached the bank on the chimp side of the exhibit, he wrapped his left arm around Joe-Joe's chest and attempted to heave him onto the island.

"He needs some help! Somebody help him! Somebody help him!" pleaded one woman. "Hurry up! This man can't do it by himself!"

Swope never once looked over at the crowd or said a word. "There was a lot of people screaming and everything, but I had my mind on business. I was in the zone," he told me.

Once Swope had rested Joe-Joe's limp form on the grassy bank, he wrapped both arms around the chimp. "I could see his eyes, and he was looking at me—his eyes were blacker than all get-out—and I squeezed him almost like a Heimlich maneuver. I pulled on him pretty hard, almost like to pump the water out of him. Then he started shaking a little bit, shaking around almost like an epileptic fit or something, probably choking on the water and what all. So I knew he was still alive."

Swope steadied Joe-Joe with one hand and started to retreat, but the chimp began to slide down the bank. The crowd began to yell again. "Oh, don't leave him! Don't leave him! Stay there!" instructed a woman.

"No! Don't leave him!" hollered another.

Swope then saw Joe-Joe's nemesis—and likely the reason the chimp jumped into the water in the first place—Chuck, knuckle-dragging toward them. "That male chimp was coming down after one of us or both of us, I don't know," said Swope, laughing at the memory. "But he was coming down."

At that moment, Joe-Joe grabbed and reached for some stalks of grass. "He was looking at me, and I was looking at him, and I was nose to nose with this thing, and I'm telling you right now, this thing was grateful I was there," he recalled. As Swope backed away into the moat, Joe-Joe shuffled a few feet up hill and then sat down and looked at the stranger who had rescued him. "He didn't in no way try to come after me or anything like that," said Swope. The possibility of being attacked by Chuck, however, remained. Swope wasn't worried. "I had the advantage because I was in the water, and I knew then that these things can't swim." He swam back across the moat and climbed the bank to the humans on the other side.

Swope had cut his legs on a cable in the moat—a cable that was supposed to help chimps rescue themselves if they fell in—and the zoo's director asked him to come to the infirmary to have his wounds cleaned. Swope was shuttled in a zoo golf cart, feeling more humiliated than triumphant. "People are looking at me 'cause I'm with security, and they think I'm some weirdo who jumped in the fountain somewhere," he said. "They took me up there and they put me in this room, and I thought to myself, You know what, I'm getting the hell out of here. So I walked out." He found his family, drove to a Target, bought new clothes and tennis shoes, and stopped at his brother's house nearby, to shower and change. He then drove home to Cement City.

The next morning, Swope stopped to pick up a newspaper on the off chance that there might be an article about the rescue. "I seen the front page of *Detroit News* with me in that water," said Swope. "Holy mackerel, I couldn't believe it." The newspaper had only learned Rick Swope's name because his wife had disclosed it to the woman who shot the videotape of the rescue. "You can blame my wife for that whole thing"—Swope laughed—"I keep my private stuff private. Believe me, I never capitalized on nothing. A guy sent me a letter with a ten-dollar bill, and said, 'Great job, the beer's on me.' So I made ten bucks."

For Swope, the public recognition peaked—to his horror—on the night of May 6, 1991, at the Beverly Hills Hilton. At Jane Goodall's invitation, he and his wife flew to Los Angeles to attend a star-studded fund-raiser for Goodall's conservation efforts. The actor Jack Lemmon served as master of ceremonies. The honorary gala chairman was pop star Michael Jackson, who infamously lived with a chimpanzee named Bubbles and had befriended Goodall because of their shared interest in chimps and children.[2] Jackson, who bought a ten-thousand-dollar table for the event, did not show, but a thousand others did, including Ted Turner and Jane Fonda, Gregory Peck, and Jimmy Stewart.[3] "The only poor people there were my wife and myself," said Swope, who, for the first time in his life, wore a tuxedo.

Near the end of the evening, Goodall turned the audience's attention to screens around the room that exhibited the video of Joe-Joe's rescue. "This chimpanzee thing, it seemed to me it got all out of proportion," said Swope. He was presented with the sculpted head of a chimpanzee in honor of his rescue.

Goodall, who would continue to tell the story of Rick Swope and Joe-Joe at her many public talks and in her writings, believed that the rescue held "truly symbolic meaning."[4] Yet she errs on many of the details. In her retelling, onlookers are yelling at Swope about the danger he is in, when in fact they were encouraging him to go on, seemingly unconcerned about his safety. Before plopping Joe-Joe on the bank, Swope heroically hoists the chimp over his shoulder and crosses a barrier, neither of which happened. But most revealing, Goodall has the zoo's director phoning Swope the next day and asking him why he did it.

"Well, I looked into his eyes," Swope supposedly said. "And it was like looking into the eyes of a man. And the message was, 'Won't anybody help me?'"

Swope could not see Joe-Joe's eyes until he had the chimpanzee in his arms on the bank.

In an article she wrote for *Science* magazine describing the incident, Goodall concluded, "Rick Swope risked his life to save a chimpanzee, a nonhuman being who sent a message that a human could understand. Now it is up to the rest of us to join in too." Like legions of legend and myth tellers, Goodall twists the facts to serve the moral of the story she wants to emphasize: more and more people are recognizing that chimpanzees are almost human, and they should be treated with similar respect, dignity, and concern. But to me, the Swope legend has precisely the opposite moral.

Rick Swope dove into the water to save Joe-Joe because chimpanzees cannot swim. What is more, humans teach other humans to swim, and chimpanzees do not teach each other anything. The mere fact that chimpanzees cannot swim means that rivers, lakes, and oceans severely restrict their geographic range, and, in turn, the food they digest, the pathogens they meet, the immune systems they develop, and the very ability they have to survive when exploring the unknown.

So while chimpanzees are almost human, in the end, they are not. And to my mind, at the beginning of the genome century, it is more accurate to flip the rubric on its head. We are almost chimpanzees.

—◦◦◦—

WHEN the Yale psychologist Robert Yerkes published *Almost Human*, his landmark book in the history of chimpanzee research, he largely told the story of the colorful Madam Rosalía Abreu, who lived in the fabulous Quinta Palatino of Havana. On the grounds of her estate, Abreu kept apes and monkeys in cages; her menagerie included chimpanzees, gibbons, orangutans, and a siamang. Yerkes spent part of a summer at Quinta Palatino, gauging the intelligence, social habits, and emotions of Abreu's primates. He also raised two chimpanzees in the United States—he named them Chim (which actually was a bonobo) and Panzee—and recorded their development. In his book, Yerkes sounded the call to fellow scientists that they had to more carefully study apes. "Where we most need reliable, systematic, detailed descriptions, we find observational fragments cemented together with guesses, some shrewd, some ridiculous," he lamented.[5]

In part, Yerkes blamed the lack of serious inquiry on the fact that apes live in tropical climes and scientists did not, leading to few accounts of wild chimpanzees. So in 1930, he established a chimpanzee colony in Orange Park, Florida—the first of its kind in the United States to study and breed chimps, and Yerkes's most lasting legacy. His research station, by design, was far removed from "an enervating tropical climate and unnecessary isolation from civilization and from centers of scientific activity," which meant nowhere near the African forests and savannas that were the natural homes of chimpanzees.[6] He directly purchased four chimps for the colony, slyly naming one of them Bill after William Jennings Bryan, the three-time candidate for U.S. president best remembered as the anti-Darwinist lead prosecutor in the Scopes "monkey" trial. Sixteen more came as a gift from the Pasteur Institute's chimp research center in New Guinea. Rosalía Abreu died in November 1930. Though she willed her chimpanzees to the Cuban government, her son allowed Yerkes to take ten of them the next year.[7]

The colony flourished, continuing to acquire some chimpanzees and breeding many others. After Yerkes died in 1956, Yale University transferred the colony to Emory University in Atlanta. The animals were moved to nearby Lawrenceville, and it soon became known as the Yerkes Regional Primate Research Center. In 1980, a mating between two of its chimps produced Clint, one of the most closely studied chimpanzees of all time.

Clint nearly died shortly after his birth when his mother, Cheri, slammed him into a wall and fractured his skull. "We made a football helmet out of cast material, and kept that on him until he healed," remembered the Yerkes veterinarian Elizabeth Strobert. "He had a rough beginning." The staff raised Clint in their nursery, and he grew into a charming adult. "He was tall, dark, and handsome," said Strobert. "I had a crush on him."

Clint went out of his way to solicit attention from humans, and Strobert admitted that they spoiled him. "He was so personable," she said. But if Clint deemed that a human did not pay him proper respect, watch out: he notoriously scooped up his feces and pelted a new president of the university during his first tour of the facility.

Lisa Parr, a Yerkes researcher who studies facial recognition in chimpanzees, said Clint was a "goofball," slightly built by chimp standards, and not exactly blessed in the brains department. Parr used him in her experiments for a time, but he proved more helpful in studies that explored handedness, hormonal influences on reproduction, and other less mentally demanding questions.

When, in 2002, a group of researchers decided to sequence the complete genome of a chimpanzee, just as had been done for humans the year before, Clint was tapped into service. Soon machines at Washington University in Saint Louis, Missouri, and the Broad Institute in Cambridge, Massachusetts, were spitting out long strings of the As, Cs, Ts, and Gs—adenine, cytosine, thymine, and guanine—the backbone of the nucleotides that made up the double helixes in Clint's DNA. The researchers' goal was to produce a finely detailed comparison of the chimpanzee and human genomes, detecting not only the similarities in genes but the vast differences in other parts of the genome, which account for most of the As, Cs, Ts, and Gs. With the analysis finished, the scientists realized that humans are not as similar to chimpanzees as commonly believed, but at the same time the two species are more closely intertwined than even evolutionary biologists conjectured.

—◦◦◦—

"BLOOD is a very special kind of juice," declares the demon Mephistopheles in *Faust*, Goethe's 1808 classic. The Cambridge biologist George Nuttall chose this inscription for his 1904 book *Blood Immunity and Blood Relationship*, which compared blood samples from 586 different species in an exhaustive attempt to fill out the branches on the evolutionary tree. Yet scientists would develop a polio vaccine, split the atom, crack the DNA code, and send humans to space before they finally did away with the niceties and called humans "apes"— and it was blood studies that pounded the gavel.

Science pieced together the place of humans and other primates on that tree slowly and contentiously. The Swedish zoologist and physician Carl Linnaeus, who in 1735 concocted the taxonomical "tree of life," listed both humans and bats under primates. He had never seen a chimpanzee, and indeed only three had been brought to Europe by 1738.[8] Johann Blumenbach, a German physiologist, referred to humans as "bimana" (two hands) to distinguish them from the four-handed apes, "quadrumana." Thomas Henry Huxley, Charles Darwin's self-described "bulldog" who vigorously made the case for evolution shortly after the publication of *On the Origin of Species*, made the link between humans and apes eight years before Darwin dared spell it out in *The Descent of Man*. As Huxley wrote in *Evidence as to Man's Place in Nature*, "Is Man so different from any of these Apes that he must form an Order by himself?"[9] He went on to argue that it was "quite certain" that man was closest to the gorilla or chimpanzee, but he could not narrow it down any further (neither could Darwin)—and he, too, separated humanity from other primates into its own suborder, Anthropoidea. In the *History of Creation*, published in 1868, the German biologist Ernst Haeckel lumped humans and other apes together as Anthropoidea, contending that the orangutan was closest to humans.

Nuttall believed that blood would resolve these debates. He realized that the attraction of antibodies to their antigens (typically, proteins or polysaccharides) would reveal the similarities and differences between species. By literally mixing the blood of species in test tubes, he could determine whether their antibodies and their antigens were closely related: mix human antibodies with chimp blood and they'll tightly bind to the chimp antigens, forming clumps; mix human antibodies with rabbit blood and fewer clumps will form. The more similar the species, the more precipitation. Nuttall's tests provided the most compelling evidence then available about the close kinship between humans and apes, but they couldn't resolve the finer question about which ape most closely shared its ancestry with humans.

So the taxonomists continued to squabble, until George Gaylord Simpson, a paleontologist at Columbia University, effectively settled the row in 1945. Simpson recognized that humans and apes likely descended from a common ancestor and thus were an evolutionary "unit," but he rejected lumping them together because humans had a superior "mentality." He believed "there was not the slightest chance that zoologists and teachers generally, however convinced of man's consanguinity with apes, will agree on the didactic or practical use of one family embracing both."[10] Indeed, since the time of Linnaeus, the greatest barrier to accepting a close tie between humans and apes had been religion. As Linnaeus wrote in a letter to a fellow scientist: "It does not please [you] that I've placed Man among the Anthropomorpha, but man learns to know himself. . . . If I would have called man a simian or vice versa, I would have brought together all the theologians against me."[11] Whereas Huxley had had so little patience for the religious hordes that he declared himself an agnostic (coining the word), Simpson outright mocked those who claimed man was a separate creation. As a purely pragmatic solution, he classified apes as Pongidae and humans as Hominidae.

By 1962, that charade could no longer be kept up. Morris Goodman, a biochemist at Wayne State University in Detroit, had refined Nuttall's methods of antibody analysis and reported that the proteins found in the blood of humans, chimpanzees, and gorillas were so similar that all belonged in Hominidae. But Goodman could not illuminate whether humans were closer to chimps or gorillas or, as many believed, equidistant to both.

That impasse was intimately connected to solving a vexing evolutionary question: When did humans branch off from their ape relatives? Paleontologists and anthropologists who dated fossils held a ridiculously wide range of estimates, saying that humans diverged from other primates somewhere between 4 million and 30 million years ago. The fossil finds used to make those estimates had been spread from Kenya to Pakistan to Nepal, which meant that an African chimp, bonobo, or gorilla, or an Asian orangutan, could be humans' closest primate kin. Two researchers at the University of California, Berkeley, Vincent Sarich and Allan Wilson, exploited the fact that close species share nearly identical proteins to figure out when humans diverged from other primate species.[12]

Because proteins accumulate neutral mutations—mutations that do not change a protein's function or alter its evolutionary fitness—at predictable rates, Sarich and Wilson surmised that each species carried a "molecular clock." They compared human serum albumin, a protein in the blood, to the version found in other apes. Their study, conducted in 1967, dated the origin of humans to about 5 million years ago, a date that immediately put into question whether some famous fossils were indeed human. "It was a molecular bombshell," wrote Ann Gibbons in her authoritative account, *The First Human*.[13] "When the article landed on the desks of paleoanthropologists, it created a decades-long rift between scientists who studied molecules and those who analyzed fossils." It also forced paleontologists to reexamine the oldest putative human fossils, which they eventually would conclude were precursors to orangutans—not human at all.[14]

Sarich and Wilson did not have the data to draw each branch on the family tree, but the molecular clock proved to be a powerful scientific tool. Rather than continuing to compare proteins, researchers dove deeper into the blood. They calibrated molecular clocks by tracking evolutionary changes in DNA, which holds the code for making the amino acids that link together and form proteins. In a 1984 tour de force, two Yale ornithologists who specialized in molecular biology, Charles Sibley and Jon Ahlquist, announced that they had tagged the timeline for each split among primates. A common species of monkeys existed 27 to 33 million years ago, from which gibbons had split off at 18 to 22 million years ago, orangutans at 13 to 16 million years ago, and gorillas at 8 to 10 million years. In the molecular breakdown, the common ancestor to humans and chimpanzees lived 6.3 to 7.7 million years ago.[15]

Still, Sibley and Ahlquist did not quell the doubters. Their technique took advantage of a remarkable feature of the As, Cs, Ts, and Gs that comprise DNA: A binds to T, and C binds to G. The double helix of DNA is made up of two complementary strands of these nucleotides, so if one stretch contains ATCG, its complement is TAGC. To assess the branching points on the evolutionary tree, Sibley and Ahlquist had separated chimp DNA and human DNA into single strands. They then mixed the single strands from each species and measured how tightly the two strands bound to each other, or hybridized, by identifying the "melting point"—the amount of heat it took to separate the strands. The more similar the DNA sequences of the two strands, the more heat it requires to melt them apart. Critics of the results rightly noted that the melting point technique did not actually spell out the DNA sequence. In essence, hybridization was one step removed from the real thing, smelling the food from across the table rather than tasting it.

In 1991, Maryellen Ruvolo, a leading Harvard anthropologist who had tried to end the fracas with yet more DNA sequence data, called the question of our closest primate kin "one of the most contentious and seemingly intractable problems in mammalian systematics."[16]

—◦◦◦—

ACCORDING to Morris Goodman, in the chimpanzee/human ancestor debate, "the final nail was put in the coffin" by Clint's sperm.

By 1997, researchers had spent seven years working on unscrambling the human genome—3.2 billion As, Cs, Gs, and Ts. Goodman and the biologist Edwin McConkey believed the time was ripe to start a human genome evolution project. As they envisioned it, the project would compare genes between species to reveal the code that allows humans to walk upright, speak in complex sentences, and grow such large brains. Studies might uncover the differences in DNA that tell genes to turn on or off. They predicted the effort would ultimately reach far beyond "an excursion into natural history." "Until we have a detailed understanding of the genetic differences between ourselves and our closest evolutionary relatives," they wrote, "we cannot really know what we are."[17]

Momentum for the idea built gradually. At the time, sequencing DNA required machines that each cost hundreds of thousands of dollars. The devices worked slowly and drank costly chemical reagents to produce accurate data. Proponents of sequencing the genomes of rats, the African clawed toad, cows, dogs, zebra fish, rice, dolphins, and cassava were also lined up at the door of the U.S. National Institutes of Health (NIH) and the Department of Energy, the main funders of the Human Genome Project. Plus, somewhat surprisingly, not many researchers actually studied chimpanzee DNA.

Several of the researchers who would emerge as the main champions for the chimpanzee sequencing project had only recently become interested in chimp DNA when Goodman and McConkey started petitioning for it. One was Evan Eichler, who was then working on his Ph.D. in molecular genetics at the Baylor College of Medicine in Houston, Texas. Eichler investigated a human disorder called fragile

X syndrome. X and Y are the two chromosomes that determine gender: women have two Xs, and men have one X and one Y. In fragile X, which can affect both genders, stretches of CGG repeat themselves in the chromosome more than sixty times, like a photocopier gone haywire, and disrupt a critical gene. People who have a bad copy of the gene can develop learning disabilities, autism, an elongated face, speech and language problems, and other symptoms. Women often show no symptoms; if it only affects one of their X chromosomes, the other functions normally. Men have no such fallback.

Eichler investigated how this disorder evolved, comparing CGG repeats in humans to those in forty-three other mammalian species, including every ape. He observed that the repeats became progressively longer over evolutionary time. It was fundamental research with profound implications about how new genes and genetic disorders emerge, as well as how chromosomes change shape.[18] Eichler eventually wanted to compare his findings in human DNA to stretches of chimp DNA in the same chromosomal regions to help determine whether the repeats he singled out were sequencing errors or the real thing. For this task, he had to sequence longer strings of chimp DNA than had been studied before.

That's when Clint came into the picture. Researchers could not just stick some chimp blood into a machine that would sequence thousands of DNA nucleotides. They had to first package chunks of the DNA into what are known as artificial chromosomes. Pieter de Jong, an investigator at the Roswell Park Cancer Institute in Buffalo, New York, specialized in making artificial chromosomes out of bacteria for that purpose, and in 1997 Eichler went to de Jong's lab to learn the technique. Sperm, according to leading thinkers of the day, provided a better source of DNA than blood, so de Jong asked the Yerkes center for a sample from one of its males. Yerkes selected Clint, an especially willing sperm donor. "We called it Clinton—that was the name on the sample tube," remembered de Jong. "It was kind of Freudian, comic relief." Yet Eichler had little success with the Clinton sperm sample, and when he left to start his own lab, de Jong decided to continue with the project, switching to blood instead of sperm, from which the team managed to make artificial chromosomes.

Across the country at the University of California, San Diego, the physician Ajit Varki was intensely interested in chimpanzee DNA but for entirely different reasons. Originally from India and trained at the Christian Medical College in Vellore, Varki specialized in hematology, and in 1984 he became intrigued by a patient diagnosed with aplastic anemia. In this autoimmune disease, the bone marrow does not sufficiently replace the body's blood cells. Standard treatment called for infusing the patient with a serum generated from horses. The serum contains antibodies made in reaction to an injection of human white blood cells into a horse, and, theoretically, those antibodies calm the patient's immune system such that it stops attacking the bone marrow. "It's actually black magic," Varki told me one Saturday morning, sitting in the standard-issue university office abutting his large lab on the La Jolla campus. "There's no explanation for how it works, but it's still used."

Varki, who often has an amused look on his face, was less intrigued by the treatment than its side effects: the woman he had treated in the 1980s developed serum sickness, an allergic reaction. More recent studies had shown that a key component to the reaction was a sugar molecule called a sialic acid. This perplexed Varki. "I said that's ridiculous; how can that be?" he told me. "Sialic acid is in all organisms; it's in all mammals, all vertebrates."

Varki did not immediately throw himself at the problem, but it stuck with him. The particular sialic acid that caused the reaction is known as Neu5Gc—short for N-glycolylneuraminic acid—and other labs had shown that it was nearly nonexistent in humans. A 1965 study further suggested that some ape blood might contain it. Varki obtained a small amount of chimpanzee blood, and a collaborator, Elaine Muchmore, discovered that it was loaded with Neu5Gc. "We got really interested," said Varki. "But you're not in the primate field, you don't know where to go. Where are the samples? How do you get these things? How do you make the contacts? It took us several years, but we finally got blood from humans and all the great apes."

Going through the samples, Varki and his colleagues identified a recent evolutionary change in humans: whereas high levels of Neu5Gc were found in the blood of nineteen chimps, bonobos, gorillas, and orangutans, none was found in the eight human samples.[19] That loss of Neu5Gc might explain differences in species' susceptibility to a wide range of pathogens. Neu5Gc helps establish infections, opening the cellular doors to influenza viruses, a type of *E. coli*, and different strains of the plasmodium that causes malaria. The loss of Neu5Gc might also alter immune responses and could play a role in the dissimilar reactions apes and humans have in response to HIV and hepatitis B viruses. These ideas were hugely speculative, Varki recognized, but they also made the case for a comparison of ape DNA to the new genetic map emerging for humans.

It was 1998, and progress in the Human Genome Project regularly commanded headlines, often describing dramatic showdowns between scientific celebrities such as the Nobel laureate James Watson and the upstart entrepreneur Craig Venter. The human project had gone to great lengths to find a diverse mix of people to sequence, creating a "consensus" genome that attempted to best represent the entire species. The nascent chimpanzee project relied on one animal, Clint, chosen because humans could easily get him to masturbate. The chimp project badly needed someone to connect the dots. Varki stepped into the vacuum, joining forces with researchers in anthropology, paleontology, neurology, linguistics, genetics, and primatology to launch the gloriously ambitious Project for Explaining the Origin of Humans.

By February 2002, Varki, Evan Eichler, and Edwin McConkey had teamed up with leading investigators in the Human Genome Project to draft a paper spelling out why the NIH should fund the sequencing of the chimp genome. "Seizing this magnificent opportunity ranks among the highest basic science priorities in all of biomedical research," they asserted. The sequence may lead to "altogether novel routes to improved treatments for human disease" and could help explain how humans developed their extraordinarily sophisticated brain and how human populations diverged as they moved out of Africa. A refined proposal that October, which enlisted support from heavy hitters including Svante Pääbo of the Max Planck Institute for Evolutionary Anthropology in Leipzig, Germany, and Eric Lander, the founder of the Whitehead Institute at the Massachusetts Institute of Technology, persuaded the NIH's National Human Genome Research Institute to pony up $42 million for the scheme.[20]

Soon, the Chimpanzee Sequencing and Analysis Consortium would challenge one of the most fundamental, oft-repeated "truths" about humans and chimpanzees: that we are 99 percent identical.

—◦◦◦—

IN a groundbreaking 1975 paper published in *Science*, the evolutionary biologist Allan Wilson of UC Berkeley and his erstwhile graduate student Mary-Claire King reviewed a host of different studies that firmly established the 1 percent genetic difference between humans and chimpanzees. "At the time, that was heretical because humans and chimpanzees were supposedly from different families," King, who went on to become a leading medical geneticist at the University of Washington in Seattle, told me. The data they amassed were convincing and came from myriad sources. Some studies compared proteins, while others looked at the nucleic acids that make up DNA. A table that ran for a dazzling three pages provided side-by-side comparisons of forty different enzymes, components of blood, muscle, and mitochondria. Again and again, there were no differences between chimps and humans, or the differences were slight. Wilson and King even went so far as to show that mice, frogs, and flies differ more from their own sibling species—meaning, for example, one type of mouse compared to another type of mouse—than do humans from chimps.

But Wilson and King's paper had a second point that received far less attention outside biology circles. Chimpanzees and humans, they stressed, differ so dramatically in anatomy and behavior that a 99 percent similarity in genes and proteins creates a paradox: the nearly identical molecular makeup does not lead to nearly identical organisms. What five-year-old human child—or five-year-old chimp—would mistake one for the other? Not only do humans have bigger brains, language, and walk upright, as King and Wilson noted, "nearly every bone in the body of a chimpanzee is readily distinguishable in shape or size from its human counterpart." Chimpanzees are hairier, sleep in trees, and don't run long distances or swim. They do not dance with each other, paint on walls of caves, naturally live outside of Africa, have blue eyes, make jewelry, or store food. King and Wilson quoted George Gaylord Simpson, who observed that *Pan*—the genus name for chimpanzees and bonobos—"is the terminus of a conservative lineage" that retains features of other apes. Chimps, in other words, had hit something of an evolutionary doldrums, while humans had caught a strong wind and continued to evolve a wide spectrum of unique features and capabilities.

Then add in the other differences that King and Wilson did not know about at the time. Chimpanzees are susceptible and resistant to different diseases than humans, they do not appear to teach each other, and their "cultural" traditions do not become more sophisticated from one generation to the next. It's not clear whether females go through menopause, and they miscarry much less frequently than humans. The males ejaculate far more sperm. Chimps do not cook, and they metabolize food differently. A 1 percent genetic difference accounts for all of this?

One percent actually is a lot when you are talking about the human genome, which contains 3.2 billion pairs of nucleotides: it is 32 million differences. But many, if not most, of these differences have no functional importance. Three nucleotides in a row code for an amino acid, and the system has a built-in redundancy that allows different trios to play the same tune. In other words, a change in one nucleotide may still produce the same amino acid. Further complicating the meaning of "difference," amino acids link together to form proteins, and even if a change in one amino acid creates a novel protein, it may have an identical function. If a violinist in an orchestra replaced a broken string (the protein) with one made from sheep rather than cow gut, the symphony would sound the same.

King and Wilson had made a startling assertion: maybe the major genetic differences between humans and chimpanzees lie not in the genes themselves but in the biological knobs, whatever they may be, that determine whether genes are turned on high, medium, or low, or completely turned off. This regulatory phenomenon, which scientists poetically call "expression," ultimately determines how much of a protein a gene produces. And quantity matters. King and Wilson argued that if a human and a chimp shared a particular gene but, say, one gene was triggered to produce loads of a given protein and the other was not, those identical genes would create two distinct organisms.

The Chimpanzee Sequencing and Analysis Consortium—which included King, Varki, Eichler, and sixty-three other researchers—

published its draft of the chimpanzee genome in the September 1, 2005, issue of the journal *Nature*, and the findings confirmed how perspicacious Wilson and King had been. Roughly one-third of proteins are identical between humans and chimps, and another third only differ by one or two amino acids. At the level of nucleotides, the difference was 1.23 percent. It dropped even lower—to 1.06 percent—when the researchers adjusted for the fact that variations exist between individuals within each species; in other words, had they looked at the genomes of hundreds of chimpanzees, they would have found many differences, and some would have matched the sequences in humans.

Just as the research team provided the best validation yet of a 1 percent difference, it also revealed the most dramatic evidence of the figure's limitations. To begin with, the number does not factor in the many stretches of DNA that have been inserted in or deleted from the genomes. The consortium calculated that these "indels" alone account for another 3 percent difference. And indels can disrupt genes and cause serious diseases, including fragile X syndrome, cystic fibrosis, Alzheimer's disease, and Parkinson's.

But more striking still, the chimpanzee genome project highlighted the importance of the biological knobs that control gene expression, which in the thirty years since King and Wilson's paper had come into a much clearer focus. The vast majority of the genome—roughly 98 percent—does not contain genes.[21] Although these noncoding regions were once called "junk DNA" because they are not expressed, researchers have discovered that they regulate the expression of other sections of DNA. It appears the noncoding regions have also evolved more quickly. Indeed, they may be chiefly responsible for what genetically makes humans human.

A nucleotide is constrained, or "conserved" in evolutionary jargon, unless changing it leads to something more beneficial—another way of saying there's a price to pay for changing it, and the new version needs to provide a useful adaptation. Katherine Pollard, a researcher from the University of California, Santa Cruz, compared the mouse, rat, and chimpanzee genomes and identified more than thirty thousand short regions that were nearly identical. She and her colleagues then overlaid these sections on the human genome and identified 202 regions that had evolved rapidly. The vast majority of these "human accelerated regions," or HARs, are located in noncoding regions; several appear to be involved with regulating gene expression. Intriguingly, the "most dramatic" difference—eighteen nucleotide divergences between chimps and humans, compared to only two between chimps and chickens—appears in a region that scientists believe regulates a gene in the neocortex of embryonic human brains.[22]

In another example, researchers working with the chimpanzee sequence described the evolution of a noncoding region that may play a central role in shaping the hands and feet. Unlike chimps, humans have relatively long thumbs that rotate toward our palms, and our feet have much shorter digits and are more rigid—two features that help us walk upright. In 2008, a team of investigators from the Lawrence Berkeley National Laboratory discovered "the most rapidly evolving human noncoding element yet identified"—thirteen nucleotides out of eighty-three that remained identical all the way from frog to chimp and yet somehow changed in humans. In a series of clever experiments that involved stitching these different DNA sequences from chimpanzees, rhesus monkeys, and humans into mice, they showed this noncoding element regulates developmental genes responsible for limb development.[23] So these mutations in a noncoding region of the human genome appear, in the language of geneticists, to have been "positively selected" because they made humans more fit by contributing to our bipedalism and uniquely dexterous hands.

In addition to the role of noncoding regions in determining how much protein a gene turns out, researchers have begun to appreciate another genomic mechanism that alters the line between humans and chimpanzees. The process of meiosis divides our chromosome number in half so that both the male and female each contribute twenty-three chromosomes to the zygote. During meiosis, accidents happen, and chromosomes routinely and randomly lose chunks of entire genes, or whole chunks of genes are duplicated. A person theoretically should have two copies of each gene, one from each parent, but because of these mistakes, genomes may have a dozen or more copies of the same gene, or maybe just one. In 2006, Matthew Hahn, a computational genomics researcher at Indiana University, reported that he and his colleagues had assessed gene gain and loss in the mouse, rat, dog, chimpanzee, and human. Human and chimpanzee gene copy num-

Human	GCAGCCTTGGGTTCCGCAAATAGGGCACCCACAGTAACACGTGTGGCGCCGACCCCGCCGTGCGCAATCGGGGCTTTATAC
Chimpanzee	A···G········T··········T····A····A····T·A·TA··············AT················G
Rhesus	A···G········T··········T····A····A····T·A·TA··············AT················G
Mouse	····GT···C··T··········T····A····CA···T·A·TA··············AT············C·G
Rat	····GT········T··········T····A····CA···T·A·TA··············AT·············C·G
Dog	····GT········T········T·G···A····A····A·TA···T······AT············C·G
Chicken	·GG·G········T·A··········T····A····A····T·A·TA···T······AT········A·T······G

The Difference Made by Noncoding DNA

A comparison across different species of the string of nucleotides from a noncoding region that regulates limb development. Thirteen nucleotides in the human string differ from the chimpanzee string, but the chimpanzee usually matches the species below it, suggesting that this highly "conserved" noncoding region changed to benefit humans.

ADAPTED FROM SHYAM PRABHAKAR, AXEL VISEL, JENNIFER A. AKIYAMA, ET AL., "HUMAN-SPECIFIC GAIN OF FUNCTION IN A DEVELOPMENTAL ENHANCER," *SCIENCE* 321:1346–50 (SEPT. 5, 2008).

bers differed by an incredible 6.4 percent. That led Hahn and his team to conclude that "gene duplication and loss may have played a greater role than nucleotide substitution in the evolution" of humans.[24]

Could researchers combine all of what's known and come up with a precise percentage difference between humans and chimpanzees? "I don't think there's any way to calculate a number," Svante Pääbo told me. "In the end, it's a political and social and cultural thing about how we see our differences."

From a Darwinian perspective, it does not matter whether we differ from chimpanzees by 1 percent or 5 percent. What's important, and stunningly clear from the data, is the relative difference between us and other species. And just as Darwin suspected, we are closer to chimps than to monkeys, closer to monkeys than to chickens, and closer to chickens than to flies.

—◆◆◆—

NICK Patterson has a cartoon on his office wall of one chimp holding a bra and looking incredulously at another chimp. "You had sex with a what?" the chimp holding up the bra asks.

Patterson works at the Broad Institute, a sleek and imposing edifice fabricated of glass and steel in a nod to its futuristic vision. The building sits across the street from MIT, which, along with nearby Harvard, backs the institute's efforts to bring together scientists from different disciplines to study genomics. Patterson embodies the cross-disciplinary ethos: he is a mathematician who once worked as a cryptographer for the British government, later developed mathematical trading models for a hugely successful Wall Street hedge fund, and then wound up helping biologists unscramble DNA sequence data. He has a genius aura, accentuated by the fact that he was born with a deformed skull that shaped the left side of his face in an unusual way and blinded his left eye. But despite his outsized intellect he has no pretensions, and he charms through a mix of mirth, mischief, and modesty.

Nine months after the chimp DNA sequence was published, Patterson and David Reich, a human population geneticist at the institute, offered a fresh theory for how and when humans and chimpanzees evolved from a common ancestor.[25] They looked at 30 million nucleotides from overlapping sequences of primate genomes—from humans, chimpanzees, gorillas, orangutans, and rhesus macaques—eight hundred times the number of nucleotides previously used in similar comparisons. "Geneticists have to work off variation: that's the name of the game," said Patterson. "If everything's the same, what are you going to say?" And based on their extensive analysis, everything was far from the same.

To peg a more precise estimate of the origin of humans, Patterson and Reich had turned to fossils they knew predated the separation of old- and new-world monkeys (*Aegyptopithecus*, dated at 33 million years ago) or predated the speciation of the first great ape, the orangutan (*Proconsul*, at 18 million years ago). They then analyzed the nucleotide differences between species, and, using the known mutation rate and the molecular clock, calculated how much time must have passed to see that divergence. Whether they used *Aegyptopithecus* or *Proconsul* as a starting point, they calculated that the split between humans and chimps must have occurred less than 6.3 million years ago—before the time frame identified by Sibley and Ahlquist in 1984.[26]

Molecular biology once again peppered question marks all over the oldest putative human fossils. The granddaddy of them all, which goes by the nickname Toumaï and formally is known as *Sahelanthropus tchadensis*, supposedly lived 6.5 to 7.4 million years ago, according to the research team that found the skull, jaw, and teeth fragments in Chad in 2001 and 2002. The two next oldest fossils, *Orrorin tugenesis* and *Ardipithecus kadabba*, date to 5.6 to 5.8 million years ago. It could be that the Toumaï fossils were younger than estimated. One member of the original team that discovered Toumaï—which means "hope of life" in the indigenous Goran language—has claimed as much, saying the fossils were found on the surface rather than in the soil that was used to date them.[27] Still, many who had invested their careers in studying fossils blithely dismissed Patterson and Reich's timeline. "Their explanation is just a hypothesis, while Toumaï is a true fossil," said Michel Brunet, a leader of the Toumaï team.[28]

A second curiosity centered on the vast amount of variation Patterson and Reich found in the genetic divergence between humans and chimps. Reich, who looks young enough to be a grad student, explained how he and Patterson focused intensely on the DNA from modern humans, chimpanzees, and gorillas, which he said hold "frozen information" from the ancient past. "We have these three lineages that allow us to go magically back in time, and this data will allow us to learn what the population, in very crude form, was just at the time of the split." On average, the chimp and human genomes differed by about 7.5 million years, but some stretches of DNA differed much more than the average and others differed much less. Some DNA regions differed by only 6.3 million years, while others differed by nearly 11 million. Much of this divergence occurred because in 18 to 29 percent of the genome, humans or chimps were closer to gorillas than they were to each other.

The most shocking finding, however, involved the X chromosome. "This was the most important observation in our paper and hardest to explain," said Reich. Their analysis showed that far less genetic divergence occurred between the chimp and human X chromosomes than the rest of the genome. When they converted the divergence into time, it represented a staggering 1.2 million years less divergence than the 7.5-million-year average seen on the other chromosomes. A comparison of human X to gorilla X revealed no such pattern. And chimp X has virtually no regions that are closer to gorilla than they are to human.

If humans separated from chimpanzees less than 6.3 million years ago, why would so much gorilla DNA be sprinkled in the human genome but not in the chimp genome? Why would chimps have gorilla DNA that is absent in humans? Why did these speciation dates conflict with the fossil record? And why was the X chromosome younger?

One day, Reich stumbled on the answer. He phoned Patterson.

"It's Haldane's rule," he said.

2

TWO BECOME ONE

A FEW MILES FROM THE ENDLESS SHOPPING MALLS AND GARISH tourist attractions that line the main highway in Myrtle Beach, South Carolina, lives a group of liger brothers. The ligers, born to a lion father and a tiger mother, are the world's largest cats, weighing up to half a ton—double the heft of either parent. They are a hybrid, never seen in the wild, and rarely seen in captivity, since most zoos don't display big mammal blends. But that's how it is with hybrids: their existence escapes detection.

Yet, when you start looking for them, they're everywhere.

Zorses, wholphins, blynxes, and pizzlies. Leopjags, zonkeys, beefalos, camas, pumapards, tigons, yakalos, and bonanzees. These are some of the natural and captive-bred hybrids that exist, and they are joined by an assortment of somewhat less exotic hybrid birds, fish, insects, and, of course, plants. With new techniques to isolate and compare DNA, more are being added to the list each year.

Hybrids evoke a mix of wonder and fear, magic and folklore, the ancient and the untested future. They raise fundamental questions that have roiled scientists all the way back to Darwin's day about what it means to call something a species and how evolution unfolds. Some people herald them for their power and odd beauty, while others dismiss them as biological curiosities that deserve no more attention than the most notorious of hybrids, the mule. However you look at it, their very existence—just like the discovery in 2007 of a planet, 20.5 light-years away, that possibly has conditions conducive to life—dramatically unsettles our concept of what's out there. And evidence suggests that humans, too, evolved through a hybrid intermediary. For those who recoil at the notion that we had a common ancestor with the chimpanzee, buckle up: we may have once mated with them, which, if true, means "humanzees" once knuckle-dragged the earth.

At The Institute for Greatly Endangered and Rare Species (TIGERS) on the outskirts of Myrtle Beach, the liger brothers Hercules, Sinbad, and Zeus share a fifty-acre spread with some eighty other big cats, bears, primates, wolves, raptors, a white crocodile, and an African elephant. The animal trainer Bhagavan Antle, forty-seven, runs TIGERS with assistants who live on the grounds, eat communal vegetarian meals, and learn how to work safely with the menagerie. "Interactive" tours allow visitors to watch the ligers and other animals up close and even hold the younger tigers, chimpanzees, and orangutans.

I first met Antle at the carefully fenced-off preserve on a cold, wet morning in January 2007. Inside a safari-themed lodge, he showed me videos of his many media appearances and movie gigs. On one wall hung a lacquered cover of *Vogue*, an Annie Leibovitz shot of the actress Drew Barrymore posed in a mock formal painting with one of Antle's lions at her side for a spread called "Beauty & the Beast." He had also provided animals for such Hollywood films as *Ace Ventura: Pet Detective*, *Forrest Gump*, *Jungle Book*, and *Dr. Doolittle*.

Antle, who had the look of a middle-aged rock star with his long ponytail, soul patch, and hoop earrings, was something of a hybrid himself. Raised on an Arizona cattle ranch, he became a disciple of Swami Satchidananda—whose claim to fame was offering the opening blessing at the Woodstock music festival—in the late 1970s. Antle's animal career began when he got a job at a health clinic affiliated with the Swami's ashram, Yogaville, in Buckingham County, Virginia. In the type of non sequitur backstory that is common among professional animal people, Antle received a tiger cub as a gift from a visitor to the clinic in 1982. Later, another visitor who worked for "tiger-in-your-tank" Exxon asked him to lecture on health, cub in tow, at a company gathering. Antle soon was toiling as a Renaissance fair magician with a fake Hungarian accent, charging people to take photos next to his tiger. He acquired and trained ever more exotic animals, breaking into Hollywood in 1986. To his astonishment, a few years later, his male lion, Arthur, successfully mated with one of his tigresses. A second liger litter arrived in 2002.

In 2004, a movie that Antle had nothing to do with made ligers famous—and raised his stature a few notches. In the geek-glorifying *Napoleon Dynamite*, Napoleon sketches a liger in his school notebook and declares, "It's pretty much my favorite animal. . . . Bred for its skills in magic." Antle, one of the few liger owners in the world, made the TV rounds, chatting with CNN's Anderson Cooper and NBC's Matt Lauer. "We had such a big splash of exposure," he said. "It was like opening a chapter of myth that had come to life."

After Antle finished showing me around, three of his assistants suddenly appeared on the lodge's deck with one of the liger brothers, Sinbad. The supersized beast had lighter stripes than a tiger and a lion-shaped head with no mane. Sinbad's limbs looked stubby, and his pectoral muscles sagged. As we watched through a glass wall, a woman offered a chunk of meat from atop a platform to make Sinbad stand and show off his twelve-foot-long frame. The assistants guided him around using two chains and a baby bottle, and then Antle invited me out to the deck for a closer look. He walked up and snuggled Sinbad's muzzle.

"Hi, bud," he cooed, pursing his lips like he was playing kissy face with a kitten.

Sinbad could have removed Antle's head with a single chomp, but I was more enchanted than scared. It was as though I was looking at a Sasquatch or a centaur in the flesh. "In our core belief, people don't want to accept the idea that two distinctly different-looking wild animals can reproduce," Antle said. "Ligers make people understand that hybridization is real."

—◦◦◦—

CHARLES Darwin understood that hybridization is real, and it deeply confused him.

In the *Origin of Species*, he devoted a chapter to hybrids, but their existence was an evolutionary riddle that he never really solved.[1] Hybrid animals such as mules, Darwin correctly noted, are usually sterile. He deemed it a "strange arrangement" that nature would afford two species the "special powers" to create hybrids but then prevent these offspring from propagating. He offered squishy theories about why this is so—nobody knew anything about genes in 1859—and how hybrids fit into his overarching theory of natural selection.

Enter Haldane's rule. The British geneticist J. B. S. Haldane in 1922 recognized that in hybrids, fertility differs between the sexes. In the sex that has mixed gametes—the males in humans, who are XY—sterility occurs much more commonly. (In mules, research suggests that the chromosomes from the horse and the donkey cannot properly line up with each other, which prevents males from making spermatozoa.) That difference between males and females is what got the lightbulb to flick on in David Reich's brain as he tried to unscramble the differences in the X chromosome found in the human and chimpanzee genomes.

What if humans and chimpanzees initially diverged as species, Reich wondered, and then, perhaps millions of years later, mated again? The hybrid males, according to Haldane's rule, might be sterile, but the female hybrids could have mated with either chimps or humans. For the sake of argument, say a hybrid female mated with a chimp and became pregnant. Her female offspring would inherit her two X chromosomes (one Xchimp and the other Xhuman). Her male offspring would only get a single X chromosome (either an Xchimp or an Xhuman). If a male inherited her Xchimp and his father's Ychimp, his mother would still pass on her other hybrid chromosomes. So he would be hybrid himself but could mate with chimps, other hybrids, and humans without sterility problems.

This scenario, which is called "complex speciation," introduces the Xchimp to humans, effectively resetting the molecular clock on the human X chromosome to match the one in chimps. Because they are hybrid, the non–sex chromosomes—the "autosomes"—retain older mutations, and a comparison to chimps shows more divergence. And gene flow would have been jumbled in many other ways. Chimps could have passed older gorilla parts of their genomes back to humans and subsequently lost them in themselves. Humans could have passed younger parts of their genomes to chimps. These would explain why David Reich and Nick Patterson found the great range of divergence in the autosomes, and why X displayed less divergence.

Hybridization, then, not only solves the X conundrum but also resolves the apparent inconsistencies in the age of human fossils and the vast divergence between chimp and human genomes. Imagine that

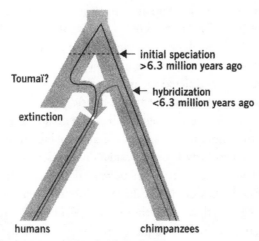

Toumaï?

initial speciation
>6.3 million years ago

hybridization
<6.3 million years ago

extinction

humans   chimpanzees

Humans, Chimpanzees, and Hybrids
David Reich and Nick Patterson's illustration of the emergence of humans and chimpanzees with a hybrid. Reading from the top, a river splits into two. This is the common ancestor speciating into humans and chimps. Reich and Patterson do not calculate precisely when this happened, only noting it was more than 6.3 million years ago. The rivers then rejoin at some point in time that they do not specify to form a hybrid. Toumaï, who may be a hybrid, too, branches off and becomes extinct. The hybrid line then becomes humans.
ADAPTED FROM NICK PATTERSON, DANIEL J. RICHTER, SANTE GNERRE, ET AL., "GENETIC EVIDENCE FOR COMPLEX SPECIATION OF HUMANS AND CHIMPANZEES," *NATURE* 441:1103–8 (JUNE 29, 2006).

---

an initial human/chimp speciation occurred, say, 8 million years ago, and then the two species began hybridizing a few million years later, finally separating for good less than 6.3 million years ago. Toumaï could be an early "hominin"—or, for that matter, a hybrid.[2] And hybridization helps explain why regions of the chimp and human genomes today have these odd sections that are either much more or much less variant than average.

The theory of a human/chimp hybrid attracted more skepticism than any other part of Reich and Patterson's analysis, which they published in *Nature* in June 2006. Jeffrey Schwartz, a physical anthropologist at the University of Pittsburgh, said the hybrid idea "pushes the limits of credulity." (Schwartz, incidentally, was one of the last researchers to contend that humans are closer to orangutans than to chimps or gorillas.) The Harvard paleoanthropologist David Pilbeam, who discovered fossils of proto-orangutans, said, "I don't buy these hybrids," contending that humans and chimps that hybridized back then likely would have differed too much to have any fertile offspring.

The most compelling dissent to the hybrid theory focuses on the 4-million-year divergence and suggests that you do not need hybrids to explain it. If the common ancestor to chimps and humans had a large population—at least forty-five thousand individuals by one estimate—there would have been a lot of genetic variation, or polymorphisms. Different individuals would have kept or lost different stretches of DNA, or even genes, that match those in the gorilla genome. If one of these common ancestors was a parent to the human lineage and the other was a parent to the chimp lineage, humans and chimps could appear closer to gorilla in some parts of their genome than to each other—even though, if you look at their entire genomes, humans and chimps clearly are more closely related to each other.

When a Harvard evolutionary biologist, John Wakeley, took Reich and Patterson to task on this point, they conceded that he was right. But they believe that polymorphisms only explain the autosomes, which still leaves open their peculiar finding about the X chromosome.[3] Only hybridization explains this, they asserted. Another critic, Nick Barton from the University of Edinburgh, told me that he thought the X data "were genuinely puzzling" but that Patterson and Reich's argument "seems contrived."[4] Reich and Patterson remain open-minded. "It's a model, it's refutable, and it's consistent with the data, " Patterson told me. "It doesn't prove it's right."

——∞——

DARWIN'S confusion about hybrids went beyond his simply pulling his beard over the mystery of the strange arrangement that they existed yet could not easily reproduce. He thought they were trouble.

In 1868, nine years after the publication of *Origin*, Darwin explored hybrids more closely in a book that examined variation in domesticated animals. Darwin asserted that hybrids—and he explicitly mentioned ligers, which the naturalist Georges Cuvier first reported were bred in 1824—might inadvertently push back the evolutionary clock, resurrecting traits that were better left behind. He used the mixing of human racial groups as an example, stating that travelers frequently noted "the degraded state and savage disposition of crossed races of man." He allowed that there were "many excellent and kind-hearted mulattos," but wrote that he was struck in South America that "men of complicated descent between Negroes, Indians, and Spaniards, seldom had, whatever the cause might be, a good expression." Although he did not actually view different races as different species, he recounted how the explorer David Livingstone had met someone on his African travels who told him, "God made white men, and God made black men, but the Devil made half-castes." Matters were even worse with mixtures of two races that were both "low in the scale." The progeny, Darwin wrote, "seems to be eminently bad."[5]

Maybe you can forgive him his racism: the United Kingdom did not abolish slavery until Darwin was twenty-five. But do not let him off so easily for impugning hybrids.

Darwin's hybrids-are-bad dictum became orthodoxy during evolutionary biology's "modern synthesis" in the 1930s and '40s, which firmly connected genetics to natural selection. The Harvard ornithologist Ernst Mayr, a leading neo-Darwinist, set the tone by dismissing hybrids as an evolutionary dead end. "The total weight of evidence contradicts the assumption that hybridization plays a major evolutionary role among higher animals," he wrote.[6]

Mayr's verdict involved a surprisingly contentious issue: What exactly is a species? Darwin had seen "species" as an arbitrary designation for animals that have similar physical features. Mayr came up with a concrete definition known as the "biological species concept." A species, he declared, is a reproductively isolated group that can interbreed. And the main force that isolated species, he argued, was geography.

By this formula, species were a fixed unit that became more fit over time through forces such as random mutation and sexual selection, not by having "gene flow"—another term for doing the nasty—with other species. "Species were rocks," explained Michael Arnold, an evolutionary biologist at the University of Georgia who studies hybridization in both plants and animals.

But maybe, in fact, they aren't. Arnold is part of a growing camp that sees species as more liquid than solid, and he rejects the idea that hybrids are always evolutionary losers just because they often cannot reproduce. "A lot of us have been hammering away on this for many years," Arnold said. "They used to call us the Mongol hordes at their gates, but now we're inside. Paradigm shifts are always painful things."

Arnold and his like-minded colleagues are pushing a profound reconception of evolution, one in which hybrids are far from bit players. Forget the tree of life, with new species neatly branching off from a common ancestor. It's a web of life, and hybrids help genes flow in unexpected directions. Species are not on a linear march toward perfection. Nature is reticulate, a network that allows "species" to profit from any genetic advantage they can find.

Reich and Patterson share the view that hybrids have not received the respect they deserve—and not just in human history. "There's a view that hybridization is not an important process in evolution, not with regard to just humans, but generally," said Reich. "Even though it seems like a weird thing to do just on a lark, it potentially can confer useful traits for an ancestral population to adapt to a new environment." Patterson was more blunt. "Ernst Mayr was a very great man, but his arguments about hybrids in the field are weak," he said. "We're talking about evolutionary time and the whole planet. Over a million years, funny stuff happens."

But what about their famous sterility? Some hybrids can reproduce, and Arnold stresses that rare events have an "overwhelming importance" in the evolutionary process. Hybrids often have desirable traits—some are actually more fit than either parent, a condition called "hybrid vigor"—and there have been instances when hybrids were able to find enough fertile hybrid partners to create a new species. This process appears to be under way right now in the United States, involving a hybrid of the Pecos pupfish and the sheepshead minnow that has greatly multiplied and expanded its range. And some scientists assert that mating between gray wolves and coyotes thousands of years ago created an entirely new species: the red wolf. More commonly, though, hybrids mate with one of their parent species, influencing the mix of what gets passed along to subsequent generations; essentially, they provide a bridge for genes to cross the species divide.

In a paper about hybridization and primate evolution that Arnold coauthored in 2006 for the journal *Zoology*, he offered several examples, including chimpanzees and their close cousins, bonobos. DNA studies suggest that these two great apes swapped genes sometime after separating about eight hundred thousand years ago. Arnold believes these ape cousins occasionally mated, but the resulting bonanzees did not establish a new species. Instead, they hooked up with either chimps or bonobos. Bonanzees ultimately vanished, but they left genetic footprints in the genomes of their descendants.

In addition to comparing genomes for evidence of unusual gene flow, scientists increasingly have used DNA analysis to confirm the existence of natural hybrids that no one previously knew existed. Their existence offers evidence that the process still occurs.

On April 16, 2006, James Martell shot what he thought was a polar bear on Banks Island in the Northwest Territories, Canada. But on closer inspection, it had orangish fur, strange claws, a humped back, and black rings around its eyes. "There was a photograph on the news, and we looked at it and thought, Boy, that's a pretty weird-looking bear," remembered David Paetkau, a population geneticist at Wildlife Genetics International in British Columbia, which has pulled DNA from the hair of some forty thousand bears in North America to help estimate the size of their population. Suspecting that it was a hybrid polar/grizzly, Canadian officials sent a sample of the bear's skin and hair to Paetkau, who isolated its DNA and then compared the results to the large genetic database his company has collected of both species. "This animal was right smack in the middle," Paetkau said. Tests also revealed that the polar bear was its mother. The custom of combining words, portmanteau, for hybrids is to use the name of the father's species first, so the bear should have been called a grolar or a grizzlar, but, for whatever reason, the name pizzly stuck. Though pizzlies had been bred in captivity, it was the first confirmed wild pizzly ever found.

In 1998, a bobcat trapper in Maine notified wildlife officials that he might have caught a lynx, which is protected by the Endangered Species Act. The officials radio-collared and released the odd-looking animal, but it soon died of starvation and they retrieved the carcass. By 2003, scientists working in the U.S. Department of Agriculture's

Forest Service had five other suspect bobcat/lynx samples from both Maine and Minnesota. DNA tests confirmed they were blynxes. Other hybrid mammals identified using DNA analysis include the forest/savanna elephant in Africa, mink/polecats in France, and a sheep/goat in Botswana.

Although somewhat more mundane, a discovery of a hybrid fruitfly was deemed so unusual that it merited a 2005 report in *Nature*. The fly's discovery emphasizes how hybrids hide in plain sight. While jogging around a golf course at Pennsylvania State University, Dietmar Schwarz noticed a honeysuckle bush, *Lonicera*. Schwarz, then an undergraduate visiting from Germany, had studied fruit flies that lived in a European honeysuckle. Sure enough, the Penn State fruit looked as though they were infested with fly larvae, so he gathered a few. Working in an entomology lab on campus, he conducted DNA and enzyme analyses and found that the larvae were the offspring of blueberry and snowberry flies.

Blueberry and snowberry flies look exactly the same, so entomologists never would have noticed this hybrid with their naked eyes. And had fly researchers not examined the genetics of thousands of species over the past century, Schwarz said he may not have detected the *Lonicera* fly's lineage, cataloging it instead as just one more overlooked insect species.

Another possible hybrid hits much closer to home: modern humans and Neandertals, who disappeared from Europe about thirty thousand years ago. Fossils of Neandertals have led some researchers to conclude that we did breed with our heavy-browed, big-boned relatives, but so far reviews of the DNA do not support this.[7] Nearly thirty years ago, the Finnish paleontologist Björn Kurtén wrote a novel about "Neandermans," *Dance of the Tiger*, in which he describes the act of hybridization.[8] After having sex with a Neandertal for the first time, the Cro-Magnon protagonist, Tiger, has one terrifying question on his mind. "What is this animal I am holding in my arms?" writes Kurtén. "And for a second he felt that he had made love to a bison or a wolf bitch."[9] That may, in the end, explain as forcefully as the genetics why ligers, zonkies, pizzlies, and other animal hybrids remain such oddities.

—◦◦◦—

IF levels of intelligence are measured by levels of curiosity, Kurt Benirschke is one of the smartest humans on the planet. And one of the questions that intrigues him the most, and that he is most expert on, is how humans evolved from chimpanzees. But when I visited him in his basement office adjacent to the morgue at a hospital run by the University of California, San Diego, he was exploring nothing of the sort.

"Come take a look at this," Benirschke said the moment I entered his office.

Benirschke, then eighty-one, was sitting at his microscope, which has a set of eyepieces extending to his side of the desk and another pair for visitors. I rolled up a chair and took a peek.

"That's 40 million years old," he said. "What do you think it is?"

"Looks like a fly," I said.

"Or a bee." He maneuvered a green electronic arrow on the insect, which was encased in amber. "I think it has four legs," he said. He guffawed, overwhelmed and amazed. He pointed the arrow to another body part. "Look!" he said, sounding like a four-year-old. "What's that?"

"A wing?"

"A wing!"

I asked him how he came to possess an ancient fly, and he told me the story of a butcher and a milkman in East Berlin who had become the world's most proficient inseminators of elephants. Benirschke went to hear the butcher and milkman speak at a conference in Germany, where he met a man who was writing a book about them, tentatively titled *The Boys from Berlin*. The would-be author gave Benirschke the amber-encased fly as a gift. "They sell it at a museum gift shop!" he bellowed.

Benirschke then shared the tale of a dentist friend who had complained that his job was all filling and drilling. Benirschke, a pathologist by training who started the Center for Reproduction of Endangered Species at the San Diego Zoo, suggested the man work on the teeth of zoo animals, and soon the dentist found himself operating on an elephant with a severed tusk. The broken tusk kept oozing fluid, so the dentist fashioned a spigot that fit over the ivory and could periodically be turned on to drain it. To prevent infection, the dentist flushed the tusk with chlorine. The animal didn't seem to mind, which piqued Benirschke's interest in the folklore of tusks: it's said that when elephants injure tusks, it's so painful they go mad. Benirschke took a piece of tusk back to his lab and hunted for nerves. "There were none!"

Kurt Benirschke is an intellectual cowboy who gallops from one idea to another, creating a near psychedelic haze of ideas. Segues are simultaneously bizarre and seamless and revolve around the two main passions in his long and hugely successful career: reproductive biology and the evolutionary forces that create new species. He has twinkly blue eyes and a sly youthfulness that men half his age have lost for good. In an amusing twist, his broad, Czech-German face is known to many, especially in San Diego, because his son Rolf, a former star field goal kicker for the San Diego Chargers, is remembered for having kicked barefoot. Kurt has the big rugged look of a football player himself, though the only obviously jockish thing about his demeanor is that he couldn't give a damn what anyone thinks of him. Then again, he's friendly and, for a pathologist, unusually personable with the living.

Benirschke is part of an endangered species himself, the physician scientist. He grimaces at the modern idea of a doctor who runs from patient to patient and has no time to stop and smell the formaldehyde. Of course patients are people to heal, but his medical degree has also given him the opportunity to regularly obtain material to study the grander questions of how humans are born and die. His autopsies of babies as a young pathologist at Harvard led to a fascination for the placenta, and then for twins—which die more frequently than singletons—and then for, of all things, armadillos, which give birth to identical quadruplets. He learned that marmoset monkeys only give birth to fraternal twins, and because Harvard didn't have a place to keep these animals, he built his own colony, keeping thirty-five of them at home in his basement. He conducted a crucial study explaining why mules are sterile. After seeing a photo in *Life* of a zebronkey—a zebra donkey hybrid, also called a zonkey—in the

Philippines, he resolved to get a piece of skin from the animal. Fortunately, he recently had trained a Filipino doctor who had returned home to Manila and whose husband was on the board of the zoo there.

Benirschke's chromosomal analysis of the zebronkey showed that the zebra parent was a different subspecies than the breeders had thought, which spurred an outcry from disbelieving taxonomists. So he decided to explore the comparative genetics of zebras kept at a game farm in the Catskills of upstate New York. He also became intrigued by an endangered American pig called the peccary, and started a breeding colony of them in the jungles of Paraguay. From an interest in placentas and twins, Benirschke had branched into comparative genetics, zoos, and endangered species, which is what led him in 1975 to start the Center for Reproduction of Endangered Species.

Way past retirement age, but with the energy of a grad student, Benirschke turned much of his attention to the mother of all questions in the study of human biology: What finally triggered the split between humans and apes? Benirschke had an answer, and he overlooked no detail. Not all of his colleagues agreed with him. But none dismissed Kurt Benirschke's version out of hand.

—◦◦◦—

TURN back the clock somewhere between 5 million and 7 million years ago. The Miocene epoch is ending, ushering in the Pliocene. A land bridge is just beginning to form between North and South America. Tree-studded rainforests in Africa are giving way to grasslands and savannas. And an ape, living in those trees and knuckle-dragging through those grasses, has evolved from the gorilla but is not yet a chimpanzee.

One day, two of these proto-chimps have what scientists daintily call a "consort." The male can tell that it is business time because the female is in estrus, making her vagina swell to the size of a grapefruit. The male mounts her, shoots in hundreds of millions of sperm, and one of them finds its way to an egg in her fallopian tube. When the sperm begins to penetrate the egg, a fateful mistake happens that, ultimately, starts the process that gives birth to the human species.

Humans have forty-six chromosomes. But this proto-chimp—like all great apes other than humans—has forty-eight. When proto-chimp egg and sperm meet, they each should therefore contribute twenty-four chromosomes to the baby. But in this egg, two of the mother's chromosomes accidentally have crashed into each other and fused into one, leaving it with twenty-three. The embryo would have only forty-seven chromosomes, one less than its ancestors. This is how Kurt Benirschke's biblical tale begins, and most scientists see eye to eye with him until this point. The academic fur begins to fly when it comes to the rest of the story.

Let's say this offspring proto-chimp with forty-seven chromosomes was Adam rather than Eve. Both chimps and humans have two copies of each chromosome, which are only divided in half in sperm and egg cells. But forty-seven cannot be evenly split. When faced with odd numbers of chromosomes, the biological cleaver that separates chromosomes—the process of meiosis—makes lots of mistakes. Some of Adam's sperm cells would have had twenty-four chromosomes, which would have allowed him to father normal proto-chimp babies. Some of the sperm would have had twenty-three, with the same recombined chromosome he had. But others might have had twenty-five, with a double of one of the twenty-four chromosomes, a dilemma that causes Down syndrome and other abnormalities in both humans and chimps, but that more often than not leads to a miscarriage. Indeed, in humans, odd numbers of chromosomes, a condition called "aneuploidy," is the main cause of miscarriage. (Some might have had twenty-two, a complement that is missing yet another chromosome, an even more lethal scenario for an embryo.)

Now Adam grows up and mates with another proto-chimp, and they have babies. One of their offspring also is born with forty-seven chromosomes and survives. The same process continues through the generations until there are many forty-sevens. As centuries pass, two individuals with forty-seven chromosomes mate, each contributing twenty-three and creating the first hominin with forty-six chromosomes, the "Neo Homo" in Benirschke's lingo. As more and more Neo Homos come into the world and mate with each other, it creates "a reproductive barrier" between humans and chimps. Basically, a forty-six that mates with a forty-eight will have even more miscarriages than a forty-seven and a forty-eight, which, after all, make normal chimp

Fusion

Chromosomal Fusion

Human chromosome 2 is a fusion of two chimpanzee chromosomes, which joined at their tips. The two chimp chromosomes look similar to their homologous chromosomes in gorillas and orangutans.

ADAPTED FROM JOSEPH G. HACIA, "GENOME OF THE APES," *TRENDS IN GENETICS* 17:637–45 (NOV. 1, 2001).

babies a relatively high percentage of the time. The chromosome-dropping Neo Homo might sound far-fetched, but back in 1972 a French research team published convincing evidence that what then were called chimp chromosomes 12 and 13 had fused to form human chromosome 2.[10] The paper, published in French, at the time received next to no attention outside of a small cadre of scientists.

The architecture of human and chimpanzee chromosomes differs in other ways that may well have driven the wedge between species deeper still. Chromosomes can break in two places, allowing a chunk of DNA to float out, flip, and reinsert itself. This is called an "inversion." Nine human chromosomes have inversions from their similar, or "homologous," chimp chromosomes. These architectural changes make it more difficult for humans and chimps to have fertile offspring. The problem occurs during the manufacture of new sperm and eggs in the offspring: homologous chromosomes from the mother and father line up with each other and swap bits of DNA. Such "rearrangements" between the homologs lead to genetic errors, many of which are lethal.[11]

When I suggested to Nick Patterson and David Reich that rearrangement—and in particular the fusion of chromosomes 12 and 13—was *the* speciation event that finally separated hybrid human/

chimpanzees and humans, to my surprise, neither seemed to have given it much thought, and neither had a particularly strong argument about how the final speciation happened. "We know very little," Patterson said. "Many of the key steps are entirely mysterious. And it may remain so for a very long time."

—∞—

IN late November 2006, on a cold Moscow day that kept switching from rain to snow, I met Kirill Rossiianov, a historian of science, to discuss the similarities between modern chimpanzees and humans, as well as an even more outrageous possibility: that they could mate. We met in a tea shop around the corner from Red Square. The topic was so outré that I felt like we were Cold War spies trading state secrets and should whisper.

I had contacted Rossiianov because in 2002 he published a remarkable account of Il'ya Ivanovich Ivanov, a Russian scientist at the turn of the twentieth century who became so intrigued by the idea of a chimp/human hybrid, a humanzee, that he wanted to breed one.[12] Ivanov, a pioneer of artificial insemination, had created hybrids of several other species, including zonkeys, zorses, and zubrons (a hybrid of a wisent and a domestic cow). He worked tirelessly to find support for his humanzee experiment.

In 1927, Ivanov's dream came true.

Over a ten-year period, Rossiianov unearthed Ivanov's diaries and lab notes from the Soviet archives. He tracked down Ivanov's correspondence with scientists and various supporters in the United States, Europe, and Cuba. Rossiianov, a shy man, told me Ivanov's work repulsed him. "What do you think about the ethical dimension of Ivanov's experiments?" he asked without waiting for my answer. "Because, I dare say, I found them disgustful. Even now I find it terrible difficult to understand."

The interest in breeding humanzees traces back to at least 1900, the year that science first recognized the unusual similarities between the blood of humans and other apes. The German physiologist and anthropologist Hans Friedenthal had taken samples of blood from an orangutan and a gibbon, a lesser ape, at the Berlin Zoo and mixed the drops with human blood. Nothing happened. When he did the same experiment with blood from monkeys, the human serum appeared to destroy the red blood cells from the monkeys. This led Friedenthal to suggest that not only was Darwin right about humans having descended from apes, but it might be possible for a human and an ape to have a child.

By 1903, Élie Metchnikoff, a father of immunology who worked at the Pasteur Institute in Paris, had decided that the overlap between humans and chimpanzees made our next of evolutionary kin an ideal model to test cures and preventive therapies for infectious diseases. Metchnikoff began importing chimpanzees from Africa to study syphilis, tuberculosis, and typhoid fever, but he never had enough animals for his needs.

The Prussian Academy of Sciences in 1912 established a chimpanzee colony in the Canary Islands, where the psychologist Wolfgang Köhler, a cofounder of gestalt therapy, conducted pioneering studies of chimp behavior. As Rossiianov documents, the colony nearly became home to the first chimp/human hybridization experiment. The German sexologist Hermann Rohleder hoped a hybrid could make the case for evolution, and pushed for breeding. He published his ideas, but the project never moved forward because Germany's economy collapsed and, with it, support for the colony (which closed in 1920).

In 1915 Rosalía Abreu had attracted international attention when her colony at Quinta Palatino produced the first chimp ever born in captivity. Metchnikoff corresponded with her to learn about the sexual habits of apes, but she had little reliable information, insisting they were monogamous. That same year, the Pasteur Institute started its own chimpanzee colony in French Guinea. The directors of the institute knew of Ivanov, who had given several talks in Paris and once worked at the Pasteur, and in 1924 they wrote him that it would be "possible and desirable" for him to conduct his humanzee experiment at their colony. They had nothing to offer him other than the chimps, though, and he could not afford the travel and other expenses on his own.

In the summer of 1925, Americans became riveted by the trial of the high school teacher John Thomas Scopes for defying Tennessee law and teaching Darwin's theory of evolution. Journalists, partisans in the evolution versus creationism debate, and huckster hordes flocked to the small town of Dayton, including two men who brought chimpanzees. One of the chimps, Jo Mendi—dressed in white spats, brown fedora, and a plaid suit—was offered by his owner to the defense team to help prove that man descended from apes. Among the media assembled was the unapologetically biased Science Service, a syndicate run by Edwin Slosson, who quietly was helping Scopes's defense lawyer, Clarence Darrow, prepare expert scientific witnesses.[13] Slosson learned about Ivanov's plans from one of the Pasteur Institute's directors and, like Rohleder, thought a humanzee would finally put an end to the skepticism surrounding evolution. Like a self-appointed press agent, Slosson contacted other publications to tell them about Ivanov. The stories attracted the attention of a Detroit lawyer, Howell S. England, who said he would raise one hundred thousand dollars to stage the breeding experiment.

England was affiliated with the American Association for the Advancement of Atheism, which joined in the cause. The *New York Times* ran a curious story on June 17, 1926, that quoted the head of the Atheism Association claiming that the Soviet government had invested ten thousand dollars on the experiment (which was true) and that it was under way at the Pasteur Institute's chimp colony in Kindia, New Guinea (which wasn't).[14] The story also included a lengthy statement from England that read, in part, "We are confident that hybrids can be produced, and, in the event we are successful, the question of the evolution of man will be established to the satisfaction of the most dogmatic anti-evolutionists." England went on to cite the theories of F. G. Crookshank, a mental asylum clinician who believed orangutans were most closely related to "the yellow race," gorillas to blacks, and chimpanzees to whites. Thus the different apes would best reproduce if paired with the proper human race. According to England, "several prominent American patrons of science" had become interested in the project, and he would soon travel to French Guinea himself to help Ivanov conduct the experiments.

At the time, Ivanov in fact was working in Paris at the Pasteur Institute, having returned from a short, disappointing trip to its "anthropoid station" in French Guinea a few months earlier. The institute's chimps had yet to breed in captivity, and it hoped Ivanov could lend his artificial insemination expertise to populating its African colony. What neither Ivanov nor the French scientists appreciated was that the chimps had all been taken as babies—they captured the animals by having hunters shoot the mothers—and none had yet reached puberty. Ivanov, who was not warmly welcomed by the locals, even went so far as to try to surgically remove sperm from a chimpanzee testicle. His first journey to French Guinea ended with nothing to show for it, and his dream of creating a humanzee remained as distant as ever.

Ivanov planned to revisit French Guinea and the hybridization experiment that fall, but he needed more money to do it. Thus, he proposed to his Atheism Association supporters that he come to the United States and hold a lecture tour to raise funds. As Rossiianov documented, Howell England warned him off, noting that it might "raise a perfect storm in our fundamentalist press, all insisting that you be deported." England suggested instead that Ivanov visit after the "first little anthropoid hybrid" existed. "We have enough scientists here to assure you not only a safe entrance into the country, but a welcome here." The one hundred thousand dollars never materialized.

At the Pasteur Institute, Ivanov began collaborating with a Soviet expatriate, Serge Voronoff, who had started an international craze with the "rejuvenation" movement. Voronoff fashioned himself a Ponce de León and believed that by slicing off pieces of chimp testicles and grafting them onto human scrotums he could turn back the hands of time. He performed the procedure on his own aging brother, who, in photos, appeared to have undergone a dramatic transformation. (Sigmund Freud signed up for a variation of the procedure, called "vasoligation," as did the poet William Butler Yeats.) Ivanov's interest in rejuvenation may have had less to do with the procedure than with Voronoff's plans to start his own chimpanzee colony on the Côte d'Azur, which Ivanov imagined might be a more hospitable place to launch his own experiment.

Ivanov then approached Doña Abreu in Cuba, who at first entertained the idea of his inseminating one of her females with human sperm. But she got cold feet when she realized that the island press would certainly find out, ruining her reputation.

He decided to go it alone. He had established good relations with the governor of French Guinea, who agreed to give him space at the botanical gardens near the capital, Conakry, and permission to hunt chimpanzees. With nothing more than the money he had received from the Soviet government and his twenty-two-year-old son, a medical student, Ivanov returned to French Guinea in November 1926 to breed a humanzee.

By the winter, Ivanov and his hunters had captured thirteen adult chimpanzees with the help of nets. He noted that two of the females, named Babette and Syvette, had menstruated, and decided to inseminate them with human sperm. He worried the experiment would upset his assistants, so he hid it from them, claiming to be helping the chimps with medical problems. On the morning of February 28, 1927, Ivanov and his son threw nets around Babette and Syvette to hold them down and managed to squirt male sperm, taken from an unidentified local man, into their vaginas. The Ivanovs each held a Browning pistol in their pockets, in case either animal became violent. As Rossiianov recounted in his paper, "The experiment was carried out by the two of them in a particularly brutal and hurried way, which made the description read like it was a rape."

Both Babette and Syvette subsequently menstruated. Ivanov had failed.

On June 25, they tried again with a chimpanzee named Black. This time, they knocked her out with ethyl chloride before performing the insemination.

Ivanov knew that inseminating three females, especially under these horrific conditions, had little likelihood of working. So to hedge his bets, he lobbied local authorities to allow him to inseminate hospitalized women with sperm taken from a chimpanzee. The women, he imagined, need not be informed about the experiment, and they would be much easier to deal with than chimpanzees. Astonishingly, his friend the governor actually entertained the idea and discussed it with local doctors. Ivanov wrote in his diary entries that it was a "bolt from the blue" when the governor said no, "a terrible blow."

When Ivanov left French Guinea in July 1927, he took Black, Babette, and Syvette with him, along with ten other chimps. Black and Syvette died before reaching the Soviet Union, and an autopsy revealed that Black was not pregnant.

The indefatigable Ivanov, however, persuaded the Soviet authorities to allow him to inseminate women with ape sperm. To get their approval, Ivanov, who had raised the hackles of the Soviet Academy of Sciences when it learned that he had tried to do a similar experiment without the consent of African women, agreed to seek volunteers who would live in isolation for a year. The only postpubescent male ape then alive in the Soviet Union was an orangutan named Tarzan, and tests showed that he had viable sperm. A woman, identified as G., corresponded with Ivanov and urged him to use her in the experiment. "With my private life in ruins, I don't see any sense in my further existence," she wrote Ivanov. "But when I think that I could do a service for science, I feel enough courage to contact you. I beg you, don't refuse me."

But fate worked against Ivanov again. In June 1929, Tarzan died of a brain hemorrhage.

In the summer of 1930, the Soviets acquired five more chimps, but the political winds had changed, and that December, secret police arrested Ivanov for supposed counterrevolutionary activities. He was sent to prison in Alma-Ata, in what is now Kazakhstan. He was released in 1931 but died on March 20, 1932, and that was the end of his quixotic quest.

—◦◦◦—

IT remains unclear whether humans and chimpanzees could successfully hybridize. Ivanov's experiments, as Rossiianov emphasized to me, say nothing: you wouldn't expect humans to conceive after only three attempts, especially if the inseminations were done while the women were wrapped in nets or drugged. So I put the question to J. Michael Bedford, an emeritus reproductive biologist at Weill Cornell Medical College who published a paper in 1977 that looked at the interaction of human sperm and the eggs of other species.[15]

Sperm must attach to an egg to fertilize it, and Bedford wanted to see if he could uncover clues to making a male contraceptive. He found that human sperm did not stick to the eggs of monkeys and more distantly related species, but did attach to the eggs of gibbons, a lesser ape. He never did the experiment with great apes, but he was confident that it would work. "My gut feeling—and this is just instinctive and not scientific—is you would be quite likely to get fertilization if sperm survived in the chimpanzee." Yet he questioned whether that embryo would come to term. "I doubt there would be very much development," said Bedford. "That degree of difference would not allow a viable form."

Nevertheless, Bedford's and Ivanov's experiments could not ultimately reveal the real evolutionary scenario. Chimpanzees and humans split off from a common ancestor at least 5 million years ago, and logically the two species were much closer to each other then than they are now. The true test would require turning back the clock. Fossils allow paleontologists to do just that, but to date only one chimpanzee remnant from the distant past has been found—three teeth, about 550,000 years old, unearthed in Kenya.[16] These teeth, however, offer no information about hybridization between the species.

For now, the only type of fossil record that addresses the question head-on is the DNA of modern humans and chimpanzees. And what it says, if Patterson and Reich are correct, is that hybrids played a larger role in human evolution than Ivanov or anyone else ever imagined.

—◦◦◦—

DOC Bhagavan Antle's many gigs include parties—an elephant parading at an Indian wedding, a tiger posing for photos with guests at a bar mitzvah—as well as the venerable Miami theme park named Parrot Jungle Island where he displays a liger and a gigantic "crocosaurus" that's a hybrid saltwater/freshwater crocodile. His work involves a lot of animal shuttling, and he invited me to join him on a road trip from Myrtle Beach to Miami, where he was exhibiting a liger at a fund-raiser for a panther sanctuary and, later, at a Super Bowl party.

We met at the Myrtle Beach compound, where Antle and his team led Hercules the liger and two tigers into a large trailer with small windows. Antle and I traveled in an RV, along with three assistants and a diapered, nine-month-old orangutan, another hybrid. In the wild, these highly endangered great apes only live on the islands of Borneo and Sumatra, which means they never interbreed and, under the definition of geographic isolation, are considered two separate species. The infant, sporting a comical mess of wild orange hair, rode in a baby sling hung around one of the assistants.

During the two-day trip to Miami, the truck-and-trailer hauling the big cats was always directly behind us. We refueled at crowded truckstops, pulled into strip malls to buy groceries, and even parked one night behind a Holiday Inn. No one noticed any of the exotic animals until we were stopped at the Florida border's agriculture station. The officer checked the paperwork and stumbled on the word "liger."

"What's that?" asked the man, a good ole boy with Brylcreem hair and sideburns.

"A mix of lion and tiger," Antle said.

"Those exist?"

Antle took out his business card, which showed him sitting beside three tigers and TV talk show host Jay Leno. "Jay Leno!" said the officer. "Well, that beats all!" He waved us on without bothering to look at the liger.

Hybrids, like the humanzee, often escape detection, even when they are right under our noses.

3

IN SICKNESS AND HEALTH

I N 2001, AJIT VARKI EMBARKED ON A SEARCH FOR THE PERFECT cup of pig spit. His bizarre culinary excursion was part of an experiment that he believed might help reveal why a long list of infectious diseases, cancers, and heart ailments much more often afflict humans than chimpanzees. And it dated back to his discovery four years earlier that humans are the only apes incapable of making the sialic acid Neu5Gc.

Shortly after detecting our limitation with this sugar molecule, Varki's lab identified the reason for it.[1] Humans can synthesize another acid, Neu5Ac, which differs from Gc by just one oxygen atom. In chimpanzees and other apes, an enzyme converts Ac to Gc. Varki and his team found that humans have a mutation in the gene, *CMAH*, that codes for that enzyme.[2] So a few million years after humans and chimpanzees split from their common ancestor, humans lost the ability to make an enzyme that every mammal from chimps to mice depends on. What had driven this evolutionary change? What was the consequence? The answer to both questions, Varki suspected, was disease.

In biology, oddities often expose clues, and Varki was struck by reports that found trace amounts of Neu5Gc in human tissues and tumors. Human blood also has antibodies to Neu5Gc, an indication that the body does not make the molecule and views it as foreign. When his lab did test-tube experiments that fed Neu5Gc to human cells, the cells incorporated the sugar into newly synthesized glycans— complex sugar molecules—on the cell surfaces.

Maybe Neu5Gc enters humans through their diet, Varki reasoned. There was one way to find out: ingest huge amounts of Neu5Gc and see what happened. But when Varki asked the institutional review board at the University of California, San Diego, that oversees human experiments for permission to ingest Neu5Gc, several of its members balked. "I was at first told that self-experimentation was not allowed anymore," Varki said. He assured the board that he could prove the sugar's presence using objective measures such as mass spectrometry, which can identify substances even if they exist in tiny amounts because their ions possess what amounts to a fingerprint. He also slyly asked if any of the committee members wanted to serve as volunteers. They gave him a green light.

To get things started, Varki's colleague Pascal Gagneux determined which foods contain the highest concentrations of Neu5Gc. Gagneux went to a supermarket and purchased duck, chicken, beef, lamb, pork, cod, salmon, and tuna. From the dairy aisle, he bought buttermilk, Gruyère, buffalo mozzarella, and goat cheese. Poultry and fish, it turned out, had little or no Neu5Gc, while red meat and dairy products were loaded with it. Goat cheese won the competition, but when Gagneux tried to extract the Neu5Gc, he found that it had loads of salt that he could not eliminate. They decided to go with the next best bet: pig.

Neu5Gc is abundant in salivary glands—"sialic" comes from the Greek word *sialos*, for saliva—so they contacted Pel-Freez Biologicals, a company that supplies labs with everything from rabbit retinas to rat tails, and ordered a few kilos of submaxillary glands from pig jaws. Gagneux and a coworker minced and homogenized the pig glands to extract the mucins, the proteins secreted by mucosal surfaces. From these mucins, they plucked off the pure Neu5Gc.

For two days before the experiment took place, Varki restricted what he put in or on his body: no red meats, milk products, or lanolin shampoos, all of which contain Neu5Gc. This would not necessarily clear his body of any lingering Neu5Gc, but it would set a baseline for future experiments. Then on the morning of February 16, 2001, he checked into a clinical research center at UCSD and slugged back 150 milligrams of the Neu5Gc dissolved in 100 milliliters of water. "It was slightly sweet and sour, slightly acidic," said Varki, pig-spit connoisseur.

Varki wasn't particularly worried that the Neu5Gc would make him sick. "It was like eating fourteen pork steaks," he noted. "People do that on July Fourth." But to be safe, the clinical center kept him under observation all day, taking blood samples every two hours. No side effects surfaced. Urine, saliva, and facial hair trimmings collected over the next week all showed increased levels of Neu5Gc. Closer analyses proved that Varki's cells had actually taken up the sugar and incorporated it onto their surfaces, as they do with other sialic acids in the synthesis of new glycans. A few months later, Gagneux and another colleague, Elaine Muchmore, went through the same self-experiment with similar results.[3] "There's no other example I know of where you eat something foreign that outfoxes the biochemical systems and becomes part of you, no different from molecules made in your body," said Varki.

Part one of the mystery was solved. Neu5Gc can enter humans through the diet. And evidence that it threatens our health would steadily build, contributing to the many profound insights about human/chimpanzee disease differences that Varki and others were beginning to amass.

—◇◇—

AMID tall saguaro cacti and endless stretches of reddish sand in the Sonoran Desert, nearly one hundred chimpanzees have lived at the Primate Foundation of Arizona for some twenty years. Run by Jo Fritz, a chain-smoking, straight-talking ex-wife of a circus performer who worked with chimps, the foundation has kept meticulous medical records on all of its animals. In February 2007, Ajit Varki spent a few weeks combing through the files, trying to figure out what ailed chimps and how their afflictions differed from those of humans.

One day during his stay, I watched as Varki and his wife, Nissi, a pathologist who directs a separate lab at UCSD, went about their work. He picked me up at the Phoenix airport, and as we drove the twenty miles past Tempe and into the sparsely populated desert, he said the investigation had been a great success. "They have amazing charts, truly amazing." He particularly was interested in the causes of death. "It's completely different from humans." This mirrored the results of an earlier review he had done of chimp medical records at what in 2002 was renamed the Yerkes National Primate Research Center.

As we pulled off the highway and drove into the desert toward Red Mountain, we passed by a sand quarry, crossed a mostly dry canal, and then pulled up to a gate topped with razor wire. We were literally in the middle of nowhere. "The address is a telephone pole," Varki noted.

Fritz, who rented the land from the Salt River Pima-Maricopa Indian Community, glided up in a golf cart and let us in. The site once had hosted a hydroelectric dam, run by the Bureau of Reclamation, and though it closed in 1914, the old concrete was still visible. Fritz lived in a low-slung house, and the chimps were kept a few hundred yards behind it, in three-story cages built in and atop the old grading for the dam. She stashed a pack of clove cigarettes in a case that also held her lighter, and like everyone else at the Primate Foundation, she wore a walkie-talkie on her hip. We sat down to talk in a prefab building that served as the main office and conference room. Paintings made by chimps hung on the walls. "No chimp has ever made anything but an abstract painting," said Varki. "You wouldn't find one that anyone would say, 'Oh, look, that's a stick figure.' And a four-year-old human child could draw a stick figure." Fritz said she thought people exaggerated how much smarter chimps were than other intelligent animals. "The more we learn about dolphins and elephants, the more we see that chimps aren't so different," she said.

Fritz is not callous about chimpanzees, and she clearly cares deeply about their fate. But she knows them, at least ones that live in captivity, as well as most anyone alive: when I visited, she had fifty-seven of them, the smallest population she had had in years. She tires of people pretending chimps are almost human. "I call them animals," she told me. She made no apologies that she shipped off many of the chimps she raised to participate in biomedical research.

Fritz became a chimp person through her ex-husband, Paul, whom she met while working at the Phoenix Zoo in the 1960s. He invited her to join him on the road with his trained animal act, which included chimps, lions, and elephants. Next thing Fritz knew, she was living la vida chimp. Three friends of Paul asked if he would take their chimps, and the Primate Foundation was born. That was 1968, and the chimps lived in a downtown Phoenix apartment. By the end of the year, she and Paul had eight chimps, mostly retired entertainers. They moved to a chicken farm in south Tempe, and more and more people kept bringing them chimpanzees. In 1973, she secured a deal to move the chimps to the old dam site. Money has always been an issue, and for a time she lived on food stamps and kept her chimps fed by passing the hat at Lions Club meetings and picking up surplus produce at grocery stores. Eventually, she won grants from the National Institutes of Health, which paid for the construction of the facility's cage structure as part of an effort to breed more chimps for HIV/AIDS research. Fritz began attending meetings of the American Society of Primatologists and sat on federal government panels to decide what to do about chimpanzees in biomedical research. "I have a high school diploma, if I could find it," she said, laughing. Chimps bred so well at the foundation that most of the population today was born there. But NIH decided to stop breeding chimps in 1995 because HIV/AIDS researchers had not used them as much as had been anticipated. Now it was getting out of the chimp research business altogether, forcing Fritz to close the foundation for good.

Loud screams suddenly erupted from the fifty-seven chimps living on the grounds. "Bread," said Fritz. It was lunchtime. Chimps in the wild of course do not eat bread. Most of their diet consists of leaves and fruit, with a little meat thrown in when they can snatch a monkey or a bush pig. Here, the main diet was Purina Monkey Chow. "These guys are doing well, but too well in some cases," said Varki. "They're getting twenty-four times what's needed in iron, and a heck of a lot of calcium. Someone at Purina made this. What the heck did they know?" Diet can have a powerful impact on health. Still, the medical records of wild chimps are few and far between, and in the chimp research business, you go with what you can get and hope that it leads to meaningful insights.

Two of Fritz's staff joined us at the table: her research director, who previously had worked with wild chimps at Mahale Mountains National Park in Tanzania, and her colony director, hired by Fritz nineteen years earlier while he was a college student. Both believed passionately in their work and thought the NIH decision was shortsighted. "It's such a waste of a facility," said Jim Murphy, the colony director. "And it's a great place. Jo demands excellence in everything we do. It's sad to shut down, but that's just how it is."

I went for a walk to see the chimps up close. They were housed in eleven social groups of three to six. A few of them spit at us as we walked by, which seemed more designed to elicit a reaction than to exhibit real aggression. The chimps were family to the staff, and several had storied pasts. Simba was in the Ice Capades and, as a baby, appeared on the Osmonds' TV show, *Donny & Marie*. Jamie Lynn was born without pupils. Next door was one-eyed Willy. And then there was Tanya, who used to ride on a motorcycle and smoke pot with her owner. It was like visiting Sherwood Anderson's Winesburg, Ohio, or listening to a Tom Waits song.

Back at the office trailer, Ajit and Nissi Varki were going through the boxes that held the medical records of the seventy-eight females and eighty-one males that had lived at the Primate Foundation of Arizona over the years. Harriet, transferred from the San Diego Zoo in 1976, died at age fifty. At that time, Nissi had autopsied the body and found end-stage renal disease, a pituitary tumor, and a follicle on her ovary, which indicated that Harriet was ready to ovulate. Ajit said Harriet's kidney looked like "an old battlefield," and that this was the most common cause of death in female chimps, which live longer than males.

Flipping through the folders, Ajit noted that he had not found a single case of rheumatoid arthritis, which occurs in about 1 percent of humans. "Anybody can diagnose it: big fat fingers," he said. "In Renaissance paintings you can see that a hand had it. South American mummies have it." And he had not found any asthma. Again, he stressed that it was an easy disease to detect. "The chimps would be breathing heavily. You just put a stethoscope to them." He was shaking his head in disbelief. "Horses get asthma," he said. He also had not found much diabetes. "But what do you use to compare all of this to? There's no decent population. I'm looking for the absence of things, and at the end of the day if you never see a myocardial infarction, what do you make of it?"

Among the male chimps, the one cause of mortality that stood out was sudden death. "It's not a human type of heart attack," said Ajit. "They get a strange cardiac muscle problem. Usually it's after an exciting event. Apparently, they throw up their hands and scream and drop dead. The only way to explain it is an arrhythmia." A note in the file of a chimp named Geronimo, a male that died in 2003 at thirty years old, described exactly such an event.

> Geronimo got upset with my new boots while I was cleaning Kobi's cages. He started to display and was throwing straw/sawdust at me. As soon as I could get past him, I ran to the changing room and took my boots off and grabbed my old shoes. I ran back in with the old shoes before putting them on. When I came back in Geronimo was lying down. I called to him. He sat up slowly and looked disoriented and then fell back down. I called Erika on the radio. While I waited for Erika, I kept calling to him trying to get his attention. This all took place in about five minutes.

Ajit suspected that Geronimo had experienced a fibrosing cardiomyopathy. A condition common in male chimpanzees and gorillas, fibrosing cardiomyopathy occurs when the heart cannot expand and contract properly because its muscle fibers become rigid.[4] "If that's the case, humans have lost some gene or genetic program that tends to lead to this," he suggested.

Jo Fritz also believed Geronimo was a chimp/bonobo hybrid because of his behavior and his webbed toes; there indeed had been one report of a bonanzee born in a French traveling circus.[5] Bonobos tend to stand upright more than chimpanzees, and Geronimo's suspected parentage led Ajit to mention that chimps don't have the back problems that afflict humans. "We are still paying the price for bipedalism today," he said. "Just imagine how bad it must have been at first. My theory is there was no choice originally. You had to be bipedal or you died. With all these other theories, there should be many bipedal mammals out there. But there's only one that stood up." Ajit was on a roll, and he let his imagination race with an idea that echoed a hypothesis popularized in *The Aquatic Ape*—a highly speculative book by a journalist that he and other scientists thought was wide of the mark. "You need an island that has dangerous predators, but the predators hate water. So the only place to live is in the shallows. Because you're standing in water, your displacement is not as difficult. And you know there's another difference between humans and apes: acne. They have lots of sebaceous glands, but they don't get acne. And the interesting thing is acne is mainly in the forehead, neck, and down to the shoulders. If you're in water, those are the only things that are exposed."

Ajit knew that he was jumping in the deep end, guilty of piling speculation upon speculation, but he was having fun, thinking out loud, not pushing an argument. "There's only one case where humans walked on all fours," he concluded. "It was two girls who lived with wolves."[6]

But how to make sense of why humans but not chimps suffer from backaches, acne, asthma, or rheumatoid arthritis? Or that a disproportionate number of female chimps die from kidney failure, and males from fibrosing cardiomyopathy?

—◦◦◦—

In 1903, Élie Metchnikoff, the researcher at the Pasteur Institute in Paris, won five thousand francs from the Madrid Medical Congress. His colleague and soon-to-become director of the Pasteur, Émile Roux, that same year won the Olfa-Ilziris Prize—one hundred thousand francs. The scientists decided to pool their unexpected income to run experiments together, and at the top of the agenda was the purchase of chimpanzees, which then cost up to a few thousand francs a head. In what would become the first biomedical experiment ever conducted using chimps, the researchers showed that the apes could catch syphilis from humans. At the time, syphilis infected 16 percent of the population in Paris. The disease does not afflict chimps in the wild, but Metchnikoff and Roux thought that if they could spread the infection from a human to a chimp, they might be able to develop a diagnostic test, discover what caused it, and identify treatments and even a way to prevent it.

First, the scientists demonstrated that they could infect a chimp penis by rubbing it with goop taken from the penis of a human who had syphilis. Next, they passed the infection from chimp to chimp. Interested in how infection set in, Metchnikoff also apparently applied syphilitic material to a chimp ear, which he cut off twenty-four hours later. The disease did not spread, and as told by Paul de Kruif in his 1926 classic, *Microbe Hunters*, Metchnikoff declared that the "germ lingers for hours at the spot where it gets into the body—now, as in men we know exactly where the virus gets in, maybe we can kill it before it ever spreads."[7] With these results in hand, he proceeded to test a localized treatment, calomel and lanolin. After putting syphilitic material onto the genitals of two chimps, he treated one but not the other. Lesions only developed on the untreated chimp.

In a daring and ethically dubious experiment even by the standards of the day, Metchnikoff and Roux in 1906 agreed to test their treatment on a medical student in Paris who had read their work and urged them to use him as a human guinea pig. They injected material from a syphilitic chancre into the student's penis and then rubbed their ointment onto his genitals for five minutes. Over the next three months, repeated exams showed no evidence of a syphilitic lesion.[8] It was a stunning success. Later that year, Metchnikoff won the Nobel Prize for his work on immunology.

Others had trouble repeating Metchnikoff and Roux's work, and in 1910, an arsenic-containing drug debuted, Salvarsan, that specifically targeted *Treponema pallidum*, the bacterium that causes the disease. Developed in the lab of another Nobel winner, Paul Ehrlich—the subject of the 1940 film *Dr. Ehrlich's Magic Bullet*—Salvarsan became the first chemotherapeutic and outsold every other drug on the market until the development of penicillin.[9]

The discovery of Salvarsan owed nothing to Metchnikoff and Roux's tests with chimpanzees; the drug had been deemed effective based on a "rabbit model" of the disease. This was a lesson that repeatedly would escape researchers who assumed that because chimpanzees were closer to humans than any other species, they made the best animal for medical experimentation.

—◁◇▷—

METCHNIKOFF and Roux were not entirely off base. *Treponema pallidum*, after all, can cause syphilis in chimpanzees. But some pathogens that wreak havoc in human populations cause no detectable disease in chimps. Further confounding attempts to understand human infectious disease by studying chimps, both species independently can evolve to dodge a bug, and bugs, in turn, can evolve to dodge the roadblocks that their would-be hosts erect. Consider the perplexing efforts to make sense of malaria.

At the start of the nineteenth century, malaria killed so many Europeans living in West Africa that the region gained a reputation as the "White Man's Grave." According to military records, which provide the best indicator of malaria mortality of foreigners then living in West Africa, between 1819 and 1836, 48 percent of newly arriving British soldiers in Sierra Leone died each year, mainly from malaria. By the turn of the twentieth century and the widespread use of quinine as a prophylactic and a treatment, that figure had dropped to 4 percent, but it was still high, and malaria research attracted intense interest.[10]

In 1920, a German researcher working in Cameroon, Eduard Reichenow, reported finding the main cause of malaria in humans, the protozoan parasite *Plasmodium falciparum*, in chimpanzees. Reichenow went so far as to suggest that because the parasite moved via the mosquito, chimps were the "reservoir" for the lurking disease. Two years later, Donald Blacklock and his student Saul Adler, who worked at the recently opened branch of the University of Liverpool School of Tropical Medicine in Freetown, Sierra Leone, set out to prove Reichenow wrong.[11]

To test their hypothesis, Blacklock and Adler first took blood from a chimp suffering from a bad case of malaria and injected the animal's parasitized blood into themselves. Luckily for them, nothing happened.

To cover their bases, the next year they took blood from a human infected with *Plasmodium falciparum* and injected it into a three-month-old chimp. Again, nothing. This led Blacklock and Adler to conclude, correctly, that a different *Plasmodium* infected chimpanzees. They named it *Plasmodium reichenowi*, a rare instance in infectious disease research where the guy who got it wrong still had the honor of having his name attached to the bug. (A similar study, reported in 1939, again found that *P. reichenowi* caused no harm to humans that were intentionally infected with it.)[12]

Although antimalarial drugs improved steadily during the twentieth century, the various plasmodia that harm humans have developed resistance to most every medicine thrown at them, and malaria remains a major health threat, each year sickening up to 500 million people and killing 1 million.[13] So researchers continue to study malaria intensively, unlike, say, polio, which has nearly disappeared because of powerful vaccines. In the 1990s, when investigators compared *reichenowi* and *falciparum* on the molecular level, the plot thickened. *P. falciparum* is one of four strains of malaria that infect humans and, as it turns out, it is closer to the chimps' *reichenowi* than are the others. *Falciparum* is also the meanest: it causes 80 percent of the malaria cases each year and 90 percent of the deaths. The strain also has more in common with *reichenowi* than with any other identified plasmodium in birds or monkeys.

The evolutionary biologist Francisco Ayala was the first to unravel the genetic similarities between *reichenowi* and *falciparum*. In 2008, he presented the most compelling explanation yet for the connection between chimp and human malaria at the opening of the new Center for Academic Research and Training in Anthropogeny (CARTA) that Ajit Varki had spearheaded at the University of California, San Diego. (Anthropogeny is the study of human origins.) Ayala told of how he had become interested in the origins of *falciparum* after reading a 1991 paper that said it was most closely linked to an avian plasmodium, *gallinaceum*, and likely moved into humans with the development of agriculture and the domestication of the chicken, about ten thousand years ago. The authors of that study argued that "burgeoning human populations altered local environments and served as a vast and stationary blood-meal source for the mosquito."[14] Ayala did not buy it. His rough calculations suggested that tens of millions of years separated *falciparum* from *gallinaceum*.

Ayala wrote the authors of the study and suggested they compare *falciparum* to *reichenowi*. "They were not interested," he recalled, chuckling. So he did the analysis himself. "We just got the sequences, lined them up, and counted the differences," he said. "It was apparent to me immediately that *falciparum* was very similar to *reichenowi*." He noted that there were far fewer differences than between *falciparum* and the other human parasites or the bird parasites.

Ayala explained that some questioned whether *falciparum* and *reichenowi* were one and the same. At the time he did his initial comparison, scientists had only one strain of *reichenowi* available for study, and he could not easily dismiss this possibility.[15] But a push to isolate *reichenowi* from ninety-four captive and wild chimps unearthed eight new strains. One of the strains came from Loukoum, a chimp from the Taï Forest in Côte d'Ivoire that died in 1999 from a baffling respiratory disease. (Pascal Gagneux, the associate director of CARTA, whispered to me, "I followed Loukoum around for a year.")[16] Ayala had conclusively solved a mystery that had sparked debate for nearly a century: *reichenowi* was different from *falciparum*, and the nine *reichenowi* strains had more diversity than in *falciparum*, a strong argument that the chimp strain came first.

Ayala and his colleagues now had the data to describe the evolution of *falciparum* in more detail than ever before. If they were right, at the time of human/chimpanzee speciation some 5 million years ago, *reichenowi* infected both species. But then a few million years ago, something happened that made humans resistant to *reichenowi*, and *falciparum* emerged. For a while, *falciparum* remained restricted by geography because of the glaciation that occurred during the most recent Ice Age; the mosquito, which serves as the vector for the disease, relies on warm climes. As the glaciers receded, mosquitoes, and the parasite, spread from Africa to the Mediterranean and the Middle East. The advent of agriculture further aided the mosquito. Only one *falciparum* made it through this genetic "bottleneck," and all current strains link to what Ayala has dubbed "Malaria's Eve."

But the history Ayala has cobbled together with the equivalent of "fossil" DNA still left a major question: What drove the resistance to *reichenowi* and created a niche for *falciparum* to evolve into one of the biggest killers of humans? Ajit Varki thought he had found the engine for the change: the *CMAH* gene mutation, which cripples the enzyme that transforms Neu5Ac to Neu5Gc.

Malaria does its damage by destroying red blood cells, or erythrocytes. A plasmodium first binds to an erythrocyte's surface and then invades the cell. Varki, working with Gagneux and others, noted that scientists had discovered that *falciparum* relies heavily on one surface molecule on erythrocytes—a molecule that contains Neu5Ac. Surface molecules on chimpanzee erythrocytes, in contrast, have loads of Neu5Gc and a sprinkling of Neu5Ac. In test-tube experiments, Varki's team showed that *falciparum* heavily favored human erythrocytes and *reichenowi* much more readily bound to the chimp red blood cells. They then fed Neu5Gc to human cells, and the acidic sugar appeared on the cell surfaces and inhibited *falciparum* binding.

The owl monkey provided still more evidence of the attraction between NeuSAc and *falciparum*. Owl monkeys are susceptible to *falciparum* and long have served as an experimental subject for studying malaria. As Varki and his fellow researchers suspected, erythrocytes from owl monkeys have NeuSAc on their surface molecules, but not NeuSGc, and much more readily bind with *falciparum* than *reichenowi*. Given that the human and owl monkey surface molecules likely differ in many ways, this furthered the case that *falciparum* had a sweet tooth for NeuSAc.

Evolutionary biologists constantly have to guard against "just so" stories, the fitting of the few facts they can accumulate about events that took place millions of years ago into coherent narratives that ultimately are too tidy. Varki realizes this, and when he wrote about his malaria findings, he dressed up his big-picture take on the evolution of *falciparum* with phrases such as "it is tempting to now propose" and "it is reasonable to suggest."[17] Setting aside the caveats, Varki believes the *CMAH* mutation that evicted NeuSGc from human red blood cells occurred about 3 million years ago and provided protection from *reichenowi*, effectively eliminating malaria from the population. But the subsequent increase in NeuSAc on the surface of these cells opened the door for *falciparum*.

The small community of researchers who study the origin of *falciparum,* as it turned out, had little time to enjoy their puzzle-solving prowess. An outsider to the field, Beatrice Hahn from the University of Alabama at Birmingham, in 2010, led a team that provided convincing evidence that gorillas, not chimps, harbored the ancestor to human *falciparam*.[18] Hahn's group, which specializes in AIDS research, studied a whopping 2,739 fecal samples from chimpanzees, gorillas, and bonobos for DNA from *Plasmodium*. They had 473 hits, all from chimps and gorillas. The chimpanzee sequences of *reichenowi* clustered together in an evolutionary tree as expected. But surprisingly, none showed a close relationship to human *P. falciparum*. A gorilla branch of the *Plasmodium* family tree neatly included all human *P. falciparum* sequences. (The nomenclature of the new sequences has yet to be sorted out, and for now they are simply referred to as gorilla *P. falciparum*.) Ayala said he was convinced that Hahn's team had it right. "It's very impressive," he told me. "This does not invalidate most of the previous work but adds considerable new information, and the picture changes in a constructive way." Varki argued that the *CMAH* mutation still drove the emergence of deadly human *falciparum*. "If I put that *falciparum* from gorillas into humans, I'd bet it wouldn't make them sick," he said.

If the gorilla story holds, it underscores a recurring problem: The close relationship between chimpanzees and humans can easily obscure relationships between us and other great apes that better explain our evolution.

Humans developed another way to ward off malaria that is one of the clearest pieces of evidence of evolution in our species: a mutation in a gene that hardens red blood cells into a sickle shape, rendering them less vulnerable to *falciparum*. This trait, which is much more common in Africans and their recent descendants, causes the disease sickle-cell anemia.

Human evolution away from malaria susceptibility—through both the sickled cell and the Neu5Gc shortcoming—exacted a steep price.

—◦◦◦—

ALTHOUGH differences between humans and chimpanzees make the two species vulnerable to different diseases, pathogens take advantage of our similarities, too, often in unexpected ways.

On June 30, 1958, three members of the Holloman Air Force Base in Alamogordo, New Mexico, arrived in Yaoundé, Cameroon, to pick up chimpanzees that a vendor had captured for them. The U.S. Air Force wanted the chimps for experiments deemed too dangerous for humans, a scheme that rested on the assumption that their similarities to us made them the best "animal model" for humans. They did not sufficiently factor in, however, that when it came to infectious diseases, those similarities could cause the human handlers harm.[19]

The Holloman team did not want the chimps for the study of disease. The men worked in the base's Aeromedical Field Laboratory, which had begun a research program with chimpanzees that aimed to prepare humans for space travel. The experiments subjected chimps to extreme situations that could cause serious injuries, even death. As a *New York Times* story from that January reported, twelve chimps had been strapped into sleds that rocket motors shot down a rail at up to 727 miles per hour. Brakes then abruptly stopped the sleds,

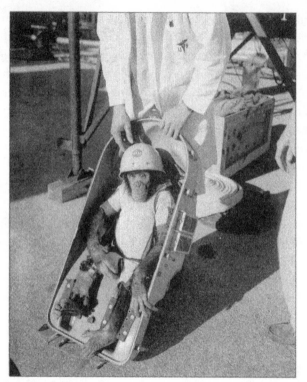

Before NASA sent humans into space, the U.S. Air Force conducted studies on chimpanzees to assess the physical and psychological impact of the weightlessness and G-forces. Ham, the first chimponaut, rode in a Mercury Redstone rocket for 157 miles without harm, but some of the chimps used in the space program died during testing. The chimps also infected their human handlers with hepatitis.
PHOTOGRAPH COURTESY OF NASA.

which "hurled the occupant against the harness with a force many times its own one-hundred-pound weight." A "G" is a force equal to ten times the occupant's weight. Chimps did not receive "any significant injuries" until 135 Gs—a force more than three times greater than what the researchers expected humans to experience from space flight. "A force of 237 Gs was required before the rocket ride proved fatal," the *Times* noted. The paper did emphasize that the chimps had been anesthetized and that the experimenters had followed procedures spelled out by the American Medical Association.[20]

While vetting subjects for their experiments, the air force team spent ten days in Yaoundé. They examined the health of the chimps, which of course required handling them. They then selected twenty-two chimps and flew with them to Holloman, transporting the animals in the passenger compartment. Two of the chimps died during their forty-five-day quarantine at Holloman, and their deaths were attributed to a combination of pneumonitis, a lung inflammation that typically is not caused by pneumonia, and unspecified parasites.

In August, one member of the team that evaluated the chimps and transported them to the base became jaundiced and was hospitalized with viral hepatitis. Three weeks later, a second team member ended up hospitalized with the same symptoms.

In 1959, four airmen twice went to Florida to collect fourteen chimps from a rare bird and game farm. All but four of the chimps died during the quarantine period, again due to parasites and pneumonitis. Two more of the group that moved the chimps developed hepatitis.

The next year, a two-man team went back to Cameroon to pick up fifteen more chimps. All of these animals had received infusions with human blood, which a local veterinarian believed would protect them from human diseases. (The other chimps from Cameroon may have received the same putative prophylaxis, but there is no record of it.) Three chimps died shortly upon their arrival in New Mexico, and, again, one of the airmen developed hepatitis. Three more people working on the base soon developed the disease, too.

All told, more than half of the staff at Holloman that intimately handled these chimpanzees developed hepatitis. Nearly half of the sixty-two newly acquired chimps died before or during quarantine. What had happened? The hepatitis outbreak occurred in people who went to Africa and in those who went to Florida, so geography alone did not explain it. And people at Holloman who did not work with chimps had no hepatitis. Studies that intentionally had tried to infect chimpanzees with hepatitis had shown that they were not susceptible to the disease. In 1961, William D. Hillis, an air force captain and medical doctor, published a study of the outbreak, concluding that the animals may have had "inapparent" infections. "It may be totally erroneous to assume that chimpanzees, if 'infected' with human hepatitis virus, will react pathologically in a fashion similar to that in human beings," Hillis wrote. "It is entirely conceivable that human hepatitis viruses may exist or even propagate in animal reservoirs without producing in their hosts evidences of disease as seen in human beings."[21]

"Hepatitis" is a catchall term indicating liver disease, and today, scientists have described five causative viruses, identified by alphabetical letters, that have little in common with one another. At the time Hillis made his analysis, no one had isolated *any* of these hepatitis viruses. When I spoke with Hillis in 2008 he supported the widely held suspicion that the Holloman outbreak was caused by what today is known as hepatitis A, which is readily transmitted through fecaloral contact. It does not naturally occur in chimps. Indeed, no hepatitis virus in chimpanzees causes the type of liver damage that is seen in humans. So chimpanzees can aid and abet pathogens that harm humans without suffering any harm themselves.

—◦◇◦—

DESPITE the fact that hepatitis viruses cause severe disease only in humans, scientists have used more chimpanzees in the study of that disease than for any other biomedical purpose. Researchers routinely herald these experiments as the strongest evidence that biomedical research with chimpanzees has great value. *Chimpanzees in Research*, a report published in 1997 under the auspices of the U.S. National Research Council, went so far as to declare that these hepatitis studies have had "enormous benefit to humankind."[22]

The first great success came against hepatitis B, a virus that spreads between humans through sex, contaminated needles, blood transfusions, and childbirth. Roughly 90 percent of the adults who become infected by hepatitis B will clear the virus, but it chronically infects some 350 million people worldwide and causes 80 percent of all liver cancers.[23] More than 250 chimpanzees were used in the development of a hepatitis B vaccine, which came to market in 1982 and today is a standard childhood immunization.[24] (Hepatitis B does infect chimpanzees in the wild, and some evidence suggests the chimp version of the virus evolved into the human one.)[25]

Chimpanzees similarly helped researchers solve the mystery of what was once called non-A, non-B hepatitis, a disease that surfaced in people who received blood transfusions but did not have either of the known hepatitis viruses. As a 1989 report in *Science* explained, the cause of the disease was yet another novel virus, hepatitis C.[26] The discovery of hepatitis C quickly led to a test that "virtually eliminated" spread of the virus from blood transfusions, a major public health achievement.[27] The World Health Organization estimates the virus infects 170 million people, and today mainly is spread by drug users sharing needles. One in four chronically infected people will develop cirrhosis, and each year, about 2 percent of them will develop liver cancer. To date, experiments have used more than five hundred chimpanzees in the search for hepatitis C treatments and vaccines.[28]

People who oppose using chimpanzees in biomedical experiments insist they no longer are needed in hepatitis B or C research, but leaving aside the merits of that argument, this avenue of investigation has, perhaps more than any other, lived up to Robert Yerkes's vision that our closest relatives might one day help humans solve their own biological problems.[29] The fact that these viruses do not cause serious disease in chimpanzees also underscores that the chimp model, imperfect as it may be, can yield great benefits to us without causing great harm to them.

—◇◇◇—

SOME human viruses that infect chimpanzees *can* cause great harm to chimps.

Jane Goodall recognized an outbreak of what looked like polio in the chimpanzees she was studying at the Gombe Stream reserve in Tanzania in 1966. Although no samples were taken to confirm that poliovirus was to blame and that it had passed from humans, six chimps died and six others were paralyzed. As a 1999 paper on the prevention of disease transmission in wild primates noted, "Poliovirus was widespread in the local human population during the 1960s, and all symptoms of the chimpanzees were identical to those found in human polio victims." Coauthored by the veterinarian D. Rick Lee (who later headed Holloman's chimp colony) and Janette Wallis (who ran chimp research at Gombe in the early 1990s), the report also documented an outbreak of respiratory disease at Gombe in 1996 that killed eleven chimps, as well as the spread of scabies there in

1997 in at least nineteen chimps that led to the death of three infants, one of which had the mite of *Sarcoptes*, which causes the skin disease scabies.[30]

Not until February 2008 did researchers publish direct proof that humans had infected wild chimpanzees.[31] The data came from a group led by Christophe Boesch, who with his wife, Hedwige, has studied wild chimps in the Taï National Park of Côte d'Ivoire since 1979. I met with Boesch in his office at the Max Planck Institute for Evolutionary Biology in Leipzig. On a table, he had a collection of sticks, each of which had a number written on it, that chimps had fashioned as tools to extract honey from bee nests. Taï chimps modified their sticks specifically for honey gathering, and it was Boesch's careful, long-term, methodical studies that allowed him to observe this fine detail, as well as other facets of chimp life and death that no other researchers had reported.

Boesch is one of the alpha males in chimp research circles, and he has the requisite muscular presence with an I-don't-suffer-fools air. Originally from Switzerland, he has wavy graying hair and a broom mustache, and reminds me of Kurt Vonnegut, minus the mirth. He first studied gorillas in Rwanda with Dian Fossey, but political turmoil in that country led him to West Africa and what became the study of several different groups of chimps in the Taï National Park.

Over twelve days in November 1992, Boesch's team of observers noted that eight of seventy-four chimps in one community disappeared. Never before had so many animals vanished at once, and they suspected disease, but no one could collect blood or tissue from any dead chimps. Two years later, ten of eighty-two chimps they were following died within two weeks of one another. This time, the scientists managed to take tissue from two corpses and identified the cause: Ebola virus, which also infected and nearly killed one of Boesch's graduate students who took a field sample. "Because of Ebola I was forced to watch out for these problems," Boesch told me. "Sadly enough, we have had more deaths because of disease." They also conclusively showed that anthrax killed six other chimps in 2001 and 2002. But Boesch could not determine the origin of the Ebola or anthrax outbreaks.

In 1999, what looked like a respiratory disease appeared to kill six of thirty-four chimpanzees in one community Boesch was following, and four infants died of starvation after their mothers disappeared. Again, they took samples, but could not find the cause. Subsequent outbreaks of what looked like respiratory disease between 2004 and 2006 killed another nine chimps, and samples sent to the lab proved equally vexing. "The problem we encounter is almost all of the pathogens we find in chimpanzees are new to science," said Boesch. "The other major problem is where do they come from? Humans? We could bring it in."

After nine years of searching through the samples, Boesch and his collaborators at the Robert Koch Institute in Berlin finally identified a set of respiratory viruses—one of which was only discovered because of newly available scientific tools—that seemed culpable. A molecular analysis of the bugs revealed that both of the two killers, respiratory syncytial virus and metapneumovirus, appeared to have come from humans.

Humans must come close to chimps to spread these viruses, and Boesch suspects that the animals likely were infected by tourists traveling with ecotourism groups, which pay the habitat for the opportunity to observe chimps in the wild, and by members of his field team. The irony does not escape him. Boesch, the founder of a conservation effort called the Wild Chimpanzee Foundation that tries to protect chimps by monitoring their populations and educating local communities, is especially vexed by the idea that his own research has caused them harm.

Boesch has used the outbreaks of Ebola and anthrax in the Taï National Park to hammer home the threat of diseases spreading between chimpanzees and humans—species close enough to pick up the same viruses as they reconfigure themselves from host to host, yet distinct enough to be missing the defenses that evolved over millions of years as protection. Nothing brings that message closer to home than HIV and AIDS.

4

OF EPIDEMIC PROPORTIONS

WHEN THE AIDS EPIDEMIC SURFACED IN 1981, RESEARCHERS around the world scrambled to figure out its cause. In May 1983, a team at the Pasteur Institute published a report in *Science* that fingered a newly discovered virus, but it would take another year to amass enough data to confirm that it was the culprit. In this confusing, frightening atmosphere, researchers turned to the chimpanzee for answers.

On August 17, 1983, scientists at the National Institute of Neurological and Communicative Disorders and Stroke injected brain tissue from a human who had AIDS encephalopathy, a degenerative brain disease, into the brain of a chimp, A207C, living in Frederick, Maryland. The lab followed a strategy picked by its leader, Carleton Gajdusek, who had won a Nobel Prize in 1976 for his work with the brain disease kuru. Gajdusek believed that kuru spread among the Fore people in Papua New Guinea because they practiced ritual cannibalism of their dead kinsmen. He had proved his theory by injecting brain tissue from humans who had died of kuru into three chimpanzees. The injections caused the chimps to contract the disease.[1] Now Gajdusek and his team wanted to find out if chimps would contract AIDS if they were injected with material from humans who had the disease.

A207C became the first chimp infected with what would be called HIV, and by May 1984, researchers collaborating with Gajdusek's group were routinely infecting chimps with the virus itself. Less than a year later, they had infected twenty-three chimps with HIV, and all remained well, except for one infant that died of undetermined causes.[2] They also showed that the virus did not infect seven different monkey species.

In humans, HIV slowly and steadily destroys the immune system, eventually allowing otherwise harmless viruses, bacteria, and fungi to run riot—the stage of the disease known as AIDS. Despite the fact that HIV did not seem to cause a similar immune destruction in chimps, the species quickly became the model of choice for AIDS vaccine researchers, and in 1986 NIH established a breeding program that included Jo Fritz's Primate Foundation of Arizona, the Yerkes center in Atlanta, Holloman (which had been renamed the Primate Research Institute), and two other chimp colonies to support these efforts. In all, 350 of the 1,500 or so chimps in facilities that conducted biomedical research and received NIH support would join the ambitious if short-lived breeding program.

As years passed, researchers who argued that HIV would damage chimp immune systems if given time fell into the minority, and many of their colleagues became disenchanted with the approach. But using chimps still made some sense for testing possible vaccines. In the typical vaccine study, an inoculated animal is "challenged" with an injection of the agent that causes the disease. Because HIV can infect chimps, experimenters could see if a vaccination triggered antibodies and other immune responses that would thwart the virus before it had a chance to establish an infection, a result called "sterilizing immunity." A similar strategy had led to the discovery a vaccine for hepatitis B, which came to market in 1982 after testing in some two hundred chimpanzees.[3]

But by the early 1990s, even the AIDS vaccine researchers had little interest in the chimp model. The staggering costs of using research chimps—then, roughly fifty thousand dollars per animal—meant that experiments uniformly had too few animals to determine whether positive results were due to chance alone. (To reach what scientists call "statistical significance," a vaccine test typically needs a minimum of four vaccinated animals, with two nonvaccinated animals as controls.) And when a few AIDS vaccine groups did report having achieved sterilizing immunity in chimps, the field largely raised its eyebrows and shrugged.[4] Scientists also had discovered a related simian immunodeficiency virus, SIV, that caused AIDS in rhesus macaques, which were much more plentiful and cheaper to use. Though the program had led to nearly four hundred live births, in 1995 the NIH instituted its breeding moratorium on chimpanzees.

The relationship between chimpanzees and HIV turns out to be more complex than researchers initially imagined. In 1986, a curious report appeared in the British medical journal *Lancet* about a chimpanzee at the Primate Research Institute. Screening the blood of ninety-four animals there revealed that one, a female named Marilyn, had tested positive for HIV antibodies. Shortly after giving birth to stillborn twins in 1985, Marilyn, twenty-six, died from pneumonia and toxins related to the birth. An autopsy did not reveal any strong evidence of the immunologic or neurological damage seen in AIDS patients. The researchers tried to isolate the virus itself but failed. Marilyn, they reasoned, could have been infected in the wild, before she was brought to Holloman in 1963 at about four years of age. Or maybe experiments that infused her with human blood caused the infection. Or she might have become infected by another chimp. Confused, they stored Marilyn's brain, liver, and spleen in a freezer and simply warned their colleagues that they risked becoming infected by chimps that had never intentionally been exposed to the virus.[5]

Three years later, a French and African team working at the International Center for Medical Research of Franceville, a primate center in a remote part of Gabon, announced that it had definitively identified two wild-born chimps that had HIV antibodies.[6] The researchers, led by Martine Peeters and her husband, Eric Delaporte, isolated a virus from one chimp and parts of the same virus from another. They dubbed it SIVchimpanzee, or SIVcpz. The virus differed from both HIV and the SIVs found in monkeys, but appeared most closely related to the human virus. In 1992, the same group, working with scientists in Europe, screened forty-four chimps that had been caught in the wild and found another SIVcpz-infected animal, Noah, housed at a zoo in Antwerp, Belgium, and originally hailing from what was then Zaire.[7]

Now there were four cases of wild-born chimps that tested positive for SIVcpz antibodies, which offered compelling support for the popular hypothesis that HIV had originated in chimpanzees and jumped into humans. But the theory did not move into the realm of accepted fact until 1999, when Beatrice Hahn from the University of Alabama at Birmingham gave the keynote lecture at the opening session of the largest annual AIDS meeting in the United States, a gathering of thousands of researchers held that year in Chicago.[8]

As Hahn explained it, five years earlier Larry Arthur of Program Resources, Inc., a contractor to the National Cancer Institute, was cleaning out his freezer and came across the autopsied remains of Marilyn. He suspected that Hahn, a prominent molecular biologist who studied HIV's evolution, might have more luck isolating a virus and sent her some samples. Hahn found SIVcpz. And when she compared it to the other SIVcpz isolates, she noticed a twist. Marilyn and the two Gabonese chimps were *Pan troglodytes troglodytes*, one of four subspecies of chimps.[9] The SIVcpz in these animals when placed on a family tree with HIV looked closely related to the main strains that infected humans. And the range of *Pan troglodytes troglodytes* closely matched areas where HIV appeared to have existed in humans for the longest time: Cameroon, Equatorial Guinea, Gabon, and the Republic of Congo. Noah, in contrast, was a *Pan troglodytes schweinfurtii*, a subspecies which lives farther east in the Democratic Republic of Congo and which, because chimps cannot cross rivers, had not mixed with *troglodytes troglodytes* for thousands of years, back when the land ranges were less divided by bodies of water. Hahn believed SIVcpz had passed into humans again and again—most likely through hunting and butchering—and only established the foothold that had caused the AIDS epidemic in recent times. "These transmissions have been going on forever and usually didn't do anything: they may have killed the person, but not gone anywhere," Hahn told me. "Then an infected person got out of West-central Africa into an urban area, Kinshasa or Brazzaville, and seeded an epidemic in a sprawling urban environment. Sex, sex and needles, who knows what accelerated the spread. Then people infected from the first case fanned out, probably up the Congo River." Reporting on Hahn's finding, a *New York Times* story went so far as to assert, "The riddle of the origin of the AIDS virus has apparently been solved."[10]

At the retrovirus meeting the next year—which was becoming something of a conference on the origins of HIV—Bette Korber unveiled a molecular clock analysis that she had conducted with Hahn and a group of other scientists.[11] By plugging in the known mutation rates of genetic sequences from old and new HIV samples, Korber calculated that an ancestor common to the virus seen in humans today dated back to about 1930, plus or minus twenty years. Her provocative finding overshadowed equally intriguing data about HIV's origins from François Simon, the chief French researcher at the primate center in Franceville, Gabon.[12] Simon described three more SIVcpz-infected chimps found in Cameroon, where he also worked. All were infants, suggesting they had been infected by their mothers at birth or through breast milk, common forms of transmission in humans. The origin of the AIDS epidemic was coming into sharper focus. And much to the relief of Korber, Hahn, and others, it was putting to rest a controversial theory that an early polio vaccine had triggered the epidemic.

Popularized in *The River*, a meticulously researched, one-thousand-page book by the journalist Edward Hooper, the polio vaccine hypothesis focused on an experiment conducted in the Democratic Republic of Congo and a few neighboring countries in the 1950s.[13] The vaccine contained a live, weakened version of poliovirus. Hooper contended that SIVcpz had contaminated the vaccine when the researchers used chimpanzee kidneys to grow the poliovirus. The team had used chimps to test the vaccine, but they denied ever having grown the virus in chimp tissues or organs. In addition, no old samples of contaminated vaccine existed, nor was there strong epidemiologic data—such as children with AIDS who had received the vaccine—to support the idea. Although Hooper challenged the new data, it effectively put an end to the speculation.[14]

Hahn in 2003 further clarified the murky waters surrounding the origins of the AIDS epidemic when she and Paul Sharp of the University of Nottingham showed how SIVcpz likely entered chimpanzees.[15] More than thirty different SIVs infect African monkeys, and a comparison of the genetic sequences of those viruses revealed that two had come together in a process called recombination to form SIVcpz. Both of the species that harbor these two SIVs, red-capped mangabeys and *Cercopithecus* monkeys, are chimpanzee prey, and Hahn and Sharp posited that hunting was the mode of transmission.

—◁◇▷—

THE backside of the verdant cane fields of the Kinyara sugar plantation in northwestern Uganda is framed by the tall, lithe trees of the Budongo Forest, which stretches across 250 square miles on the western edge of the Great Rift Valley that runs from Syria to Mozambique. More than five hundred chimpanzees live in Budongo, and in 1962, the British anthropologist Vernon Reynolds and his wife observed them for eight months, uncovering a fission and fusion system through which chimps repeatedly separated from a group and then rejoined it.[16] Reynolds, who became a professor at Oxford University, had hoped to develop a long-term research site in Uganda, but political unrest kept him out of the country, and he was not able to return to Budongo until 1990. Since then, Reynolds and other researchers have monitored several of the communities in great detail. In May 2008, I spent a day in the forest with Zinta Zommers, a young Canadian Rhodes Scholar studying the transmission of disease from chimpanzees to humans as part of her Ph.D. thesis.

Early in the morning, we stopped to watch a group of chimpanzees screaming high in the trees. Zommers and her field assistant tried to determine which individuals were making the racket and why. In the distance, we could hear the hoots and grunts of other chimps. Suddenly, a torrent of feces and urine came flying at us.

"Ugggh," groaned Zommers, jerking back to avoid being splattered. "They're going over to the group that just called, and often they pee and poo prior to leaving. So that's the time you've got to watch out."

Zommers was wearing a T-shirt from WILDCRU, Oxford University's Wildlife Conservation Research Unit, and she checked her shoulder for damage. She had successfully dodged the barrage, but her bundle of gear on the ground had taken a direct hit. "This is why you have an extra shirt with you," she said, breaking off a leaf and scraping chimp feces from a pocket of a long-sleeved shirt she'd left on the ground. Opportunity overtook whatever disgust she felt, and she asked her assistant to collect some of the gunk. "Sometimes the feces will land on the assistants' heads, and we collect from their heads directly," she said.

As her assistant snapped on latex gloves, scooped samples using a tongue depressor, and scraped them into what looked like pill bottles, Zommers took out a small pipette and collected chimp urine from the nearby leaves, dripping it into her own vials. She later would ship the vials and bottles to the same lab at the Robert Koch Institute that analyzed samples for Christophe Boesch. But unlike a decade earlier when Boesch needed tissue from dead chimps to determine what pathogens they carried, new technology allows researchers to pluck out viruses and bacteria from urine and feces. Although no SIVcpz had yet been found in Budongo, this noninvasive technique of screening wild chimps was radically altering the understanding of the virus's spread.

Beatrice Hahn's group again blazed the trail. In 2000, Hahn reported that her lab reliably could detect SIVcpz antibodies and viral genetic material in urine and feces. Two years later, the team for the first time found SIVcpz in the urine and feces of a wild chimpanzee, a twenty-three-year-old male in Gombe.[17] Before then, the only known SIVcpz infections were in Marilyn, Noah, and the Gabonese and Cameroonian chimps; the infections had been detected because these chimps either lived in captivity or were infants rescued after bushmeat hunters killed their mothers. Researchers routinely drew blood from captive chimps to monitor their health, but ethical concerns stopped them from bleeding wild chimps: if field scientists darted the chimps at ground level with tranquilizers, the animals might scamper up a tree before becoming sedated, and then, when the medicine kicked in, they could fall, causing serious injury or death. Similarly, netting chimps on the ground, like the Man in the Yellow Hat from *Curious George*, could cause harm. By January 2003, Hahn's team had found six more SIVcpz-infected Gombe chimps after screening fecal and urine samples from more than seventy-five individuals, but, interestingly, hundreds of samples from seventy chimps at two sites in Uganda showed no evidence of the virus.[18] The next month, they collected ninety-seven fecal samples from around Kisangani, in the Democratic Republic of Congo, and found one more SIVcpz-positive chimp.[19]

Within a few years, new findings recast the view that SIVcpz had a low prevalence in chimps. By the time of the 2006 retrovirus conference, the lab had collected more than six hundred fecal samples from ten different field sites in Cameroon. Nearly 450 turned out to be chimp (many of the others were gorilla), and sixteen individuals were infected. Most boggling of all, when they compared infected to uninfected chimps in each community, they found an SIVcpz prevalence that ranged from 4.4 percent up to a whopping 35.3 percent. "Our eyeballs popped out of our heads when we saw this," said Brandon Keele, a postdoc in Hahn's lab. Although the numbers of individuals sampled from each community were small—nine to forty-six—it still indicated that SIVcpz was not a rare occurrence and, in the most extreme case, mirrored the hardest-hit human epidemics in the world. The French team headed by Martine Peeters analyzed SIVcpz's isolated from these samples and found that two of them more closely matched HIV than any viruses yet found in chimps.

No evidence existed that SIVcpz had harmed any captive or wild chimps. But Hahn was open to the possibility. Six of the eight infected captive chimps had died, yet only Marilyn had been autopsied, and a few had causes of death, such as severe diarrhea and pneumonia, that are linked to AIDS. The wild chimps that they could identify appeared in good health—but then a decade or more can pass before HIV causes AIDS in humans. The field researchers would now keep a particularly close eye on their well-being.

Science had constructed a convincing, detailed narrative of the origins of AIDS. Chimpanzees hunted SIV-infected monkeys that passed two different viruses to the chimps, and the SIVs recombined to form SIVcpz. Hunting and butchering of chimps by humans—or maybe handling them as pets—had infected us with SIVcpz, which mutated into what we know as HIV-1, the main virus that caused the AIDS epidemic, the most devastating infectious disease to emerge in the twentieth century. (HIV-2, a relatively rare infection in humans, comes from SIV in sooty mangabeys, a species of monkeys also found in western Africa.)

Our connection to chimpanzees reaches all the way from past to future. As the story of the AIDS epidemic underscores, we still remain in the dark about many aspects of that relationship. Viruses traffic between us and them in every which way imaginable, moving from them to us (SIVcpz), us to them (respiratory diseases and possibly polio), and us to them to us (hepatitis A at Holloman). Our similarities sometimes dangerously intertwine us; but we are not them, and by paying close attention to our differences, we may come to understand the mechanisms that make humans ill.

—◦◦◦—

TUCKED away in a bucolic, forested neighborhood of stately brick houses about a fifteen-minute drive from downtown Atlanta sits the Yerkes National Primate Research Center, home to the oldest colony of chimpanzees in the United States. The U.S. Centers for Disease Control and Prevention is a few blocks away, and since the beginning of the AIDS epidemic, the CDC and Yerkes have led the way in trying to make sense of the disease and contain its spread.

Yerkes researchers conducted some of the first intentional infections of chimpanzees with HIV, and, as elsewhere, the virus did not seem to have much effect on the animals.[20] Then came Jerom (formally known as C499). Yerkes researchers infected him with HIV in April 1985, when he was three years old. At first his body seemed indifferent to the virus, the same as other chimps. Over the next two years, they injected two other strains of HIV into Jerom to see whether the virus could "superinfect." If it could, the researchers would know that whatever immunity had developed from the original infection was not robust enough to ward off different strains of the virus. Not only did Jerom become superinfected, but by 1993 his immune system was showing evidence of serious damage. HIV selectively targets and destroys white blood cells that have receptors on their surface, called CD4, and AIDS in humans is defined as having fewer than two hundred of these lymphocytes per microliter of blood. In 1993, Jerom had 390 CD4s, and the number dropped to 300 the next year.

By September 13, 1995, Jerom's CD4 count had slid to ten and he developed a severe case of diarrhea. The researchers suspected that the HIV in Jerom had somehow adapted to chimpanzees and become pathogenic. Worried that he would soon die, two days later they transfused forty milliliters of his blood into another chimp, Nathan, to see if it would again cause disease. Within two weeks, Nathan's CD4s plummeted from 1,240 to 320. At fourteen weeks, Nathan's count fell to 10 CD4s.

In February 1996, the Yerkes researchers euthanized Jerom.[21]

Nathan lived for six more years. Some questioned whether the transfusion he received might have harbored other pathogens or foreign cells from Jerom that led to his immune decline, so the researchers isolated Nathan's HIV, and in May 1997 infected two more chimps, Manuel and Tika. Manuel had been infected with another HIV strain in 1984 that had caused no harm; Tika was HIV free. Tika rapidly lost CD4s, dropping to fewer than one hundred cells within the span of four weeks. Manuel had developed immune responses from his earlier infection and fared better, but eighteen months after becoming superinfected he had only 196 CD4s. Manuel died of AIDS-related causes in 2001, while Tika was still alive in 2009 and living at Chimp Haven, a national chimpanzee sanctuary in Louisiana established for "retired" research animals.[22] "Pretty much the work has come to a halt," Frank Novembre, one of the lead researchers on the studies, told me. "For years and years and years, everyone was searching for a virus that would cause AIDS in chimps. We finally found one, and nobody wanted to use it."

In 1998, a prominent AIDS vaccine researcher, Norman Letvin of Harvard Medical School, reviewed the state of the field. Letvin noted that the virus that had been passed from Jerom (dubbed HIV_{JC}, for Jerome Chimpanzee) to Nathan (then called HIV_{NC}) "would provide an important new tool for testing vaccine approaches."[23] Alfred Prince, a scientist based at the New York Blood Center which maintained a colony of chimps in Liberia, and his colleague Linda Andrus shot off a strong rebuttal, signaling the beginning of the end for the use of chimpanzees in AIDS studies.[24] "To many of us who work with chimpanzees, the prospect of causing a rapidly progressive and fatal disease in this near-human species is abhorrent," they wrote. Letvin acknowledged that this did present a "difficult ethical dilemma," but with an estimated 40 million humans infected by HIV, he said that finding a vaccine was "an absolute priority." Prince and Andrus volleyed back with a letter signed by nine others, including Jane Goodall. This time, they also challenged the scientific merit of studying the trajectory of HIV_{NC} in chimps.[25]

The scientific debate focused on the question of what an AIDS vaccine should do. This might seem obvious—prevent AIDS—but many vaccine researchers worried about falling into the Voltairian predicament of the perfect being the enemy of the good. The perfect AIDS vaccine would completely prevent an infection with HIV by providing sterilizing immunity; a good vaccine would allow an infection but prevent the onset of disease. AIDS vaccine researchers, frustrated by their failure to create sterilizing immunity in both monkey and chimp experiments, had lowered their sights and begun to hope for a vaccine that simply taught the immune system how to contain a chronic infection and prevent the virus from harming the immune system—or at least slow the disease course. Prince and Andrus argued that an AIDS vaccine should be perfect. "Prevention of disease is not relevant," they wrote. For their money, an HIV vaccine had no worth unless it offered sterilizing immunity. Given that perspective, it made sense to keep using chimps for AIDS vaccine studies, and they could use infectious challenge viruses that were incapable of causing disease.

The chimpanzee researcher Patricia Fultz, who had overseen the infections of Jerom and several other chimps at Yerkes and subsequently moved to the University of Alabama, strongly objected to Prince and Andrus's argument. "That's one of the stupidest statements I've ever heard," she told me at the time. (Fultz had infected three chimps with a virus derived from Jerom and had some evidence that it caused immune abnormalities in two of them, but the studies never yielded clear evidence of harm.)[26] Norman Letvin rejected the contention, too. "If we have a vaccine that can make people live decades longer, we need to know that," Letvin told me.[27]

The ethical debate had less nuance to it. In the eyes of Letvin, Fultz, and other scientists, a few chimps unfortunately had to suffer to help humans combat one of the worst epidemics we have ever faced. Yet I heard no discussion about whether the experiments were repugnant. They were. Anyone who doubts this should read the account of a primate care technician who worked at Yerkes's Chimpanzee Infectious Disease Unit during Jerom's tenure there.

—∞—

RACHEL Weiss began working at Yerkes in 1994 immediately after completing her undergraduate degree. An engaged and eager young animal caretaker, she quit the job in disgust after watching Jerom's demise and death.[28] Weiss later described the facility as though it were part of the Gulag Archipelago. The windowless building, constructed of concrete slab, hosted a chimp room at its center that reminded her of a dungeon. It held eleven cages—which she called "cells"—each about ten cubic feet, and the thirteen chimps were separated into five groups. "The atmosphere in the building was bleak and timeless," wrote Weiss. "Unwashed stains from countless feces wars covered the ceiling; almost every one of the water devices in each cell either didn't run, or ran incessantly."

The chimps at first gave Weiss the standard chimp welcome, pelting her protective Tyvek suit with feces and sliming her hairnet with saliva. But they soon warmed to her, and Weiss got to know the individuals with caring specificity. Jerom, "dark in fur and face," had a bad case of diarrhea when she first met him in August 1995, and he spent much of his time on his bed board. "I'd been told that he was a jerk, and liked starting trouble especially with Manuel and being generally unfriendly with caretakers," she wrote. AIDS had begun to exact its physical tolls, leaving him with "wild, sunken, staring eyes" and an emaciated body. "Like the humans I've seen affected by wasting, I could see Jerom's skull under his skin."

Nathan had "a huge grin and a great sense of humor," and loved attention. Freckle-faced Manuel was more "aloof" and had a protruding bottom lip that gave him the air of "a pouty little boy." Tika, of slight build, was gregarious and developed the habit of "yelling into the telephone at whomever I happened to be talking to."

As Jerom's health deteriorated, Weiss's heart broke. "He was so severely weakened by the wasting that he had a difficult time holding his head up," she wrote. "He would sit with his knees drawn up and held his chin in his hand: he had to manually turn his head in the direction he wished to face. At times he would hang his head and sob quietly; other times he would climb down from his bed board and curl up in a fetal position on the floor in front of me. He never let me touch him then, but I desperately wanted to go in and hold him. My supervisor and I took turns sitting with him feeding him liter after liter after liter of electrolyte solution."

On the day that Jerom's blood was tranfused to Nathan, Weiss had agreed to help with the preparations but asked that she not be required to attend the actual procedure. As it turned out, everything happened so quickly that she never took her chance to leave. After anesthetizing Jerom and Nathan with medicine shot from a dart gun, a veterinarian took blood from Jerom and injected it into Nathan's arm. "I was horrified and I tried to walk away from the scene, but I kept coming back to watch. I fantasized about tackling the vet as she worked over Nathan."

Jerom had begun to receive antibiotics for his diarrhea, and he rebounded the first few weeks after the transfusion. Weiss became closer to him, grooming and tickling him during her visits. "I learned that Jerom was ticklish in his ribs, and that he was fairly shy," she recalled. "He smiled but it always looked as if he was trying to take it back, and when he laughed he bit down on one of his knuckles and turned away, so I couldn't see that I'd made him lose his cool." She successfully lobbied for him to have supervised visits with another chimp, but then his health declined again and they were cut off. She complained, but, in her account, was "admonished" and told to "stop caring so much about these chimpanzees."

As a substitute for chimp companionship, they wheeled a TV up to Jerom's cage and showed him videos, including episodes of *Oprah* and *Power Rangers* and the documentary *People of the Forest*, made by Jane Goodall and her ex-husband, Hugo van Lawick, about the chimps of Gombe. Jerom often put his thumbs in his ears and screamed at the TV. His health and moods continued to roller-coaster, and in early February 1996, the research team decided to euthanize him. Weiss gave him treats the day before, a Butterfinger and powdered donuts.

The morning he was put down, Weiss visited with Jerom for a while and then watched as they loaded him onto a gurney, draping him in white sheets. "I kissed Jerom goodbye as they took him away," she wrote. Weiss resigned two weeks later.

Weiss had a particularly harsh judgment of the ethics and scientific merits of the experiment, as well as of many of her coworkers. She is careful and thoughtful, and I have no reason to doubt the accuracy of her account. But I have a somewhat more generous take on the experiments. They unquestionably were brutal to the chimpanzees. Yet they did yield some insights. And I have met many people who do invasive research with chimpanzees at Yerkes and elsewhere, and none have left me with the sense that they are uncaring or cruel. To a person, they believe that their studies are for a greater good, and several have paid a steep price for their involvement in the work. Frank Novembre was one of several researchers who received correspondence laced with razor blades. Animal-rights activists also assailed him in letters placed in the mailboxes of his neighbors. "They told everybody I was this butcher of chimpanzees and that I did barbaric research," said the soft-spoken Novembre. "It's not our intention to just go out and harm animals. That's not what we're about. It's basic research. We're trying to find clues to disease and cures."

—◦◦◦—

ON November 5, 1999, the U.S. National Institutes of Health held a "consultation" to discuss a request to test an AIDS vaccine made by one of its investigators with a strain of HIV that could cause disease in chimps. The meeting in a Bethesda, Maryland, hotel conference room attracted many leading researchers in the debate, and they hashed out the scientific pros and cons, spending little time on the ethics. At the end of the day, the sentiment clearly was against the proposal, in part because many did not think this particular vaccine looked promising enough to warrant it. But as Harvard's Norman Letvin said, he still believed that it would be "ethically defensible" for the right experiment, and he urged his colleagues "not to consign a possible [sic] very powerful model . . . to the trash heap of history."[29]

As it turned out, the meeting for all intents and purposes did exactly that. Novembre, for one, said the decision to stop conducting vaccine experiments on chimpanzees did not bother him. "I think it's fine," he told me. "I wasn't the one that wanted to exploit that model. I was just providing a resource for those who wanted to exploit it. And I don't know if it was particularly good because of the lack of the number of animals you could use. I don't think it's a bad thing that it isn't going on. I don't miss the attention."

Yet Novembre shook his head in disbelief when I pointed out that common wisdom holds that HIV does not cause AIDS in chimpanzees. "Chimps *do* have the capacity to develop AIDS," he said.

—◦◦◦—

FROM the start of the search for an AIDS vaccine, investigators have lamented that they were flying without a compass. While scientists know the end goal—teaching the immune system to prevent HIV from establishing an infection, or failing that, how to limit the damage that HIV causes—they do not know which specific immune responses can do the job. Discovering these so-called correlates of protection has become the field's Holy Grail, and without it, researchers have been left to design vaccines based simply on educated guesswork. Frank Novembre and his colleagues in 2008 published a study of their HIV-infected chimps that they hope will offer at least a few points on the compass.[30]

For many viral diseases, antibodies play both first- and second-line defense, thwarting infection and, if that fails, preventing or ameliorating disease. In the case of HIV, the virus copies itself like the bunny rabbit of the viral world and makes many errors each time, spawning millions of mutants. Most antibodies directed at HIV are effective against very few of these mutants, and no antibody or cocktail of antibodies has yet been discovered that, in the poetry of science, is "broadly neutralizing"—in other words, works against the majority of HIVs. Indeed, if researchers had identified a broadly neutralizing antibody at the start of the AIDS vaccine search and found the piece of HIV that could prod the immune system to make it, a product likely would have helped avert the epidemics in South Africa, Botswana, Zimbabwe, and the other countries of southern Africa that today are the hardest hit in the world.

The antibody response that people mount against HIV *does* help knock down the level of virus. But too many mutants dodge antibodies and keep infecting and killing CD4 cells. It is these very CD4s, in turn, that tell other cells to launch attacks with antibodies and other immune warriors. That is the essence of HIV's viciousness: there's a feeble immune response from day one, and the virus plays bully, relentlessly punching the weakling until it completely collapses.

Given the limits of antibodies, AIDS vaccine researchers have tried to exploit what's called the "cellular" arm of the immune system. Antibodies latch onto the virus and prevent it from attaching to a CD4 cell's surface. The cellular arm of the system does a mop-up job, identifying the cells the virus has managed to infect and selectively eliminating them. The main weapon in the cellular arm is the killer cell, aka the cytotoxic lymphocyte, or CTL. Most leading AIDS vaccine strategies currently aim to teach the immune system to make both antibodies and CTLs against HIV, hoping that the one-two punch will prove powerful enough.

Novembre's team analyzed the immune responses of ten HIV-infected chimpanzees, comparing the five animals that remained unharmed by the virus, called the "nonprogressors," to Nathan, Manuel, Tika, and two other chimps whose virus progressed to cause AIDS. One of the key differences they found reflected an observation also made in humans. When CD4 cells bump into HIV, they should start copying themselves—"proliferating"—and begin orchestrating the production of antibodies, CTLs, and the like. Test-tube studies show that the CD4s from the nonprogressor chimps proliferated like mad when exposed to pieces of HIV, but CD4s from the progressors snoozed. Theoretically, then, a successful vaccine could be aimed at preserving CD4 proliferation. The chimp study does not reveal how to make such a vaccine, but it adds to the data from humans that proliferation is one of the compass's points. Ultimately, researchers could use that knowledge to evaluate a vaccine's potential efficacy.

The chimp studies supported another intriguing mechanism of the disease seen in both humans and monkeys. HIV directly kills the CD4 cells it infects, but it also causes mayhem by constantly kicking uninfected CD4s into action, telling them to proliferate more than necessary. When CD4s copy themselves, they die. So excessive immune activation, as it is known, leads to the death of many "innocent bystanders"—CD4s that HIV has not infected. The chimps that progressed to AIDS had much higher levels of immune activation than the nonprogressors. Again, this seemed to be a possible compass point to guide vaccine testers.

One similarity in the progressors and the nonprogressors offered a final cautionary note. Both groups of chimps mounted similar killer cell responses to HIV. Therefore, CTLs, at least in the chimpanzee, do not provide protection. That does not mean CTLs are unnecessary or bad. But it does question whether they should retain their lofty status in HIV studies, some of which have deemed them the primary measure of a vaccine's worth.

AIDS vaccine researchers may yet discover a way to stimulate broadly neutralizing antibodies or superpotent CTLs. Or maybe the first successful vaccine will, as many in the field believe, combine both approaches. But after more than two decades of failures in identifying a vaccine, an increasing number of investigators realize that the correlates of protection may not be found among the usual suspects.

—◦∞◦—

AJIT Varki has examined the correlates-of-protection question from another angle. He sees the chimps that progressed to AIDS as one-offs, exceptions to the rule. Rather than comparing the likes of Nathan and Manuel to unharmed HIV-infected chimps, Varki ran a comparison that assumed from the outset that humans were progressors and chimps nonprogressors. In keeping with his other work, Varki looked at the issue through the lens of the sialic acids, and specifically Neu5Gc.

Varki contends that when the *CMAH* gene mutated in humans some 3 million years ago and they lost the ability to synthesize Neu5Gc, they experienced what he calls a "sialoquake," shaking up several genes involved with sialic acid biology. Such a quake would have led to changes in the molecules on cell surfaces that bind sialic acids. His research has focused intensively on a family of these molecules called Siglecs, short for sialic-acid-recognizing Ig-superfamily lectins, and he has found they help to explain the difference between the way that humans and chimpanzees handle infection with HIV.

CD4s are T cells, so designated because they go to finishing school in the thymus. Siglecs are abundant on human immune cells—except for T cells. Working with Pascal Gagneux and others, Varki compared the expression of Siglecs on the T cells of humans, chimpanzees, bonobos, and gorillas. The great ape cells had tons of Siglecs, while the human cells had hardly any or none at all.[31]

Siglecs in T cells turn down the knobs that control immune responses. This led Varki to put forward the theory that genetically engineering human T cells to express high levels of one of these Siglecs would make them less prone to start the cascade of events known as activation. That is precisely what happened when he tested the theory in his lab. Conversely, he cranked up activation of chimp cells by using an antibody to block that same Siglec on them. "I think what's happening is that Siglecs are providing a brake in ape T cells," Varki told me. "Human T cells seem to have lost these brakes."

As Varki sees it, humans, lacking these Siglecs, have become "hyperreactive." When HIV confronts chimp T cells, they take appropriate action and contain the infection. Human T cells, in contrast, over-rev their engines and die premature deaths.

I ran Varki's results by Anthony Fauci, a prominent HIV/AIDS immunologist who heads the U.S. National Institute of Allergy and Infectious Diseases. Fauci pointed out that activation not only indirectly kills CD4 cells but, by creating more of them, provides HIV with more targets. "If you didn't have that robust activation, you wouldn't provide a fertile environment for HIV to replicate," said Fauci. "Varki makes an interesting story."

Varki's "story" may not turn out to be *the* explanation for the divergence between the ways humans and chimpanzees handle HIV. One subsequent study by a separate lab challenged his methodology, going so far as to claim that Siglecs were "unlikely" to be responsible for the differences in activation seen between human and chimp T cells—and that no one had yet devised a reliable test-tube model for activation in chimpanzee cells.[32] Still, in science, the question often is as important as the answer. Varki was probing a topic that AIDS researchers by and large had ignored: How, precisely, did human and chimpanzee immune systems differ so that HIV played tiger in one host and pussycat in the other? The answer might provide another compass point on the road to the holiest of grails, the correlates of protection.

—◦◦◦—

IN February 2009, Beatrice Hahn yet again revamped the thinking about the relationship between AIDS, humans, and chimpanzees. At that year's retrovirus meeting, Hahn's lab reported for the first time that wild chimpanzees infected with SIVcpz likely did develop AIDS.

The lab team had analyzed 1,099 fecal samples collected between 2000 and 2008 from chimpanzees living in Gombe Stream National

Park. They found evidence of SIVcpz in eighteen chimps. Seven of the eighteen infected chimps died during the study period, compared to ten of seventy-six uninfected animals. When they corrected for age and other variables, the scientists found that the SIVcpz-infected chimps had a fifteenfold higher risk of death than did virus-free apes, meaning the virus poses nearly as great a risk as HIV-1 does to humans.

Studies of lymph nodes from two of the infected chimps that had died showed the type of immunologic destruction seen in HIV-infected humans. And these chimps had low levels of CD4 cells, the main targets of SIVcpz and HIV-1. The researchers did not have tissues from the other dead chimps to do similar analyses, so they could not conclude that SIVcpz had harmed them. But Hahn and some primatologists were convinced. "You put two and two together and say it's probably the virus that's responsible for their mortality," Hahn told me.

The finding stunned researchers who study wild chimpanzees. "We were shocked at the initial discovery of SIVcpz in the Gombe chimps, and even more dismayed when we established that it seems to be pathogenic," said the behavioral ecologist Anne Pusey, who once was the research director at Gombe and now works at Duke University. "It must be the case that some of the mortality over the last decades has been due to SIV." During the study period, the SIVcpz prevalence ranged from 9 percent to 18 percent, similar to the devastating levels of infection seen in human populations in sub-Saharan Africa. At that level, SIVcpz may have had a significant contribution to the decline in wild chimpanzee populations.

In retrospect, it is not surprising that SIVcpz harms chimps more readily than HIV-1: viruses typically have a difficult time when they move from one host species to another. They must adapt to the new host, figuring out a way to efficiently infect cells and copy themselves. One way to adapt is to move through different individuals, the process of passaging, which gives the virus a chance to make subtle adjustments that increase its fitness in its new host. SIVcpz, as Hahn had argued earlier, likely made little headway in humans until urbanization—that is, when denser living conditions in African countries allowed it to passage quickly from one person to the next. One study put a finer point on this, showing that colonial health reports from Africa at the turn of the twentieth century documented a sharp spike in genital ulcer diseases including syphilis, which, especially in these typically uncircumcised populations, probably allowed SIVcpz to establish infections in humans more easily.[33] The growth of cities effectively would have thrown rich fertilizer on the already budding epidemic.

As much attention as passaging had received in attempts to unravel the cause of the AIDS epidemic, I never heard anyone suggest that the human virus might struggle in our ape cousins because it had not adapted to them through passaging. Researchers had assumed that HIV-1 could not harm chimpanzees because their immune systems had evolved to outwit it.[34] They based this assumption on the fact that though more than forty SIVs infect African primates, they rarely cause disease. No one had identified a chimpanzee in which SIVcpz had caused AIDS. HIV-1$_{JC}$ and HIV-1$_{NC}$ led to disease, the thinking went, only because the virus had combined two strains that together packed more punch than garden-variety HIV-1 or, presumably, SIVcpz. Hahn's new data showing that SIVcpz likely caused AIDS in chimps forced researchers to do a double take. And it also made plain that the overemphasis on our similarities came at real costs to both chimpanzees and humans: it misled scientists trying to solve one of the most catastrophic health dilemmas of our time.

The field had built many of its previous assumptions on the skimpiest of evidence. No one had ever studied the fate of SIVcpz-infected wild chimpanzees before, which was only possible at a site like Gombe that carefully tracks habituated animals over many years. Of the few chimps found to be infected with SIVcpz at primate centers and zoos, most had died shortly after being discovered and never had a proper autopsy: several were infants, which frequently do not survive when they come into captivity, leaving little reason to suspect that they had AIDS.

If SIVcpz does routinely cause AIDS in chimpanzees, it raises several staggering questions that could point to rich, unexplored terrain in the search for an AIDS vaccine. How does SIVcpz compare to HIV-1$_{JC}$ and HIV-1$_{NC}$ as well as other HIV-1 isolates? Has HIV-1 lost or gained something that makes it more deadly in humans? Or does the immune system of chimpanzees react in markedly different ways to HIV-1 and SIVcpz? How would SIVcpz behave in humans? How does it compare to HIV-2, the other human AIDS virus?

Some experiments, like intentionally infecting humans with SIVcpz or capturing SIV-infected wild chimps and taking blood samples, are unethical, and it is now hard for scientists to gain funding for any HIV studies involving chimpanzees. Even Hahn, who had received international attention for her breakthroughs, complained that she had trouble getting grants to continue her SIVcpz fieldwork. Indeed, the only researcher I knew who immediately pounced on the news was an outsider to the field, Ajit Varki, who planned to conduct test-tube experiments with SIVcpz in human and chimpanzee cells to revisit his hyperactivation hypothesis. But there was one peculiar, long-term experiment under way with SIVcpz that promised a unique new perspective.

After the deaths of Jerom, Nathan, and Manuel from AIDS, intentionally infecting a chimp with SIVcpz became so unconscionable an idea that no researchers even suggest it as a thought experiment. But just such an infection had taken place in 1995 with Noah, the chimp living at the Antwerp Zoo that was the fourth isolate of SIVcpz identified.

Noah had come to the zoo under the most unusual of circumstances. When the Belgian king Baudouin visited Zaire in 1986, then president Mobutu Sese Seko gave him Noah and another male chimp, Niko, as a gift. International treaties prohibited the shipment of wild chimps from Africa, but Baudouin could not politely refuse, so Noah and Niko were flown to Brussels, where customs officials confiscated them and sent the young apes to the Antwerp Zoo.[35] They were estimated to be two years old, and when researchers detected Noah's SIVcpz infection three years later during a large survey of captive chimps, he was in fine health with no sign of AIDS.

Following worries that Noah presented a risk to zoo staff, he was moved to the Biomedical Primate Research Center, in Rijswijk, the Netherlands. Niko, his cage mate, went, too; given their long relationship, the staff deemed it inhumane to separate them. Researchers from the primate center and the Institute of Tropical Medicine in Antwerp monitored both chimps closely, assuming that Noah eventually would infect Niko. When they became sexually mature, they began to roughhouse more, drawing blood, and the researchers in 1995 decided to intentionally infect Niko so they could carefully study the kinetics of the SIVcpz infection.[36]

Two years later, Noah and Niko had normal CD4 counts. They both had more activated immune systems than uninfected chimpanzees, but "markedly less" activation than found in HIV-infected humans.[37] In 2001, both remained healthy despite having higher levels of virus in their blood than HIV-1-infected chimpanzees, and the researchers said Noah and Niko closely resembled the small percentage of HIV-1-infected humans known as long-term nonprogressors.

The investigators closely examined Noah's and Niko's blood to see if they could discover what the chimps' immune systems were doing right. In healthy people or chimps, CD4 cells routinely are destroyed and replaced. HIV-infected humans cannot replace these cells as quickly as they're eliminated. Noah and Niko, in contrast, could keep up with the destruction caused by the SIVcpz.

The test-tube experiments could not pinpoint how this occurred, but the researchers discovered a few hints. The much-celebrated CTL, the killer cells, had not protected Noah and Niko from SIVcpz—just as had been the case when chimps were infected with HIV-1. Drilling deeper, they found that cells with CD8 receptors on their surfaces—the very cells that make CTLs—suppressed SIVcpz through a mysterious mechanism, possibly by secreting a biochemical they could not yet identify.

In 2006, researchers drew blood from Niko and tried to transmit SIVcpz, using a variety of different routes, to six chimps that earlier had been infected with HIV-1. Intravenous inoculations of the virus readily infected two of the chimps, and two others easily became infected rectally. But when SIVcpz was put in the vaginas of two animals, the virus did not take. The investigators tried a higher dose, but again, nothing happened. A chimp uninfected by HIV, however, easily became infected with SIVcpz through the vagina.

After one year, none of the SIVcpz-infected animals showed a decline in CD4 cells. Niko and Noah still remained healthy, although Noah had a bout of severe thrombocytopenia, a low level of the platelets that form blood clots, a condition that frequently develops in HIV-infected humans.

The vaginal transmission experiment strongly suggested that a prior infection with HIV-1 protected the two females from SIVcpz. If researchers could figure out why that was, it could help them identify the long-sought correlates of protection. The study's results also indicated that the disease-causing power of SIVcpz might more closely resemble HIV-2 in humans, which infects about 1 million people in West Africa. Roughly 75 percent of the HIV-2-infected population are long-term nonprogressors, able to live with the virus without AIDS setting in.[38] It could be that SIVcpz only leads to AIDS in some chimps, or that the course of disease is much slower than HIV in humans, which itself takes a lengthy ten years on average to cause symptoms. While relatively few infected chimpanzees have been studied over time, and few are in captivity and subject to close analysis of their immune systems and day-to-day well-being, the experiment quashed any lingering notions that the virus is harmless to chimpanzees.

The day after Hahn's group reported the Gombe data, I e-mailed Jonathan Heeney, a lead researcher involved in the SIVcpz studies at the Netherland's Biomedical Primate Research Center. "I have always stated that chimpanzees appear to be 'relatively resistant' to AIDS compared to humans, but clearly not overtly resistant to AIDS," Heeney replied. He by then had left the Netherlands for the University of Cambridge: the Dutch government in 2004 outlawed research on chimpanzees, and Noah, Niko, and the other chimps at the center had been relocated to a sanctuary. Heeney continued to try to monitor their health, but the sanctuary staff did not entirely trust researchers. (Niko died after receiving an anesthetic for a routine health exam in 2005.)[39] It proved difficult for him to even obtain blood samples. "SIVcpz could have subtle deleterious effects over time which may impact on certain individuals' long-term health and most certainly could influence their survival in the wild," said Heeney. "We have some data to this effect and are having a closer look at tissues from animals in cohorts we have studied. However, we urge caution, as we feel that the data based on a few cases is too borderline to make conclusive statements."

If scientists definitively determine how SIVcpz harms chimps, that will not by itself lead to the discovery of an AIDS vaccine—for either us or them. But the studies of SIVcpz-infected chimpanzees, precisely because so little is known about them, hold out the hope of clearing some of the fog in this protracted and urgent search. AIDS, more than any single biological factor other than DNA, poignantly illustrates that the fates of humans and chimpanzees are intermingled, and maybe that can be for better as well as for worse.

TWO

BRAINS

Where, then, is the difference between brute and man?
What is it that man can do, and of which we find no
signs, no rudiments, in the whole brute world? I answer
without hesitation: the one great barrier between the
brute and man is Language. Man speaks, and no brute
has ever uttered a word. Language is our Rubicon, and
no brute will dare to cross it.

—FRIEDRICH MAX MÜLLER,
lecture on the "science of language" and its origins, 1861

—◁◦▷—

Man is a kind of miscarriage of an ape, endowed with
profound intelligence and capable of great progress.

—ÉLIE METCHNIKOFF,
The Nature of Man, 1903

5

TALKING APES

I FIRST "SPOKE" WITH APES IN MARCH 2008. THEY WERE BOTH bonobos, twenty-seven-year-old Kanzi and his twenty-two-year-old half-sister, Panbanisha, and they lived at the Great Ape Trust outside Des Moines, Iowa.

The Great Ape Trust is the last bastion of "ape language research," a controversial scientific endeavor that began more than a century ago and has won tremendous attention in the press, spawned many books, and even became fodder for a Hollywood movie starring an actor who later became president of the United States. Sitting on 230 acres in a rural neighborhood that includes a man-made lake, the trust owes its existence to hot dogs, in keeping with the often bizarre nature of ape language research. Ted Townsend, a local heir to the fortune his father made from the Frankomatic machine, donated $25 million to create the state-of-the-art facility, which when I visited housed seven bonobos and, separately, three orangutans. The trust's researchers had made Kanzi the most famous bonobo in the world by showcasing his extraordinary communication skills. Before I met with him, William Fields, who directed the bonobo research, let me in on a surprising secret about his less known half-sister.

We sat in Fields's office in the thirteen-thousand-square-foot bonobo building. A hallway and a concrete block wall separated us from the area where the bonobos lived. Fields explained to me that there had only been three scientific articles published about Panbanisha's communication skills so far, and as the researchers accrue more data, "Kanzi will have to share the limelight with the real ape of genius."

I was taken aback. Fields's former boss, Sue Savage-Rumbaugh, who initiated the work with the sibling bonobos, had once said, "Kanzi has been shown to have acquired linguistic and cognitive skills far beyond those achieved by any other nonhuman animal in previous research."[1]

"You think Panbanisha is as skilled if not more skilled than Kanzi at language?" I asked Fields.

"Yes," he whispered. "And they understand parts of this. They don't understand everything."

"They're listening to us right now?"

"I'm certain," said Fields.

"How can they hear us right now? They're on the other side of the wall right there."

"They can't hear everything, but they hear certain things," he said. "Oh, yes, they can hear, and I'm very careful about what I say. And the reason that I'm careful is because they don't understand everything we say. They hear parts of it. But I try to make sure that I'm careful. You never know."

"I'm sorry to be skeptical. But there's a concrete block wall right there."

"That's what we thought," said Fields, laughing. "I don't have the scientific data to prove it to you. But Sue and I know when we have specific critical conversations about certain things we have to make sure we're at a certain distance or they're going to hit you with it."

"The kids are at the door with their ears on it."

"They are," he grumbled. "Eavesdroppers. It's disgusting."

From the beginning, humans who have made bold claims about the communication skills of our closest relatives have met with harsh criticism. Fields and other proponents of ape language research argue that the intensity of the criticism reflects how threatened humans can be by the prospect that another species might, as the German philologist Friedrich Max Müller declared, dare cross that Rubicon.

Charles Darwin shared Müller's assessment, asserting in *The Descent of Man* that language "has justly been considered as one of the chief distinctions between man and the lower animals." Darwin recognized that dogs, parrots, and monkeys use sounds, gestures, and facial expressions to communicate, but "the habitual use of articulate language," he stressed, "was peculiar to man." While he noted that humans had to learn language, he wrote that "man has an instinctive tendency to speak, as we see in the babble of our young children; whilst no child has an instinctive tendency to brew, bake, or write."[2] What made "the habitual use of articulate language" peculiar to man? Darwin could only wave his hands. "The fact of the higher apes not using their vocal organs for speech, no doubt depends on their intelligence not having been sufficiently advanced," he concluded. "If it be asked why apes have not had their intellects developed to the same degree as that of man, general causes only can be assigned in answer, and it is unreasonable to expect any thing more definite, considering our ignorance with respect to the successive stages of development through which each creature has passed."

Ape language research has attracted top scientific thinkers, including the linguist Noam Chomsky, the psychologist Steven Pinker, and the astronomer Carl Sagan, and the field has elevated Kanzi the bonobo, Koko the gorilla, and Washoe the chimp to celebrity status. It has often inspired cacophonous and hot-tempered debates. Adding to that drama, one researcher turned rogue: in the late 1970s, a leading proponent of ape language research who had taught American Sign Language to a chimp archly named Nim Chimpsky all but demolished the entire field.

The Great Ape Trust today yearns for scientific respectability, and it has credible, thoughtful investigators on its staff. Yet ape language research makes many scientists snicker and roll their eyes, a problem the field can trace all the way back to its grandly ambitious progenitor, Richard Lynch Garner, who at the end of the nineteenth century lived in a cage in Gabon to observe chimps communicating with each other and claimed to have deciphered some of their "words."

—◦◦◦—

RICHARD Lynch Garner frequently made the headlines during his lifetime. An adventurer and biologist who championed Darwin and spent years among the apes of Africa, he was lauded to have "actually acquired their rude speech and is now able to understand practically all of the communications to each other."[3] He crossed paths with Robert Yerkes, Thomas Edison, Alexander Melville Bell (the telephone inventor's father), and U.S. president Grover Cleveland. But Garner is largely forgotten today and remains unknown to many primatologists, despite his pathbreaking—if scientifically suspect—work.

Garner published three popular books between 1892 and 1900—*The Speech of Monkeys, Gorillas & Chimpanzees,* and *Apes and Monkeys: Their Life and Language*—as well as articles in such well-read publications as the *New York Times, McClure's,* and *Cosmopolitan.*[4] If not for the ethnologist and linguist John Peabody Harrington, however, much of Garner's history likely would have vanished. Peabody, an authority on Native American languages who worked with the Smithsonian Institution, at a young age became fascinated by Garner's books and the tales of his exploits in Africa. Peabody wrote a slim biography of Garner, *He Spoke,* and encouraged Garner's son to donate his father's papers and lantern slides to the Smithsonian, which stores them in a facility in Suitland, Maryland, a D.C. suburb that is also home to the headquarters of the Census Bureau. The archive is stuffed with Garner's detailed diaries, excited letters, and goofy poems, as well as with ornately illustrated newspaper clippings of Garner's time in Africa. He was a courageous, industrious, and incessantly curious man who wrote with brio and, at times, literary flourish. But Garner also could be self-pitying ("I have heard it said that we are to be born again; but I really do not want to be. It was bad enough this time"), self-glorifying ("The vast majority of men who follow science as an occupation are mere mechanics and they are the ones that I offend"), vituperative toward those who crossed him ("She is undoubtedly the most treacherous, ungrateful and unscrupulous little miscarriage that ever got loose on society"), and, by the end of his life, something of a misanthrope ("I must admit that I have acquired a certain contempt for the greater portion of the genus homo . . . the perfidy, the unblushing selfishness, the arrogant egotism of the human species are such an appalling and unexpected revelation to me").[5]

Born in 1848 in Abingdon, Virginia, Garner at age fourteen joined the Confederate army and was soon captured and imprisoned at Baltimore's Fort McHenry. As recounted in *He Spoke,* Garner was so young that a sentry mistook him for a local lad who had snuck into the fort, "and assisted by a good heavy brogan," kicked him out, allowing Garner to return to the war. But lucky breaks were few and far between in Garner's life. Not only was he repeatedly caught by Union soldiers and sent to a dozen different prisons, he spent much of his subsequent professional life scrambling to find patrons to support his outlandishly ambitious plans, and he moaned incessantly in his correspondence about not being paid for his work.

Immediately after the Civil War, Garner attended a few years of college and then became a schoolteacher, married, and had a son. Hungry for money, he took extra jobs, including battling Apaches and bringing wild ponies to market. He became interested in animal language by chance. Garner shunned religion from a young age, when he had become enamored with Darwin and the theory of evolution. At a seminar for teachers on phonetics, he suggested that human speech might have evolved from the sounds of animals. "This was instantly and violently assailed as rank heresy and I barely escaped being immolated as an infidel," he later wrote. "In defence of my assertion, I promised to adduce the evidence at some future meeting and stimulated by this tirade of opposition I resolved to study those sounds in a methodic manner and try to learn the speech of animals."[6]

In 1884, as Garner explains in *The Speech of Monkeys*, a trip to the Cincinnati Zoological Garden rekindled his desire to decode animal language. Garner became "deeply impressed" by monkeys of an unspecified species, sharing a cage with a mandrill, which has massive canines and a multicolored face that gives it the look of a tribal warrior with face paint. The cage had a dividing wall with a small passageway cut into it, and the smaller monkeys "instantly reported" to the other smaller monkeys on the other side the mandrill's every move. "I watched them for hours and felt assured that they had a form of speech," wrote Garner, who claimed that he could tell from their vocalizations whether the mandrill was asleep or moving about. "Having interpreted one or two of these sounds, I felt inspired with the belief that I could learn them, and felt that the 'key to the secret chamber' was within my grasp."

Thomas Edison in 1888 unveiled his "perfected phonograph," which had much improved sound from the one he developed a decade earlier and which ran off a battery rather than a crank. The phonograph could record as well as play sound, and Garner soon acquired one for his study of monkey speech. With phonograph in tow, he recorded monkeys at zoos in Chicago, Philadelphia, Washington, D.C., and New York City, and also sought out organ grinders and people who kept monkeys as pets. As he studied their sounds, he tried to decipher words, sometimes playing the recordings to monkeys to see if he could discern the meaning from their actions. In a sentence so long it could compete with the prose of Faulkner, Garner explained in a magazine essay, "The Simian Tongue": "I am aware that it is heresy to doubt the dogmas of science as well as some religious sects; but sustained by proofs too strong to be ignored, I am willing to incur the ridicule of the wise and the sneer of bigots, and assert that 'articulate speech' prevails among the lower primates, and that their speech contains the rudiments from which the tongues of mankind could easily develop; and to me it seems quite possible to find proofs to show that such is the origin of human speech."[7]

With the article's publication, Garner became famous, owing largely to his having hired a savvy literary agent, according to Gregory Radick, a historian of science at the University of Leeds and the world's foremost Garner buff. Radick, who carefully mined the primate philologist's archives for a compelling account of the "professor's" life, notes how the *New York World* reported that "the entire scientific world" knew of Garner's work and that it had "been written about in every known language save in the simian tongue itself."[8]

Garner contends in *The Speech of Monkeys* that monkeys have words for "drink" and "food," "love" and "alarm," and the state of the weather. But he stresses that he does not think monkeys are "capable of shaping sentences into narrative or giving any detail in a complaint" and that they "do not generally carry on a connected conversation." He by then had only had the chance to study two chimps and could reach no conclusion about their language, but said they would be "the chief objects of [his] studies in tropical Africa." Garner the self-proclaimed professor and scientist was about to become an intrepid explorer of the Dark Continent—Indiana Jones meets Dr. Doolittle—to excavate the truth about animal language.

Garner did not have the means to conduct studies in Africa. He solicited one-hundred-dollar contributions so that he could "give to the world the secret with which to pass the gates of speech," promising to share half his future income from lectures and writings with

Richard Lynch Garner proudly took a photograph of himself in front of his cage, Fort Gorilla, in requisite pith helmet and with spear-toting native at his side. This is an artist's rendering of the photograph that he included in his book Apes and Monkeys.

REPRINTED FROM RICHARD LYNCH GARNER, *APES AND MONKEYS: THEIR LIFE AND LANGUAGE* (BOSTON: GINN, 1900).

investors.[9] Alexander Melville Bell, Grover Cleveland, and Edison—who had corresponded with Garner about designing a phonograph specifically for the expedition—all opened their pocketbooks, and in October 1892 Garner arrived in Gabon.

He traveled heavy, most remarkably lugging along twenty-four panels, each three feet, three inches square, of what resembled chain-link fencing. When he reached his destination deep in the forest, he assembled the panels into a cage, which he named Fort Gorilla. The cage was not for gorillas or any other animals, though; it was a home for Garner, who thought it would afford him the best opportunity to view apes in their natural habitat while providing "a certain immunity from being surprised by the fierce and stealthy beasts of the jungle."[10] His cage turned the zoo concept inside out. According to Garner, it was his home for 112 days, and it allowed him to observe many a chimpanzee and gorilla.

Although he had hoped the phonograph would serve as his main scientific tool to study speech, in the end he did not take one with him. He blamed the problem on patent law and greed, but as his biographer Radick notes, Garner was making the expedition on a shoestring and likely could not afford one. He also had cultivated an awkward relationship with Edison. Garner had planned to visit the inventor before he left for Africa, and he outlined the custom phonograph he wanted built. But the project never moved forward. Then Garner's literary agent sent Edison a letter requesting a phonograph *after* Garner had arrived in Gabon. At the bottom of the letter, Edison wrote: "Garner is the most impracticable man extant. He never arranged to get a phonograph. He should have spent 2 or 3 days here & got one & learned its peculiarities instead of that he simply talked & did nothing."[11]

Even without a phonograph, Garner claimed to have deciphered some ten chimpanzee words, and he estimated that there were at least ten more. Shortly after arriving, he purchased a young chimp from a trader. The chimp, which he named Moses, lived in a cage next to Garner's, and the would-be scientist learned to imitate his subject's sounds. "In talking to Moses I used his own language mostly and was surprised at times to see how readily we understood each other," Garner wrote. He also attempted to teach Moses four words that Garner thought would be relatively easy for a chimp: "mamma," *feu* ("fire" in French), *wie* ("howl" in German), and *nkgwe* ("mother" in Nkami). Moses made some progress over three months but then died from a respiratory disease. "Moses will live in history," Garner pronounced. "He deserves to do so, because he was the first of his race that ever spoke a word of human speech; because he was the first that ever conversed in his own language with a human being . . . and Fame will not deny him a niche in her temple among the heroes who have led the races of the world."[12]

Much of what Garner learned about the speech of chimpanzees on his first trip to Africa is of no lasting import. As Radick details, Garner's critics later accused him of having vastly exaggerated his derring-do, even charging that he spent many nights not in his cage but at a nearby Catholic mission, dining well and drinking claret. Garner also was not the most careful observer, wrongly asserting that chimps gestated for only three months (it's closer to eight), did not sleep in nests, and had humanlike families with males behaving like "husbands" and "fathers" (females in estrus mate with many males, and there is scant evidence the males know their offspring). Still, he kept returning to Africa to do more studies, making six more trips, some lasting several years, before he died.

As Garner attracted increasing coverage, journalists imparted the minutiae of his journeys and exploits. The *New York Times* on November 16, 1893, reported that he had arrived in England with two chimpanzees "with which he is able to communicate," and then a month later that "one of his African comrades" had died. His return to New York on March 25, 1894, the next day merited a front-page *Times* story that went to some length to discredit news of a supposed letter to his brother in Australia that falsely claimed he was "a lexicographer of the gorilla" and that he had tried to hypnotize monkeys with a mirror. A serious, if far-fetched, follow-up that June suggested he would soon master communication with chimpanzees, which would open the door for them to become excellent laborers. "Brought into daily association with mankind, they would soon acquire our language and habits, and after a generation or two of culture it is more than likely that they would become qualified to hold domestic positions." During a journey to Gabon from 1904 to 1910, a mere letter to his son prompted another front-page *Times* story ("Says Monkeys Answer Him: Prof. Garner Declares They Reply to His Calls in the Jungle").[13]

When Garner returned to the United States in 1910, he brought with him a chimpanzee he named Susie, and the two became something perilously close to a circus act. Now the *New York Times* almost ridiculed him in an October 14 article that year, describing how two press agents promoted Garner's trip to the Bronx Park Zoo with Susie, then nine months old, to meet officials and other primates there. Garner and Susie—who supposedly knew one hundred English words—arrived at the zoo in a taxicab. "She did not demonstrate that she had any grip on the English language whatever," the *Times* reporter noted. "There was a moment during the visit when she seemed about to laugh, but it got no further than a grin and a hissing sound, which Prof. Garner asserted was a laugh." The *Times* also ran a vicious spoof that December 4 of Garner and his "trained monkanzee" Susie, penned by the satirist Wallace Irwin, who wrote that Garner had taught more than a million "monkeys" to speak and was sending them to the United States to vote for Republicans; he explicitly trained males, Irwin noted, as only they could vote.

I found an apparently unpublished essay in Garner's archives, "Kindergarden [*sic*] Studies of a Chimpanzee," that describes in detail Susie's language skills when she was one year and three months old. She recognized five different colors, five distinct geometric forms, and could distinguish "apple" from "orange," "knife" from "spoon," and "kiss" from "spank." She knew the difference between one, two, and three, and could open and close a door when asked. "She is now just beginning to manifest the first movements towards human speech," he wrote. "This is the result of some lessons by an expert in linguistics, who asserts that if he could have her under his tuition, until she is two years of age, he would certainly teach her to speak a dozen if not a score of words of human speech and know the meaning of them," he continued. "The same professor declares that within two years, he would teach her to actually read and write, and not simply thump on a type machine." In closing, Garner thundered, "I will venture to predict that with such opportunities she would startle the world as no other animal of any race or kind has ever done before."

Not long after he wrote this, Garner sold Susie to the New York Zoological Park—which, a *Times* story noted on May 16, 1911, trotted her out to sup at a members' day tea, resplendent in a harem skirt. At the time of Garner's death in 1920, newspapers had published countless more articles by or about him and the "very amazing achievement of the eminent exponent of the simian tongue," as an *American-Examiner* headline in 1910 decorously put it. But the fact was his success never went much farther than Susie.

Grandiose, wrong, and even pitiful as Garner often was, he was far ahead of his time. His phonographic techniques would lay the foundation for what today is called the playback experiment, which has helped show conclusively that primates indeed have a few specific sounds that equate to words. Apes, others would demonstrate, can be taught dozens of "words" to communicate with humans. Perhaps most striking, apes can learn to recognize a few hundred words.

Then again, so can dogs.

But to be fair to Garner, his greatest insight of all, which sometimes was at odds with his enthusiasm about their "speech" and potential, was his repeated insistence that primates have very real communication limitations.

—◦◦◦—

IN *Almost Human*, Robert Yerkes devoted a chapter to "anthropoid speech," and he tipped his hat to Garner. "As Garner was not adequately trained for his difficult research and failed to command the scientific resources of his time, his results have not been accepted generally by scientific authorities," wrote Yerkes. "It is nevertheless true that many of his observations have been substantially verified, while some have been proven incorrect. Probably his enthusiasm led him to exaggerate the degree of intelligence, and the power of vocal communication, of his subjects. But the writer humbly confesses that the more he learns about the great apes and the lesser primates by direct observation as contrasted with reading, the more facts and valuable suggestions he discovers in Garner's writings."[14]

Yerkes, a trained scientist, parsed information much more carefully than Garner. He recognized that the chimpanzee had special communication skills, maybe because it is so sociable. "It always seems as though this creature had something to say, and every now and then he makes one feel that he actually is saying it." But, he noted, ants also had a sophisticated way of communicating. And he doubted Garner's claim that chimps had speech or language. "There is no obvious reason why the chimpanzee and the other great apes should not talk, but it seems to be the consensus of opinion among expert observers, as well as those who know the animals only casually, that they do not do so," he wrote.[15]

Yerkes went on to wonder whether chimpanzees could be taught to talk—which he tried to accomplish with his bonobo, Chim, and his chimp, Panzee, without success. He blamed himself for the failure and noted Garner's reported success with Moses.

Yerkes also described the work of William Furness III, a Philadelphia doctor who between 1909 and 1911 had brought two orangutans back from Borneo and bought two chimpanzees from an animal dealer in England. As Furness explained to the American Philosophical Society in April 1916, he tried to teach them to speak but made little progress.[16] "If these animals have a language it is restricted to a very few sounds of a general emotional signification," he said. "Articulate speech they have none and communication with one another is accomplished by vocal sounds to no greater extent than it is by dogs, with a growl, a whine, or a bark. They are, however, capable to a surprising degree of acquiring an understanding of human speech."

Furness went to comical extremes to train one of the orangutans to say a few English words. "The first move in teaching her to say cup was to push her tongue back in her throat as if she were to make the sound 'ka,'" he explained. "This was done by means of a bone spatula with which I pressed lightly on the center of her tongue. When I saw that she had taken a full breath I placed my finger over her nose to make her try to breathe through her mouth. The spatula was then quickly withdrawn and inevitably she made the sound 'ka.' All the while facing her I held my mouth open with my tongue in the same position as hers so that her observation, curiosity and powers of imitation might aid her, and I said *ka* with her emphatically as I released her tongue." To add the "p," he had to close her lips with his fingers.

Imagine if humans had to do this to teach their children to speak each word.

Furness was refreshingly blunt about his failures and the limits of the orangutan and chimpanzee minds. "The crudest scrawls of the cave dwellers are hundreds of centuries ahead of the simian thought," he said. "They undoubtedly can be taught, owing to their physical resemblance, to imitate human actions to a remarkable degree, but their highest notch of mentality after four or five years of training is hardly comparable to that of a human child of a year and a half."

Yerkes did not give up all hope that apes could be taught language. "I am inclined to conclude from the various evidences that the great apes have plenty to talk about, but no gift for the use of sounds to represent individual, as contrasted with racial, feelings or ideas," he wrote. "Perhaps they can be taught to use their fingers, somewhat as does the deaf and dumb person, and thus helped to acquire a simple, nonvocal, 'sign language.' "[17]

—◁◦▷—

In October 1920, according to the Reverend J. A. L. Singh, two "strange" children were removed from a wolf den and came to live in the orphanage he ran in Midnapur, India. The "wolf girls" garnered much publicity, and in 1927 a psychologist from the University of Vermont published his correspondence with Singh to provide fresh details about their wild origins and reintroduction to humans. Singh estimated the two girls were ages eight and two when they were "excavated" from the den. He said they crawled on all fours and cried or howled in a voice that resembled neither animal nor human. The younger sister, Amala, died within a year, but the older one, Kamala, slowly learned to speak some words. "At the present time, Kamala can utter about forty words," wrote Singh. "She is able to form a few sentences, each sentence containing two, or at most, three words. She never talks unless spoken to; and when spoken to she may or may not reply."[18]

That year, Winthrop Kellogg was studying for a Ph.D. in psychology at Columbia University. Kellogg became intrigued by Singh's analysis and other tales of supposedly feral children. Were they, as some asserted, mentally damaged before they wound up living with animals in forests and jungles? Or were they normal at birth and shaped by their environments? These questions were fundamental to understanding whether nature or nurture made humans *human*.

After he earned his degree, Kellogg and his wife, Luella, carried out an unprecedented experiment to try to address these questions, which they would later describe in a book for lay readers, *The Ape and the Child*.[19] They well realized, as they wrote, that it would be "legally dangerous and morally outrageous" to force a human infant to grow up in "uncivilized surroundings." But they could ethically do the reverse, raising a wild animal in "a typical human environment."

The wild animal they chose was a chimpanzee, and Robert Yerkes in 1931 loaned them an infant, Gua, a seven-and-a-half-month-old female he recently had brought to his Florida colony from Doña

Abreu's collection in Havana. The Kelloggs would gauge Gua's development against their own son Donald, who was ten months old at the start of the experiment. Both Gua and Donald would be raised in their home and treated as similarly as possible.

The experiment ran for nine months, and as the Kelloggs explained in *The Ape and the Child*, they carefully compared everything from how the two ate and slept to their dexterity, senses, play, and intellectual development. But what would receive the most attention were their explorations of language and communication. Donald from the start toyed with making sounds. "It was as if the child, like other normal humans of similar age, was *practicing* the formation of new vowels and consonants." With Gua, in contrast, "no additional sounds were ever observed beyond those which she already possessed when we first made her acquaintance." Gua, unlike Donald, only seemed to vocalize when provoked by some stimulus, usually of "an emotional character." The Kelloggs narrowed Gua's vocalizations down to four groups: the bark, the food-bark, the screech or scream, and the "oo-oo" cry. While Donald attempted to make many of these sounds himself, Gua never tried to imitate Donald.

The Kelloggs spent several months trying to teach Gua to say "papa," and their technique mirrored the ludicrous lip manipulations Furness had used with his orangutan. "Although the possibility may still remain, we feel safe in predicting, as a result of our intimate association with Gua, that it is unlikely any anthropoid ape will ever be taught to say more than half a dozen words, if indeed it should accomplish this remarkable feat," they wrote.

When it came to comprehension of language, like Garner and others who had worked closely with apes, the Kelloggs recognized that Gua understood several words and phrases they said to her ("no," "kiss, kiss," "close the door," "hug Donald"). "Surprising as it may seem, it was very clear during the first few months that the ape was considerably superior to the child in responding to human words." But by the sixth month of the experiment, Donald won the competition, and at the end he had a vocabulary of 107 comprehended words and phrases to Gua's 95.

Reaction to the Kelloggs' experiment ranged from criticism for using their child in research to suspicion of trying to advance Darwin's theory of evolution.[20] A *Time* magazine review of their book called their work a "curious stunt."[21] But the Kelloggs' account presented strong evidence that, from infancy, humans and chimpanzees process and use sounds differently. Only human infants babble. And yes, chimpanzees can produce a few words and understand dozens, but they hit a wall precisely when humans typically experience a "vocabulary explosion": two-year-old children learn about two words per day, jumping up to three words per day the next year, and skyrocketing to as high as twelve words per day by eight years old. A high school graduate, on average, knows sixty thousand words.[22]

—◇◇◇—

FIFTEEN years after Gua left the Kelloggs' home, another husband-and-wife team, Keith and Cathy Hayes, tried a similar experiment with a baby chimp from Yerkes's Florida colony. Keith was a newly hired behavioral psychologist at what was then called the Yerkes Laboratory, and in September 1947 they "adopted" a three-day-old chimp, Viki, and she would soon come to live in their home.

The Hayeses were poking into the same nature-versus-nurture territory explored by the Kelloggs, although they were comparing Viki to caged apes, not a human infant, and language played a more central role in their project. "The art of language cannot require very much intelligence, we argued, considering that the human child of one year is already beginning to master it," wrote Cathy Hayes in *The Ape in Our House*, a 1951 book about their first three years working with Viki.[23] (The book, which shares stories of Viki's wacky antics, appears to have been the inspiration for the movie *Bedtime for Bonzo*, which starred Ronald Reagan as a professor who wants to test the nature/nurture hypothesis.)

At the start of the experiment, Cathy was convinced that Viki would be speaking in short order. "Why should not an ape, raised in a completely human fashion, acquire human speech?" She noted that Furness had taught an orang to say "papa" and "cup," avoiding his devastating final assessment of the ape's language prospects. "Why couldn't we teach an ape enough language to communicate its needs and feelings? Why should this not happen easily and spontaneously under the proper conditions?"

The Hayeses quickly realized that Viki babbled "very little," and by four months, she became quieter still. "There were a few encouraging little flare-ups, like the day she went Hawaiian with remarks like 'ah ha wha he' and 'ah wha he o,'" wrote Cathy. "But these exceptions only made the next day's silence more discouraging."

When she was five months old, they began actively trying to teach Viki to speak, using food as an enticement. They made little progress, though she did begin to vocalize "a rasping, tortured 'ahhhhh'" when she wanted food, which they called her "asking sound."

This "astonishing" advance encouraged the Hayeses to teach Viki words. "All we needed to do, we said blithely, was to hold her lips in the proper position, tell her to speak, and out would come human syllables," Cathy wrote with her own tongue in cheek. At fourteen months, they began training her to say "mama," and a few weeks later, she could say it without someone moving her lips. In time, she also learned "papa," "cup," and "up."

The Hayeses recognized that they had been overly optimistic about Viki's ability to learn language, and during a road trip—which included visiting Yerkes at his home in Connecticut and the Kelloggs in Indiana—Cathy took the two-and-a-half-year-old chimp to a speech clinic at the University of Michigan. "What would you do if we brought a child with Viki's symptoms to you?" she asked. After listening to her few words and examining her, the linguists concluded that Viki did not suffer from brain damage, as in cases of human aphasia, where the impairment of language skills is usually traced to trauma, a tumor, or another neurological disorder. "In chimpanzees, it is probably due to the lack of certain brain parts which have never evolved in that species," they told her.

Prescient assessment.

In *The Ape in Our House*, Cathy also addressed the other side of the language coin, comprehension. "Of all the questions we are asked by new acquaintances, the one we dread most to hear is: 'How many words does Viki understand?'" wrote Cathy. She explained that it was "next to impossible" to answer the question, because of several variables. Gestures and the pitch of a voice could influence Viki's response. The chimp also had a strong contrarian nature, and if she failed to follow a command, it didn't mean she failed to understand it. What's more, Viki would apparently comprehend something one day and not the next. Or she would learn routines, such as going to the bathroom or having cocoa before bed, that only made it seem as though she were following a series of commands. She could not, despite extensive training, consistently identify her body parts. She would kiss on command, but "kiss Mama" often nudged her to kiss Papa. Cathy surmised that Viki was "generally poor at understanding words," and estimated that her comprehension, at best, was limited to fifty word groupings. "The significance of Viki's speech training lies not in the fact that she has learned a few words, but rather in her great difficulty in doing so, and in keeping them straight afterward," she concluded.

Viki lived with the Hayeses until May 1954, when the six-year-old chimp died from viral encephalitis.

—◈—

The same year Viki died, David Premack, another psychologist who began his career at the Yerkes Laboratory, would move the field in a direction that Yerkes himself had suggested in *Almost Human*: he attempted to teach chimpanzees a "nonvocal" language.

As later research would verify, chimpanzees do not have the vocal apparatus of humans; basically, chimps have a different-shaped larynx (ours produces vowels) and a different-shaped pharyngeal region (which includes the tongue and palate) above it, limiting their ability to make human sounds.[24] But anatomical differences had nothing to do with Premack's decision to move away from the approaches the Hayeses, Kelloggs, Furness, and Garner had tried.

Premack was completing his Ph.D. in psychology and philosophy, and his real motivation was to study what separated humans from other species. Language was a tool to investigate the issue. "I had no interest in chimpanzees or language," said Premack, who is blunt to the point of being cranky. "There was no definition of language: it was mumbo jumbo from the philosophers of the day. The only way I could ever understand it was to get an appropriate nonhuman and cut language into pieces and install them one by one."

Premack's "words" in his invented language consisted of colored pieces of plastic that had different shapes. Chimpanzees had to place these metal-backed tokens on magnetic boards, effectively writing sentences. As Premack and his wife and collaborator, Ann James Premack, explain in the book *The Mind of an Ape*, over fifteen years they had four chimp language students, the most accomplished of which was Sarah.[25]

Not only did Sarah learn to answer simple queries ("color of apple?"), she could make and carry out requests. But she never asked any questions herself. "The ape's failure is due to its inability to recognize deficiencies in its own knowledge," the Premacks wrote. Unlike their predecessors, the Premacks had used language studies to peer inside an ape's mind, and what they found led them to believe that chimpanzees were not capable of inferring the mental states of others— what David Premack called a "Theory of Mind." That in itself severely limited their ability to communicate.

The nuts and bolts of language—grammar and its components, such as syntax—also escaped the chimps. As the Premacks wrote, "The evidence we have makes it clear that even the brightest ape can acquire not even so much as the weak grammatical system exhibited by very young children." In the end, they emphasized that more than language separates the human and chimpanzee minds. "Adding a human larynx to the ape would not make of it a human, nor would subtracting language from the human make of it an ape," they asserted.

When I spoke with David Premack in September 2007 he had become even more dismissive of what his chimpanzees accomplished than he was in 1983, the year *Mind of an Ape* was published. "I quit that sort of stuff a long time ago," he told me of his ape language research.

Premack's chimpanzees learned strings of words. But that was far from language. "It became clear to me that these strings of words— some of them got to be reasonably long—were going nowhere," he said. "There were no embedded phrases. They never became 'The apple that is red was eaten by Sarah late in the morning.'" In Premack's opinion, this process of embedding ideas within ideas, called recursion, is a bright, shining line that divides us from them. "There is no evidence for recursion in animals," he told me. He stressed that chimpanzees' inability to grasp the simplest rules of grammar meant that they did not even have a nonrecursive language. And the only words they learned had to have a sensory component, things that could be touched; metaphors and abstract concepts, like "time," had no meaning to them.

Premack argued that chimpanzees would not evolve a language for a simple reason: they do not teach one another, which is how human children develop vocabulary. When considering the reports that later researchers had taught language to chimpanzees, bonobos, and even parrots, Premack characteristically was to the point. "It all boils down to next to nothing," he said. "It's all nonsense."

—∞—

WHILE Premack taught Sarah and his other chimpanzees to communicate with tokens, another husband-and-wife psychologist team, R. Allen and Beatrix Gardner, began a project that more closely matched Yerkes's vision of teaching apes to use their hands to communicate. The Gardners, who worked at the University of Nevada, Reno, taught American Sign Language to a chimpanzee they named Washoe. Their work would advance ape language research into a realm of scientific respectability it had never enjoyed.

Washoe lived in a trailer behind the Gardners' house, and they enlisted four other humans into their "foster family" to care for her around the clock. They assiduously spoke only ASL when around her, hoping to avoid an environment that relegated signing to "nursery talk" and vocal communication to "adult" conversation. As the Gardners reported in *Science* in 1969—the first time a top-notch scientific journal with a wide audience had published ape language research—Washoe had learned thirty-four signs during her first twenty-two months of training. She even could use a few signs in combinations that "resemble short sentences," they wrote.[26] Yet the Gardners danced around the language issue: "From time to time we have been asked questions such as, 'Do you think Washoe has language?' or 'At what point will you be able to say that Washoe has language?' We find it difficult to respond to these questions as they are altogether foreign to the spirit of our research." It was an odd assertion in an article titled "Teaching Sign Language to a Chimpanzee." And ASL, as they rightly stress, is a language. In their experiment, the Gardners had discovered there was a blurry, and sometimes hard to find, line separating "communicative behavior" from language. They simply chose to dodge the issue, setting aside the rules of grammar and ideas like recursion that help to establish that a line does exist.

The Gardners thought that one weakness in their study was Washoe's age: she was a wild-caught chimpanzee estimated to be at least eight months old when she had arrived in Reno. They contended that they might have made more progress had they started language training earlier, and so in 1972, they began a second "cross-fostering" project with four baby chimpanzees they obtained a few days after birth. In 1975, *Science* published their results working with the first two of these students, each of which had mastered more than a dozen signs by six months of age. Washoe, in contrast, only knew "come-gimme" and "more" after six months of training.[27]

Animal language research flourished in the 1970s like never before. In Georgia, Duane Rumbaugh—who would marry the Great Ape Trust's Sue Savage—taught a chimpanzee named Lana a language dubbed Yerkish (for Yerkes) that used lexigrams, symbols of words that do not necessarily resemble the object they represent. Penny Patterson taught ASL to Koko the gorilla, featured on the cover of *National Geographic* in October 1978 taking a photo of himself. Language, said Patterson, "is no longer the exclusive domain of man." Irene Pepperberg began the Avian Learning Experiment with a parrot named after the project, Alex, that would win scads of press for his extraordinary communication skills.

But in 1979, the leader of yet another experiment to teach a chimp American Sign Language brought the entire field tumbling down. For nearly four years, the psychologist Herbert Terrace and a team of students had lived and worked with Nim Chimpsky. Terrace had studied at Harvard with B. F. Skinner, who was famous for his theories about operant conditioning, which involved changing behavior through rewards. Skinner two decades earlier had written a book, *Verbal Behavior*, that the linguist Noam Chomsky had savagely reviewed. Chomsky lambasted Skinner's notion that children required reinforcement to learn a language, arguing that language was innate in humans, akin to a bodily organ. Chomsky maintained that language acquisition was guided by a "universal grammar," rules that speakers of every language knew without being taught. If Terrace could teach Nim to sign, he would undermine Chomsky.

Terrace believed that Nim had made progress and even seemed to be capable of stringing together sentences of sorts, frequently using twenty-five different three-sign combinations like "banana me eat" and "hug me Nim." But as Terrace and his colleagues explained in a 1979 *Science* article, "Can an Ape Create a Sentence?" they had amassed abundant evidence that the answer was no.[28] In children, the mean word-length of their utterances steadily increases. Over one nineteen-month period, Nim's did not. More damning still, a close review of videos showed that Nim's signs frequently imitated what his teachers had just uttered. Plus, Nim, unlike children, never learned that a conversation required taking turns giving and receiving information.

Terrace and his team went further and analyzed films of Washoe and Koko, determining that "they showed a consistent tendency for the teacher to initiate signing and for the signing ape to mirror the immediately prior signing of the teacher." Apes, they concluded, had a "severely restricted" ability to learn more than "isolated symbols," and no study had shown "unequivocal evidence of mastering the conversational, semantic or syntactic organization of language."

Critics of Terrace would claim that Nim was not properly cross-fostered—he had some sixty handlers, who themselves had varied proficiency at signing—and that the project relied too heavily on operant conditioning.[29] Still, they were no match for the skeptics who long had suspected ape language research was hooey. At a meeting held by the New York Academy of Sciences in 1980, the linguist Thomas Sebeok, the organizer, told the audience about Clever Hans, a horse that the *New York Times* in 1904 reported could reason. Hans and his owner convinced onlookers that the horse was able to solve math problems. But when a doubting psychologist had two different people each whisper a number into one of the animal's ears, Hans no longer could add. Hans's talent was actually a finely tuned ability to read his master's head movements, which unconsciously signaled to the horse when he had clapped his hoof enough times to have arrived at the correct answer.

As the reporter Nicholas Wade reported in a waggish article about the meeting, Terrace's analysis triggered "a series of mutual criticisms among ape language researchers which have made the War of the Roses look like a teddy-bears' picnic by comparison."[30] The meeting included talks by circus experts, debunkers of 1950s dolphin communication studies, and a show by the magician James Randi who, to emphasize how easily humans can be fooled, performed some Houdini-ish breakouts after audience members knotted his hands with rope.

What little funding was available for ape language research soon dried up, and that, combined with increasing scrutiny of all studies involving apes, relegated the work to the scientific graveyard. The field became a butt of jokes. "It's about as likely that an ape will prove to have a language ability as there is an island somewhere with a species of flightless birds waiting for humans to teach them to fly," said Chomsky.[31] The psychologist Steven Pinker gave the corpse a few swift kicks in his 1994 best seller, *The Language Instinct*. "Even putting aside vocabulary, phonology, morphology and syntax, what impresses one the most about chimpanzee signing is that fundamentally, deep down, chimps just don't 'get it,'" Pinker wrote. Humans were "hectoring" these chimps to learn the way we communicate, he chided. "The chimpanzees' resistance is no shame on them; a human would surely do no better if trained to hoot and shriek like a chimp, a symmetrical project that makes about as much scientific sense."[32]

—◆—

AT the Great Ape Trust in Des Moines, the "eavesdropping" Panbanisha, Kanzi, and other bonobos and orangutans have learned to communicate with humans using lexigrams, with a cup represented, for example, by a bull's-eye. They can also identify photos of objects.

I first watched Rob Shumaker work with Azy, a thirty-one-year-old orangutan. Shumaker, an evolutionary biologist by training, began a language project with Azy in 1995 at the National Zoo in Washington, D.C. In the beginning, Shumaker said, it took Azy about six hundred trials to learn a new noun. Today, he can learn a noun in thirty. Shumaker has taught Azy about seventy nouns, verbs, and adjectives, and Azy has mastered a few dozen of them.

Azy and two other orangutans lived inside a concrete building, with a three-story-high outdoor cage attached to it and a fenced, four-acre forest behind, when I visited them in 2008. The language education took place in the front room of the building, which has an oversized touch-screen computer monitor behind an inch-and-a-half glass wall. Azy lumbered into the space and folded his massive arms onto a ledge that sits under the computer screen and then plopped his pan-shaped face atop his hands to wait for Shumaker. He had a stoic, languid demeanor, and his long, matted orange hair and beard gave him the look of a mountain man or an aging hippie.

Shumaker held up a bunch of grapes, and different lexigrams appeared on Azy's screen. Azy quickly touched the correct one. He then did the same when the monitor displayed photos of grapes. Shumaker repeated the exercise with a cup, carrots, apple slices, and popcorn. Azy got everything right.

Shumaker, who is in his midforties, was still in high school when Herbert Terrace upended the field of ape language research. "At the time, the question was framed as: Is it language or not, yes or no?" said Shumaker. "In my opinion the answer is yes." But he had a measured assessment of what that "yes" meant. Apes do not acquire language to the same degree as "normal functioning humans at maturity," he said, noting that they have a limited vocabulary and do not string words together into long sentences. "And apes are not using all of the elements of language—appropriate prepositions, tense." He also allowed that some of his predecessors in the field let their imaginations run wild about what they might accomplish, wondering whether they might discuss evolution with a chimp or take a talking chimp to Africa to help decipher the behavior of wild relatives. "We now have some reasonable expectations," he said.

The field, or at least what was left of it, had matured, noted Shumaker, and he wanted to move past the bad old days. "What's the next step?" he asked. "Do we get out of this mode of saying, 'Are apes capable of language or not?' and having these big fiery debates about it? I hope so. . . . What I would much rather do is focus on the capabilities we have. Are we going to argue forever about the word 'language'? Let's get on with it. Let's really study what's going on mentally for these guys." That is the key for Shumaker: ape language research opens a window into ape minds. "I think language allows us to do things that you simply could not do otherwise, and allows us to ask and answer questions that you could not approach with an ape in any other circumstance," he said. "It has tremendous value. Language is a great platform for exploring these greater cognitive skills." I asked Shumaker whether he felt besieged by all the criticism of ape language research. No, he replied, because he worked with a group of smart people who saw "the real legitimacy" to the work. "They don't think I'm out in left field; they don't think I'm crazy," he said. The Great Ape Trust had created a home for a group of orphan investigators who fed off one another. "It's not really a sanctuary for the apes," he told me. "It's a sanctuary for the researchers."

Shumaker made no great claims about Azy's language production skills and none whatsoever for his ability to comprehend words. But there was an entirely different story across the street at the bonobo building.

Kanzi and Panbanisha, the most skilled of the seven bonobos, had mastered nearly three hundred lexigrams. More remarkable still, a carefully designed study published as a 220-page monograph in 1993 compared Kanzi's language comprehension at eight years of age to that of a two-year-old girl whose mother was also one of Kanzi's caretakers. Both were asked to respond to hundreds of sentences like "Take the ball to the bedroom," "Put the rubber bands in the plastic bag," and "Put the telephone on the TV." In 653 trials, Kanzi responded correctly 72 percent of the time, which was slightly better than the girl's performance. And in two-thirds of these trials, the person speaking the sentence was hidden behind one-way glass to avoid the "Clever Hans" effect.[33]

Sue Savage-Rumbaugh and the man who took over her work with the bonobos at the Great Ape Trust, William Fields, believed that Kanzi had made such remarkable progress with language because he acquired it at a young age and, like human babies, was exposed to it rather than taught it. As Savage-Rumbaugh and colleagues wrote in their monograph, "We suggest that, as Kanzi grew up hearing others speak and observing the consequences/sequelae thereof, enduring changes occurred in the neurological networks of his brain that most closely approximate those that were basic to the evolution of language in humans."

Kanzi's mother, Matata, originally was part of the language training program, but she had an extremely difficult time mastering lexigrams. "What she learned one day, she quickly forgot the next," Fields told me. Matata, the matriarch of the Great Ape Trust bonobos, after thirty thousand trials knew fewer than a dozen lexigrams. "She doesn't have language—you can't talk to her," Fields said. "When I need her to do something, I say, Kanzi, can you get Matata to do this or that?"

Kanzi, who lived full-time with his mother until he was two and a half, was exposed to her lexigram training from birth, and he "spontaneously" learned the symbols. He also became deft at understanding spoken English. "Words can be learned, and learning a vocabulary can be learned, but acquiring the language function is something that happens at a certain plastic period of development of an organism," Fields said. "And the only organisms capable of acquiring language on this planet are babies. Certainly with apes once you pass a certain point there is no possibility."

In the 1980s, Savage-Rumbaugh and her fellow researchers had published few studies about Kanzi in peer-reviewed journals.[34] The most detailed of them had led to a withering assault from two scientists who had taken part in Project Nim. They noted that Savage-Rumbaugh's team essentially documented similarities between Kanzi's behavior and language, and they argued that this strategy was not "falsifiable," a key to scientific studies. "Consider an art dealer trying to evaluate whether a picture was painted by Picasso or not," they explained. "He could acquire a huge amount of evidence supporting the conclusion that Picasso was the artist (by noting similarities between the painting and other works by Picasso) even if the work is a fake. In the case of ape language research, the atheoretical pursuit of similarities has led to fruitless debates of the 'is the glass half empty or half full' sort." The critique allowed that Savage-Rumbaugh's work was a "great advance over previous research." But the authors concluded that Kanzi "does not know that lexigrams designate, represent, symbolize, or name objects and events; rather, he knows how to use them in order to effect desired outcomes such as obtaining objects, being allowed to engage in favored activities, or receiving the approval of his trainers."[35]

Savage-Rumbaugh wrote a lengthy rebuttal, accusing them of having "incorrectly overgeneralized from their experience with Nim."[36]

She agreed that most of Kanzi's utterances were requests, but contended that this was more a function of having to use a lexigram board, which required him to stop what he was doing, move to the board, and touch a symbol.

At the end of the day, the evidence did little to persuade the scientific community at large that Kanzi had acquired language, and when the exhaustive monograph appeared that compared his skills to those of the child, the silence was deafening.[37]

When I first met Kanzi, he was standing behind a heavily screened door. He had a big belly and looked nothing like the svelte bonobos I had spent many hours watching at the San Diego Zoo. I introduced myself and showed him my digital recorder, suggesting that if he made a sound I'd play it back for him. He showed no reaction.

I then stood behind a glass window and watched Kanzi run through a touch-screen exercise that had him selecting from one of three options. If he chose the correct symbol, a machine at the bottom corner of the screen fed him an M&M. I could not figure out the goal of the program, and when I asked his handler to explain it to me, she said, "If I explain the rules to you right now, Kanzi is going to hear it, Kanzi's going to understand, and the experiment is kind of useless."

I continued to try and speak to him. I felt foolish, but the handler claimed not only that he understood her, but that several of his squeaks were actually English words ("right now," "M&M," "yeah").

"He may not want to speak with you because you're a new guy," she said to me at one point.

"That's right," Kanzi said, according to the handler (to my ears, it was two squeaks).

"You don't want to answer because I'm the new guy?"

"Yeah," squeaked Kanzi.

I asked Kanzi to show me a yellow M&M. He ignored me.

"The obvious questions are the ones they don't always like to answer," his handler explained.

Kanzi then moved to a glass-walled room at the entrance to the bonobo facility, which features another touch-screen computer that serves up lexigrams. Kanzi climbed onto a table in front of the screen and draped one leg over a basketball he had brought with him, a bonobo with attitude.

Panbanisha, a bonobo at the Great Ape Trust in Des Moines, Iowa, communicates with humans by means of several hundred lexigrams, which are symbols of words.
PHOTOGRAPH BY MALCOLM LINTON.

I watched Kanzi touch a series of lexigrams ("kiwi," "donuts," "swelling," "shot," "peas," "mouth," "car," "orangutan," "hamburger," "Panbanisha"), and then after a brief delay the screen would flash three photographs of the object. Kanzi selected the correct photo two dozen times in a row. In a second demonstration, the computer displayed hundreds of lexigram symbols. I read a word into a microphone ("banana," "onions," "ball"), and when Kanzi heard my voice piped into his room, he selected a lexigram. He quickly tired of this and took a pee, which he typically did not do in this room. The handlers decided to bring in Panbanisha, another Weight Watchers candidate. I continued reading words into the microphone, and she correctly selected half a dozen lexigrams.

Liz Pugh, Savage-Rumbaugh's sister and another of the bonobo handlers, invited me to join her and Panbanisha on a walk outside around a fenced area that abuts a geese-filled lake. Pugh fitted a collar and leash on Panbanisha, and we walked around the winter-brown grounds as the birds honked. Pugh had a plastic lexigram board made up of three panels, and Panbanisha touched the symbol for fire. Panbanisha then ambled over to some fallen logs and collected twigs, which she stacked together in a clearing. Pugh handed her a box of matches, and Panbanisha lit the twigs on fire.

"Who taught you how to make fire?" I asked Panbanisha.

Like Kanzi, she had ignored everything I said to her.

"She doesn't talk to strangers, don't you know that?" Pugh laughed.

Panbanisha touched the lexigram for hot dogs.

"What do you like better, hot dogs or marshmallows?" I asked.

She touched the lexigram for marshmallows. That was the extent of our conversation.

Pugh then had an assistant bring out marshmallows on a stick, and Panbanisha roasted and ate them.

I am convinced that Panbanisha and Kanzi can identify many lexigrams from an array of hundreds. True, watching Panbanisha make a fire and roast marshmallows was a remarkable sight. But my attempts to speak with the bonobos led nowhere. Maybe it was because I was a stranger. Maybe it was because they generally distrust males, or maybe my questions were too obvious or I did not talk about things that interested them. Whatever. If they have language, I did not witness it. If a three-year-old human showed as little response to what I said, I would think the child had hearing problems or was psychologically impaired.

The skills the bonobos did demonstrate reminded me of Rico, a border collie that could identify two hundred objects.[38] The researchers working with Rico showed that he could learn a new word's meaning after one exposure—a "fast mapping" ability comparable to that of a three-year-old child. But they also noted that a toddler has a much broader understanding of the meaning of words, and they asserted that his "seemingly complex linguistic skill" may consist of simple cognitive building blocks also seen in language studies with apes, dolphins, parrots, and even sea lions. These building blocks include the idea that objects have labels, the ability to store knowledge in memory, and learning by exclusion (if there are two objects and you know the name of one, a new word likely applies to the other). Indeed, Rico's team believed their findings might help elucidate the evolution of language: the existence of these building blocks in other species suggests that the ability to comprehend speech was "already in place before early humans began to talk."

Rob Shumaker's view that language studies can offer a unique understanding of ape minds makes good sense. But language ability, like all extraordinary claims in science, requires extraordinary proof. Part of the problem may be, as Fields asserted to me, that the Great Ape Trust has not published much of the data it has gathered. Part of the problem may have to do with what its researchers have published, and not simply because the data aren't fully convincing. In 2007, Savage-Rumbaugh listed Panbanisha, Kanzi, and another bonobo member of the "Wamba" family as *coauthors* on a paper about the welfare of captive apes. As Savage-Rumbaugh explained in the paper, she asked the bonobos yes/no questions, and they agreed on twelve items that were important for their welfare, including "receiving recognition, from the humans who keep them in captivity, of their level of linguistic competency and their ability to self-determine and self-express through language." The notion that the Wambas declared that their welfare depended on a recognition of their language skills is worthy of Richard Lynch Garner.

Of course animals do communicate with humans, and I am convinced that Kanzi and Panbanisha can do so to an unusual degree. But my sense after my visit to the Great Ape Trust, and after reading extensively about Lana, Washoe, Nim, Sarah, Gua, Viki, Moses, and Susie, is that Friedrich Max Müller had it right: language is our Rubicon, and no brute will dare to cross it. Granted, apes can be taught to use and recognize words and symbols, even to utter a few words. But as Pinker said, they just don't get it. And the reason why is steadily becoming clearer as a different group of scientists probes ever more deeply into the brains and genes of both chimpanzees and humans, discovering biological barriers for them and evolutionary leaps for us that promise to one day have the final word in the seemingly endless ape language debate.

6

THE FOX IN THE CHIMP HOUSE

I N 1987, THREE SIBLINGS WHO ATTENDED THE SAME SCHOOL CAME
to the Institute of Child Health in London at the suggestion of their
headmistress, who was concerned that a profound speech and language disorder afflicted them. "Of course they were aware of it,"
explained Faraneh Vargha-Khadem, a cognitive neuroscientist at the
institute who was asked to evaluate one of the children's problems.
"They recognized it was affecting their siblings. They were aware of
the fact that their cousins also had this difficulty. But it wasn't something they had done anything about."

Other researchers had shown that speech and language disorders
could run in families, but no one had pinpointed a responsible gene.
The clinical geneticists at the institute took blood samples from several members of the family to build a pedigree chart, hoping to trace
the trait as it moved through three generations. They carefully documented who in the family was affected and who was unaffected. In
all, fourteen members of what the researchers referred to as the KE
family had the disorder, while an equal number did not. It was a pattern of classic Mendelian genetics known as "autosomal dominant":
affected people had inherited one mutant gene, which caused the disorder, and had a fifty-fifty chance of passing it on to their children.

A gene linked to speech and language, if it could be found, offered
a new angle to understand what made the habitual use of articulate
language, as Darwin had said, peculiar to man. Clinicians had found
regions of the brain that, when damaged, compromised or completely
shut down the ability to speak or process what was heard. Language
problems also had ties to handedness, dyslexia, even schizophrenia.

Clues gleaned from those studies had helped clarify the anatomy of speech and language. But they said nothing about how it came to be in the first place. Only genes could speak to that.

When I met with Vargha-Khadem at the well-worn Institute of Child Health in December 2006, nearly twenty years had passed since the KE family entered her orbit. To her chagrin, the KE family's problem initially led to ridiculous, reductionist propositions that their disorder would reveal why only humans had articulate speech. More frustrating, the gene that caused their disorder did not easily give up its identity. But with world-class sleuthing and more than a dollop of luck, the gene eventually was found, and the discovery attracted international notice—with one family member even discussing his unusual condition on a video he posted on YouTube. By then, the limited effects of the gene had clarified that it was but one of several needed to create speech and language. Yet, in test-tube studies it served as something of a magnet to pull together the other genes in the language network, and helped scientists extract a startling, nuanced difference between humans and chimpanzees.

I asked Vargha-Khadem why she thought this gene, one of twenty thousand in humans, had received such intense scrutiny. "We share so many other aspects of our behavior and our genome with other species, and in particular the chimpanzee, and yet speech and language is what distinguishes us," she said. "Of course it captures the imagination because it appears as though speech and language has really played an important role in terms of bringing us out of the wild."

—◦◦◦—

In 1861, Pierre Paul Broca arrived at a meeting of the Anthropological Society in Paris with another man's brain. Broca, a surgeon and neurologist who helped found the society, had retrieved the brain from an unusual patient who had died a few months earlier. The patient had been hospitalized for thirty years and was known as Tan because he would answer "tan, tan" to any question put to him. He eventually altogether lost his ability to speak, although his doctors could tell from his responses that Tan understood most everything. Broca met Tan just five days before his death, when he was transferred to the surgery unit because of a massive, gangrenous infection that ran from his right heel to his buttocks. Broca showed the society how the most extensive and oldest lesions in Tan's brain occurred in the middle part of the frontal lobe in his left hemisphere. Broca believed these lesions had caused Tan's loss of speech.[1]

Thirteen years later, the German physician Carl Wernicke described the brain taken from a deceased stroke patient who had been able to speak but could not understand what was said to him. Again, a lesion in the left hemisphere stood out, although this one was located farther back, near the intersection of the temporal and parietal lobes.[2]

Subsequent work showed that Broca's area controls the ability to produce language, and Wernicke's area is essential in comprehending words. When these areas are damaged, people like Tan and Wernicke's patient suffer from aphasia.

The KE family members experienced dysphasia, meaning they had only a partially compromised ability to speak and process words. Regardless, after their pilgrimage to the Institute of Child Health, the KE family was featured in a documentary made by a group of parents whose children had aphasias and who thought the KE family's story might help them lobby for special programs to improve the schooling of others with similar problems. The documentary aired on British television and was seen by Myrna Gopnik, a linguist at McGill University in Montreal, who happened to be visiting the United Kingdom. Gopnik contacted the KE family and began her own investigation.

Gopnik contended that the family had a grammar-specific problem, having particular difficulty with tenses and plurals. When she first shared her idea with Vargha-Khadem and her colleagues, they cautioned that their own studies had led them to a more complicated theory. "We had been investigating them at that time for two years, and we knew that the problem affected many other areas of language, understanding, language production, and other aspects of cognition as well," Vargha-Khadem told me. She spoke quietly, precisely, and with more exasperation than anger. "So this putative gene was not selectively interfering with one aspect of language, being grammar. We pointed that out to her, but she was not very keen on our alternative explanation that this was a more widespread disorder. In particular, I don't think she was keen on linking the fact that they actually had a speech disorder, and yet the speech disorder is so profound that it overrides the language problems because they cannot express themselves."

In April 1990, Vargha-Khadem's colleagues published a study in *Developmental Medicine and Child Neurology*, a respected if little read specialty journal, that showed the inheritance pattern of the disorder in the KE family.[3] That same month, Gopnik upped the ante with a report in *Nature*, one of the most widely read scientific publications, that laid out her grammar hypothesis.[4] Vargha-Khadem and her fellow researchers fired off letters to the editor that called the work "misleading" and "inaccurate."[5] Gopnik remained steadfast, arguing in a reply that the gene likely affected "the capacity to acquire language."[6] The next year, the prominent psychologist and language maven Steven Pinker gave Gopnik something of an endorsement in an article, "Rules of Language," asserting that her work provided "evidence that some aspects of use of grammar have a genetic basis."[7] The mainstream media ignored the reports about the KE family, as well as the academic spat, until 1992, when Gopnik spoke at the annual meeting of the American Association for the Advancement of Science (the publisher of *Science*). "Genetic biologists have identified the grammar gene," wrote the syndicated columnist James J. Kilpatrick. "Researchers say inability to form past tenses or plurals may be inherited," read a *Los Angeles Times* headline. "Poor Grammar? It Are in the Genes" read another.[8] At that point, though, the hunt for a gene to explain the disorder had come up empty-handed.

Looking back more than fifteen years later, Vargha-Khadem was frustrated that Gopnik received so much "premature publicity" and that she did not mention the family's speech problem. But she did not blast Gopnik. "It was just so attractive as a hypothesis," she told me. "And it was doubly attractive to have the family members as the medium through which the hypothesis could be tested." Vargha-Khadem shook her head. She played for me an interview a colleague had conducted with a KE family member, Laura.

RESEARCHER: What's your name?
LAURA: LOW RAH.
RESEARCHER: Where do you live, Laura?
LAURA: Schwe ho oof oof hi efhoo two oof.

RESEARCHER: And how old are you?

LAURA: Foe. En af.

RESEARCHER: Where do you live?

LAURA: Ehry.

RESEARCHER: And how old are you?

LAURA: Fie. An they ar short ah eh best boht an they are six. Buh now it's over. That it how long. Ah the howar was six. Not eight. And eight I am done. Any cow odeh in an eight and I am ice likely.

"*That's* a grammar problem?" Vargha-Khadem asked me, referring to Gopnik's hypothesis. "You see, if she had played this, nobody would have *ever* believed her. *Nobody* would have *ever* believed her."

When Vargha-Khadem and her colleagues finally published a detailed description of the KE family in 1995, they put forward compelling evidence that the disorder extended far beyond grammar. The researchers, who had identified fifteen affected and sixteen unaffected family members, asked the study participants to bite and lick their lips, click their tongues, close one eye, and make animal noises. The affected group had great difficulty with these seemingly simple actions. As Vargha-Khadem told me, "The disorder actually arises from a central control of oro-facial musculature." IQ tests also detected significantly lower scores in this group. So whatever the gene was, it led to speech and language problems as well as "severe extralinguistic difficulties."[9]

At that point, the investigators had carefully characterized how the disorder affected the family members—its phenotype—and even done brain-imaging studies that highlighted affected regions, but the gene eluded them. They were having difficulty getting funding to do the work, which in part explained their slow progress. But even if they had had the money, linking a phenotype to a genotype is a bit like solving a complex crime, and they did not have a genetic Scotland Yard. Anthony Monaco at nearby Oxford University, however, did.

—◦∞◦—

MONACO, who had won much acclaim for fingering a gene that caused a type of muscular dystrophy, worked at the Wellcome Trust Centre for Human Genetics, which was established to screen large numbers of families to find the genes and genetic networks linked to common diseases. One of the problems Monaco studied was "specific language impairment," SLI, another name for dysphasia. Monaco and the Institute of Child Health in 1996 launched a collaboration to see if they could tease out the cause of the KE family's problems.

Monaco handed the job to Simon Fisher, who had just completed his Ph.D. and was eager to take on the work. "There was huge power just taking this one family to localize where the gene was," Fisher told me one morning in November 2006 at the sparkling, well-equipped research center. "Power," in statistical terms, means the ability to state with high confidence that an observed link between a genotype and a phenotype is real—a critical distinction, since many apparent associations are due to chance.

Fisher took the DNA from each of the affected and unaffected family members and, one chromosome at a time, sifted for a difference only found in people with the disorder. "We got the linkage in a couple of months," said Fisher. "It didn't take long at all."

But the Oxford group had not found the gene. Rather, they had identified a region on chromosome 7 that only the affected family members shared.[10] "All that tells you is that somewhere in this big chunk of the genome there is a mutation," explained Fisher. "Finding the mutation is where the fun comes about."

They designated the region SPCH1, which has roughly one hundred genes in it. It was 1998, and the human genome had yet to be decoded, making it extremely laborious to locate these genes—and even less was known about what the genes actually did. So they began hunting for other families that might have the disorder, hoping that would help narrow down the search. No luck. Next, they trolled through databases for kids who had speech and language problems and known chromosomal oddities like deletions or recombinations. Nothing panned out.

Shortly after the Oxford team published its findings about SPCH1, Jane Hurst, a clinical geneticist who worked a few minutes' walk from the Wellcome Trust Centre at Oxford's Churchill Hospital, phoned Fisher. Years earlier, while still doing her training, Hurst had bled the KE family, winning her the coveted first author slot on the first paper ever published about them. Now she was "a bit miffed" that the new paper did not even have her name on it. She phoned Fisher to give him a bit of good-natured grief.

Then, a few weeks later, a pediatrician contacted Hurst about one of her old cases. In 1994, Hurst had seen a pregnant woman who had abnormal amniocentesis results. The fetus had a balanced translocation, pieces of two different chromosomes that had swapped places, which only sometimes causes problems. The woman decided to continue with the pregnancy, and Hurst saw the baby six weeks after he was born. "He looked absolutely perfect," remembered Hurst. "I didn't think much more about it." Then came the call from the pediatrician, who said the boy had speech and language delays. "Do you think this could have anything to do with his chromosome change?" he asked her.

Hurst looked through her notes. "I thought, oh gosh, the break point is exactly the same place that the linkage is showing." In other words, the boy's chromosome 7 had been broken in the same region of chromosome 7 that delineated *SPCH1*.

Hurst again phoned Fisher. "Simon, I've found you the patient who is going to get you your gene," she told him.

Hurst explained that the medical charts of the boy, CS, read as though he were a member of the KE family. Fisher decided to give it a close look, and sure enough, CS's break point mapped to a location in *SPCH1* that contained a gene called forkhead box P2, or *FOXP2*.

Originally discovered in flies, "forkhead" refers to the shape of the terminal end of the embryo. *FOX* genes regulate other genes. Specifically, they code for proteins that tell other genes when to transcribe their DNA into messenger RNA that, in turn, makes amino acids; the *FOX* family of transcription factors have a starring role in embryonic development.

Hurst marveled at the serendipity of it all. If CS had not had his chromosomes analyzed as a fetus, his speech and language problem likely would not have led him to a geneticist. "His mum just happened to have that test in pregnancy because she was relatively old," said Hurst.

Vargha-Khadem told me she was floored when she learned about the CS-KE connection. "I thought, my god, this is close to a miracle," she said. "It's a huge population in the world that could have had such a translocation. For this to have occurred in Britain, but not only that, to have occurred in such a way that the same clinician who saw the family would see this child and see the connection between them . . ." The chances of it still left her speechless.

In October 2001, Fisher, his assistant Cecilia Lai, Monaco, Vargha-Khadem, and Hurst announced in *Nature* that *FOXP2* was the cause of the KE family's problems.[11] Affected family members exhibited the same point mutation—a single nucleotide—in one of their two *FOXP2* genes. That nucleotide was essential to the gene's function, as indicated by the fact that a survey of several other species showed that it remained unchanged all the way down to the lowly yeast. CS's translocation caused a different disruption of the gene and was not inherited, but it made an ironclad case that, fourteen years after researchers began studying the KE family, the mystery had been solved.

For the first time, a gene had been conclusively tied to speech and language, kicking open a door that neuroscientists had been knocking on for over a century. The finding made news worldwide, including a page-one story in the *New York Times*.[12] In conjunction with the paper itself, *Nature* published an editorial by Steven Pinker, who heralded the discovery as the "dawn of cognitive genetics" and said if it did prove necessary for the development of language, "one can imagine unprecedented lines of future research."[13]

—◆◇◆—

IN early 2001, shortly after submitting their paper to *Nature*, Tony Monaco and Simon Fisher met with Svante Pääbo, the evolutionary biologist based at the Max Planck Institute for Evolutionary Anthropology in Leipzig. Pääbo asked whether Monaco's search for genes linked to autism had uncovered any strong leads. Pääbo was interested in "Theory of Mind," the term coined by the psychologist David Premack as a result of his chimpanzee language research. Self-absorbed as many humans seem to be, we have a knack for reading minds. When a woman with a stack of books in her hands tries to open a door with one knee, even a five-year-old child will infer what she wants and open the door for her. If we receive praise for a halfhearted job, we might reason that the person is lying and wants something from us. A stranger who holds your gaze could lead you to think the person wants to meet you. People with autism have difficulty understanding the wants and beliefs of others, leading one researcher to call them "mindblind."[14] Chimpanzees have a similar mindblindness. Pääbo wanted to compare a gene linked to autism to the same gene in chimpanzees.

Monaco's nascent autism studies had yet to fish out any interesting genes, but he told Pääbo about their evidence that a mutated *FOXP2* caused the KE family's speech and language disorder. Fisher had done a preliminary comparison of *FOXP2* in humans with the orthologous gene—which means they shared a common ancestor—in mice and found the genes were nearly identical. "I wasn't sure it would be that interesting," recalled Fisher.

"Trust us," said Pääbo. "Let's have a look at it."

Pääbo's lab compared the 715 amino acids in the human FOXP2 protein to amino acids in the same protein in the mouse, monkeys, and great apes. As Fisher had indicated, humans and mice were remarkably similar: only three differences existed, none of which were related to the KE family's mutation. Given that humans and mice shared an ancestor roughly 70 million years ago, this high degree of conservation backed Fisher and Monaco's earlier assertion that the *FOXP2* gene performed a crucial developmental function and could not tolerate much change. "You need two functional copies of *FOXP2* to have proper speech and language, but that does not mean at all that *FOXP2* changed in order to evolve speech and language," Wolfgang Enard, who led the work in Pääbo's lab, told me when I met him at the Max Planck Institute in December 2006. "That's a very different question." The comparison to chimpanzees, rhesus monkeys, and gorillas appeared to answer that.

These three species had identical FOXP2 proteins, and they differed by one amino acid from the mouse and by two from the human. So two amino changes in humans had occurred after they separated from chimpanzees. "We saw pretty convincing evidence for positive selection of *FOXP2*," said Enard. "That was the exciting thing."

Most mutations that occur are neutral—they have no impact on a gene. Positive selection refers to mutations that lead to a fitness advantage, so much so that a "selective sweep" takes place as individuals with the mutation thrive and ones without it die off. Enard said several strands of evidence supported the contention that *FOXP2* in humans went through a selective sweep. "You have these suspicious two amino acid changes from the chimpanzee, which are more than you would expect because there's only one between a chimp and a mouse," he noted. Second, the gene is "fixed" in humans today: the researchers analyzed 226 chromosomes from unrelated humans who came from all over the world and found those two amino acid variants in each of them. Finally, the human *FOXP2* is significantly younger than other genes. Human genes on average have a common ancestor that's about seven hundred thousand years old; our *FOXP2*, based on molecular clock calculations, is at most two hundred thousand years old—all consistent with a selective sweep. Enard predicted that the *FOXP2* in Neandertals would be more chimplike, as they had split off from modern humans more than half a million years ago, supporting the idea that language made us superior and the popular notion that our heavy-jawed, big-browed relatives were lug nuts.

What if one or both of those amino acids gave modern humans the oro-facial control that the KE family had lost? Pääbo, Monaco, Fisher, and their colleagues raised this possibility, stressing that it was a "speculation," which in the world of science is a flag that an idea is cheap, even trampy. But if the speculation proved accurate, they reasoned, then the amino acid changes would at least have contributed to the evolution of human language.

It was a well-planted flag. In September 2007 another group of scientists reported that bats showed more variation in the equivalent gene to *FOXP2* than did any other species, discounting the theory that modern humans tolerated the changes in the gene because it directly led to language and a fitness advantage.[15] The next month, Pääbo's group effectively demolished their own initial theory with a report that they had partially sequenced *FOXP2* in Neandertals.[16]

Pääbo had retrieved DNA from two bones found in the El Sidrón Cave in Asturias, a principality in Spain. Neandertals, to the surprise of Pääbo and Enard, had the same two changes in their *FOXP2* as modern humans. They considered the possibility that the gene had entered Neandertals when they mated with modern humans, but Pääbo and Enard rejected this scenario because they saw no other persuasive evidence for gene flow between the species. Similarly, a modern human handling the Neandertal bones could have contaminated the DNA samples, but they thought this highly unlikely, for several reasons: the bones were removed from the cave by researchers wearing face masks, lab coveralls, and sterile gloves; in Leipzig, a special room was dedicated to handling ancient DNA; and the analysis included several controls to detect human contamination if it existed. That left them contemplating another idea: maybe the change occurred in the common ancestor to modern humans and Neandertals, say *Homo heidelbergensis*. The assumptions about Neandertals' limited language skills could have been based, as Pääbo put it, on "oversimplifications of reality."[17]

In making their original, and now dismissed, claims about a selective sweep in modern humans and the evolutionary advantage that language would provide, Pääbo and Enard were making a plausible argument given the data they had at the time, unlike, say, Gopnik's "grammar gene" assertion. Scientific knowledge is provisional, constantly being subjected to revision as new facts surface. But more to the point, while evolutionary biology and modern genetics make a beautiful couple, theirs is a rocky relationship. The reality of how modern humans came to be and the function in that process of genes—genes, more often than not, we barely understand—is so multifaceted that, by definition, oversimplification occurs frequently. Far too many "big discoveries" evaporate. Data that do not fit a sexy theory too often receive short shrift. Occam's razor, the maxim that encourages scientists to embrace the simplest explanation, is cherished too fiercely. And nothing screams this message more loudly than language itself.

Even if the *FOXP2* mutation was not the result of a selective sweep because language made modern humans more able to thrive and survive, the gene still had something to do with language, and the fact that it differed in chimpanzees likely helped explain their lexical limits. Now the challenge was to figure out where, exactly, *FOXP2* fit into the network of genes that controlled language. Researchers started to look, with more rigor than ever before, at how chimps naturally communicated with each other—not with humans trying to teach them to speak some version of English. They hoped that a more sophisticated understanding of the chimp's pant-hoots and grunts would more visibly locate the border that divides communication from language, and, by inference, how *FOXP2* and other genes draw that line.

—◦∞◦—

As chimpanzees in the trees above her head screeched at one another, Cathy Crockford lifted her walkie-talkie to her mouth.

"Yes, yes, hello, hello?" she said.

A voice on the other end asked, "Are you taking off? Over." It was Crockford's husband and colleague, Roman Wittig, and each was working with an assistant this morning in Uganda's Budongo Forest, documenting the movements and vocalizations of one of the most studied groups of wild chimpanzees in the world.

"No, we're with Musa, sort of," she said. "And I think most of the other chimps are in the tree in between."

"Okay, then we're coming over there, too. Over," said Wittig.

In the dense cluster of tall trees, the chimpanzees easily follow one another's whereabouts with pant-hoots, grunts, and screams. But humans who want to track wild chimpanzees—which researchers call "chimping"—do best if they work in teams and exploit modern communication tools including walkie-talkies, cell phones, and global positioning system devices. Crockford and Wittig also carried the traditional chimping aid, binoculars, which allowed them to identify the chimpanzees they spotted and monitor them as they swung from tree to tree in search of fruit, groomed each other, and jumped down to the ground for a stroll or a nap.

The Budongo researchers had given each chimpanzee in the seventy-member community a name, and this brisk morning in May 2008 the drama revolved around the return of Zefa, a second-ranking male that had reappeared after leaving the community for several months, leading some to think he had died. As one battle after another erupted around Zefa and alliances shifted, Crockford, Wittig, and their assistants documented the plot twists. "It's like a daily soap," Wittig said.

But Crockford and Wittig had not come to the forest simply to observe the hierarchy battles in this chimpanzee community. They wanted to make sense of what our closest relatives were saying to one another, and they wanted to go beyond deciphering their gestures and facial expressions, which others contend are the root of language.[18] Crockford and Wittig aimed to connect their vocalizations, in all their intricacies, to equally complex social interactions. To that end, they carted equipment that few chimpanzee investigators have taken into the wild. A strap over each of their shoulders held a high-end Marantz digital voice recorder attached to an ultrasensitive microphone covered with a foam windscreen. They hoped to use these sophisticated gizmos to analyze more closely than ever before the vocalizations made by wild chimps. "Vocalizations tell you a lot more than what you can see," Crockford explained to me.

Hoots, screeches, moans, grunts, and barks poured out from the trees in many directions, and, even with binoculars, I had difficulty figuring out which vocalization was coming from which chimp. But Crockford, Wittig, and their assistants quickly pulled the signals from the noise. "That was a submissive pant-grunt," Crockford told me, pointing toward the canopy of a tree behind us. "It's probably that Nick is beginning to move down out of the fruit tree." Nick, aka Nick the Prick, is the alpha male in the Sonso community, and Zefa's return had kept him extra busy, reminding friends and foes alike that he owns this forest and will kick your sorry chimp ass if you dare question his greatness. "He'll probably come over here and find Bwoba is here, and he'll have to display. Then we'll have some action," she said.

Crockford pressed a button to begin recording. "Pant-grunts and screams coming from where Nick just drummed," she said into the microphone. She then pointed the mic at the source of the sound and remained still and silent until it faded. "Okay, Zefa's climbing down," she noted. She walked closer to Zefa and recorded a loud outburst. "Pant-hoot and possibly a grunt," she said.

From such recordings, Crockford and Wittig are assembling a "sound catalog" that itemizes distinct calls and the context in which they were made. "These vocalizations are very context-tied," explained Crockford. "So an individual far away can probably work out quite a lot of what's happening, but I don't think that necessarily means it's symbolic, which it would have to be to be equivalent to words."

Chimpanzees, as far as scientists can tell, only vocalize about the here and now. They do not talk about yesterday or tomorrow, their dreams or fears, loves lost or sought—all of which would require using words as symbols. They are Zen. In the moment. This prevents their communications with one another, complex as they often are, from reaching the status of language. "They don't combine thoughts," said Crockford, echoing the observation that captive chimps are incapable of recursion. "But what I'm very interested in is they combine calls. It's not very common with primates."

Crockford and Wittig, who are in their late thirties, are two of the leading primate field researchers of their generation, and it would not surprise me if one day a movie is made about them. Crockford has blonde-brown hair knotted into a ponytail, and that, along with the fact that she is British and has a lanky build, makes comparisons to a young Jane Goodall inescapable. Wittig, a German, has a hint of Brad Pitt, but swap the jovial swagger for a studied circumspection that makes his closed-mouth smile almost look like a wince. Not only are they good-looking and smart; they also have a storied pedigree in the primate world.

Crockford and Wittig came to Budongo after chimping in Côte d'Ivoire with Christophe Boesch in the Taï National Park. Together, they studied conflict management between the three groups of habituated chimpanzees, the focus of Wittig's Ph.D., and Crockford separately analyzed vocal communications for her doctoral dissertation. Crockford's work most directly informed their current studies. Between 1998 and 2000 she recorded pant-hoots made in the context of travel or food from adult males from the three different Taï groups, designated as North, South, and Middle. She then analyzed the acoustic characteristics of the different vocalizations and found that each community had distinct pant-hoots. As she and her coauthors wrote:

> For travel, a typical South community pant hoot was characterized by having higher pitched exhaled build-up elements with the energy in the climax scream being distributed over a lower range than a typical Middle or North community pant hoot. A typical Middle community pant hoot was characterized by having more introduction elements, higher pitched inhaled build-up elements with a faster rate of acceleration and a longer climax scream than a typical North or South community pant hoot. For food, the North community pant

hoots were characterized by having lower pitched exhaled build-up elements and energy in the climax scream starting from a higher frequency than the Middle and South communities. The Middle community pant hoots were characterized by the climax screams having a less tonal quality, more introduction elements and a slower rate of production of build-up elements than the North and South communities.[19]

These are exquisitely detailed findings, which led Crockford and her colleagues to conclude that chimpanzees—like parrots, finches, dolphins, killer whales, and humans—are capable of vocal learning. The results backed an account of captive chimps in which one individual had introduced a pant-hoot followed by a raspberry, the Bronx cheer that involves sticking out the tongue and blowing to make a farting sound.[20] Another report from captive chimps showed that they made different rough grunts for foods they highly preferred (including bread, apples, and mangoes).[21]

In 2005, Crockford and Wittig went to Botswana to study baboons with Dorothy Cheney and Robert Seyfarth. Cheney and Seyfarth became well known in the late 1970s when they began a detailed experiment involving vervet monkey vocalizations in Kenya. Like Richard Lynch Garner nearly a century earlier, the husband-and-wife team recorded the monkeys' vocalizations and then played them back to vervets to try and suss out their meaning. They showed that the vervet monkeys in Kenya's Amboseli National Park made discrete alarm calls upon seeing a python, a martial eagle, or a leopard. Earlier researchers had noticed this behavior, but Cheney and Seyfarth proved it. First, they recorded the distinct alarm calls, which ranged from what they described as high-pitched chutters to low-pitched staccato grunts. They then concealed a speaker in the park and played back each sound. When monkeys on the ground heard the leopard sound, they scattered up trees. An eagle alarm made them look up or rush into the bush to hide. The snake sound led their gaze down.[22]

Crockford and Wittig helped their mentors do similar studies with baboons in Botswana's Okavango Delta, working out the social dynamics among females. Female baboons have a strict hierarchy, and they inherit their rank from their mothers. When a high-ranking female threatens a subordinate with a lunge, chase, or outright fisticuff with her, studies have shown that 13 percent of the time, the aggressor will grunt to her opponent a few minutes later, signaling that all is forgiven. But the researchers suspected that, as with other monkey species, a close relative of the aggressor—a third party—could also mediate the peace process with a "reconciliatory grunt."

After recording the reconciliatory grunts of individual chacma baboons in the Moremi Game Reserve, Crockford and Wittig waited to observe a fight between two females. Within five minutes of the fight's end, they set up a loudspeaker in the area where the aggressor was last seen, and then played the reconciliatory grunt from a mother, daughter, or sister. As a control experiment, they played the grunt of another high-ranking female from a different matriline. They found that when victims heard the reconciliatory grunt from a kin of the aggressor, they much more quickly tolerated the presence of the aggressor. In one third of the tests, the victim even groomed or embraced her aggressor.[23]

Now in Budongo, Crockford and Wittig wanted to use what they had learned to conduct a study they hoped would be the most thorough ever done of wild chimpanzee communications.[24] "Quite a lot of this sort of work has been done with monkeys, quite a lot of it has been done with captive apes, but very little has been done with wild apes," said Crockford. "People have tried to look at it for quite some years, and the results have not been very encouraging. I think because in the forest it is so hard to see what is happening."

Wittig explained to me that their ambition was to help lay bare "the basis for language." We were standing in a clearing, and the only sounds were the mellifluous birds. "What we are interested in finding is what are the social components? What do chimps know about others in the group? What do they know about their social environment? And what do they communicate about their social environment? This is the bottom line for us. We think that the big human brain is very likely a result of a huge complexity." At that moment, the chimps in the trees above drowned out our conversation, erupting into a barrage of pant-hoots and screams. We stopped and stared up at the canopy as it continued for fifteen seconds. The chimps were responding to vocalizations from another group of chimps they heard in the distance. They started to come down from the trees and mobilize, heading toward what they perceived as intruders.

"So this is exactly what you're talking about," I said, as Wittig entered his coordinates into a handheld GPS.

"Exactly."

Later, when the racket quieted down, I asked Wittig whether he thought chimpanzees had language. "I don't know," he said. "For me it's not so much a question of whether they do or don't. I'd like to know whether they use their vocalizations to transfer certain information, and whether they use listening to eavesdrop on interactions of others, to deduce what happens around them, and to compile information that they can use themselves to compete better."

"Don't you already know that they do these things?" I asked.

"We actually know much too little about it," he said, laughing. "There are many, many anecdotes where people say chimps can do this, chimps can do that, but social intelligence in apes has never been really studied in the wild, where we might expect the broadest use of those third-party knowledges for chimpanzees because they're not fed by a zookeeper, they're not taken care of by a vet, they're meeting other groups, they're meeting dangerous situations—all those things where information about it will be very valuable and give them advantages."

Given that researchers had spent more than four decades studying wild chimpanzees, I was surprised that so many fundamental matters of chimpanzee communication remained inscrutable. "It's a big puzzle," he said. "We are not able to go there and ask the chimps direct questions, and their communities are very complex. So it won't be possible to just observe and see what's happening. Observation is a very important tool to get an idea of what *might* happen, and then we have somehow to prove it. I think it's time to take one step further and create a playback experiment where chimps can show what their knowledge is about things."

Ultimately, Wittig and Crockford believe the complex social interactions of chimpanzees led to the evolution of bigger brains. This "social intelligence hypothesis" is an idea that stretches back to 1966, to Alison Jolly, who studies wild lemurs, and was refined ten years later by the psychologist Nicholas Humphrey, known for his hypothesis about the development of the human mind through the construction of an "Inner Eye."[25] The basic proposition of social intelligence is that there is a lot to keep track of when you're living in a community with dozens of individuals with different ranks, alliances, and temperaments. Hominins, with still more complicated social dynamics, kept selecting for an expanded brain, and language cropped up as part of that process. "To me it seems like language is an acquisition that comes on top of a big brain," said Wittig. "There was first a big brain and then language was developed, at least language in the form that we would refer to as human language." A chimp in the background started to pant-hoot. "It might be the social component is the pre-position for acquiring language."

—◦◦◦—

DANIEL Geschwind reached up to his office bookshelf and took down a toy, a 3-D puzzle of the human brain. Geschwind, a neurogeneticist at the University of California, Los Angeles, began to snap together the plastic pieces to help with his explanation of the parts of the brain that control speech and language, but for the life of him, he couldn't figure out how the left and right hemispheres attached. "I'm really bad spatially, so don't make fun of me," pleaded Geschwind, as we sat in his office in September 2007. "It's like I'm having a little stroke or something. I'll get it together, and then I'll figure it out."

The plastic model might have challenged Geschwind, but when it comes to embracing the complexity of the genes that drive the brain's control of speech and language, he excels at figuring out how the pieces fit together. To peer into the human brain and see how it typically stores, uses, and comprehends words, Geschwind examines not only normal human brains but also cases where the process goes awry, studying the genes of families afflicted by autism, dyslexia, schizophrenia, and other conditions that often coexist with speech and language disorders. "Studying disease is really a fundamental way to understand normal function," Geschwind told me. "Disease has given us extraordinary insight to understand how the brain works or might not work."

Geschwind's forays into speech and language also have led him to analyze DNA from the brains of chimpanzees, monkeys, and even songbirds. "A lot of people think our lab is all over the place," said Geschwind. "It's actually pretty integrated. Language is complex, and the only way we're going to have a hit is when two or three findings point to the same place."

He ran his hand over the left hemisphere of the plastic brain he eventually assembled and pointed to Broca's and Wernicke's areas. The Sylvian fissure, a Grand Canyon of sorts, runs between Broca's and Wernicke's areas. And the area around that fissure, the perisylvian region, is the most precious real estate in the land of speech and language.

We know this in part because of the research done by the neurologist Norman Geschwind, a cousin of Daniel's father who studied aphasias at Harvard Medical School. In 1968, Norman published a landmark paper that examined one hundred brains of people who had died and found that most had a pronounced enlargement, visible to the naked eye, in the perisylvian region on the left side.[26] He showed that the brain typically is asymmetrical: the left hemisphere is usually larger than the right. The preponderance of humans with larger left perisylvian regions was directly related to the fact that most humans are right-handed.[27] Decades later, Daniel would spell out how this surprising correlation was determined by genes.

Roughly 90 percent of humans are right-handed, and while many species, including chimpanzees, show a hand preference, we uniquely favor one type of handedness over the other. In right-handed people, 97 percent use the left perisylvian region for speech and language. In lefties, this plummets to about 60 percent, with another 30 percent equally using both hemispheres of the brain for language, and 10 percent using only the right.

Curious about the genetics of cerebral asymmetry and handedness, Daniel Geschwind studied the question in seventy-two pairs of identical twins. He compared them to sixty-seven pairs of fraternal twins, who could help tease out the effects of environmental factors—things like what babies are exposed to in utero or how they are raised—in contrast to genes. All were men, and on average they were seventy-one years old and had taken part in a long-term twin study organized by the NIH.[28] Geschwind used magnetic resonance imaging to assess the size of different brain regions, and found strong evidence for what is known as the right-shift gene theory. This hypothesis contends that genes predispose people to being right-handed and to using the left hemisphere of the brain for language. Left-handedness indicates that a person did not inherit these genes, in which case random forces determine where in the brain language resides.

I asked Geschwind why he thought a right-shift had evolved. "We don't know," he said. "But I would surmise it has to do with the idea that there's some computational benefit to the kind of processing that's going on in language—which is extremely rapid processing—to keep everything in one circuit in one hemisphere."

Humans can understand about 150 spoken words per minute all day long. To appreciate how quickly our minds must decode that onslaught of verbs, nouns, adjectives, and adverbs, consider that a Morse code operator can, at best, transcribe fifty words per minute. Operators usually do not remember the contents of the messages they send, and they quickly become fatigued, requiring a break every hour or so.[29] So there is something special about our ability to process the language we hear. Geschwind noted that there are compelling clues that this processing works best in one hemisphere. "People who are more likely to have bihemispheric language are left-handers who are susceptible as a population to language disorders like dyslexia or autism," he said.

Proving that genetics drove handedness and brain asymmetries is a far cry from finding the responsible genes—and further still from explaining how genes create speech and language. "*FOXP2* is paradigmatic: it's this beacon and the first proof that this area of research might lead to great insights about human beings and evolution," said Geschwind. "Yet it can't be more than a small piece of the story. How do we move from gene-by-gene studies to understanding systems? This plagues everything we do."

—◁∞▷—

FOR genetic detectives, the easiest case to crack is a single mutation in a single gene that afflicts half the living family members from several generations. Collaring *FOXP2*, then, was a relatively straightforward task. Yet the Institute of Child Health in London could not do it alone, and even the sophisticated operation at Oxford required spectacular luck to hear the signal from the responsible gene. Now consider that most disorders, diseases, and traits come about because of an orchestra of genes playing a complicated score. Gene hunting is a messy business—despite the advent of sophisticated techniques such as the polymerase chain reaction that can amplify miniscule amounts of DNA, microarray chips that can reveal the expression of hundreds of genes at once, and whole genome analysis that can detect interactions between proteins and DNA. Many leads turn into dead ends. And even the bona fide identification of players in the orchestra can overwhelm researchers, who effectively hear a cacophony when they put them together.

Biologists once naively thought genomics would separate the signal from the noise, but as information became available about the complete sequence of DNA in the chromosomes of humans and other species, it became clear that they had what the Nobel Prize winner Sydney Brenner described to me as a White Pages rather than a Yellow Pages: they knew where things lived but had no idea what they did. A field called functional genomics sprung up, which was aided by the specialized techniques of proteomics, transcriptomics, metabolomics, and loads of other "omics" that promised to map the entire universe of proteins, transcripts, metabolism, and everything else imaginable. Then came "systems biology," a grander attempt to analyze complex data sets from many sources.

To appreciate what a particularly confusing realm speech and language geneticists inhabit, consider a study conducted jointly by Geschwind, Monaco, Fisher, and thirty-nine researchers from nine other countries who discovered an unusual gene involved with brain asymmetry, schizophrenia, and handedness.[30] The researchers analyzed the genes in 222 reading-disabled siblings, 105 pairs of left-handed brothers, and 1,002 families that had schizophrenic members. As the investigators reported in the July 2007 issue of *Molecular Psychiatry*, the study identified a gene, *LRRTM1* (leucine-rich repeat transmembrane neuronal protein 1), that influenced handedness in dyslexics. A subsequent examination of the DNA in families of schizophrenics uncovered a similar connection.

Unlike the *FOXP2* story, the affected people did not simply have a mutant version of one of their *LRRTM1* genes. Rather, they appeared to have lower levels of the LRRTM1 protein that contributed to, but did not outright cause, their trait. The researchers further complicated matters by proposing that an unusual mechanism led to the reduced levels of the protein.

We normally inherit one gene from each parent, and this trait appeared to be inherited from the father. This led the team to the realization that the *LRRTM1* gene on the maternally inherited chromosome is either completely inactivated or turned down to a low level, a phenomenon called imprinting, which has only been found to be involved in approximately forty human genes. But the low levels of LRRTM1 did not simply reflect the muffled gene in the mother. In the affected individuals, something in their paternal DNA must have turned down the knobs on the father's *LRRTM1* during fetal neuronal development, the researchers contended. Normally, the mother's *LRRTM1* could compensate if the body detected that low levels of the protein were being made, but the imprinting effectively put it out of business. Evidence for this knotty scenario included a postmortem examination of a brain from a fifteen-week-old fetus. Studies of chimpanzees contributed, too: they have an identical LRRTM1 protein to humans, and they showed no evidence of imprinting in two brains obtained after the chimps' deaths.

All of this is maddeningly complicated, and, on top of that, Geschwind and the others had yet to find a direct link between *LRRTM1* and speech and language. Little wonder that scientists searching for the genetic underpinnings to speech and language veritably danced a jig over the finding that an aberrant *FOXP2* caused the KE family's problem.

FOXP2 offers two different potential paths to understanding the biology behind human speech and language. *FOXP2* tangos with many other genes, and one strategy looks for its dancing partners— the genes that are coexpressed with *FOXP2*—and then asks whether they independently have connections to speech and language. Another approach attempts to better understand the effects of the protein in humans and other species.

The protein turns on or off other genes by binding to them. This binding allows the protein to serve as a hook to fish out related genes in test-tube and animal experiments. Geschwind's lab identified 285

FOXP2 targets in the fetal brain. Another study that Geschwind conducted with Fisher's lab found many of these same targets in neuronal cells that originally came from a four-year-old girl as well as in mouse brains, adding credence to the hunch that these targets actually matter. Before the investigators had made sense of this intriguing data, separate research on the development of the perisylvian region, autism, and children with specific language impairment led them to yet another *FOXP2* target that their analysis for technical reasons had completely missed.

Although speech and language depend on the perisylvian region of the brain, scientists know little about the molecular forces that create this circuitry. So Geschwind and his team probed eight brains removed from fetuses that were seventeen to twenty-two gestational weeks old and looked for genes that were expressed in abundance in the perisylvian region but not elsewhere. One of the genes that surfaced was Contactin-associated protein-like 2, known in shorthand as *CNTNAP2*. When they examined mice and rats, the gene was splashed all over their brains and had no connection to the perisylvian region.[31]

At the same time Geschwind's group was conducting these studies, a report appeared in the *New England Journal of Medicine* that a mutation in *CNTNAP2* caused epilepsy, autistic behavior, brain abnormalities, and language problems in four children from an Old Order Amish community in Pennsylvania.[32] Geschwind had also found mounting evidence that tied a variation in *CNTNAP2* to autism and a delay of a child using a first word from the average of one year of age to nearly three.[33] "Here's where everything comes together in a really bizarre way," Geschwind told me.

—◦∞◦—

BLONTERSTAPING. Perplisteronk. Contramponist. People who have trouble remembering and repeating these nonsense words have specific language impairment, SLI, which affects up to 7 percent of children. And although SLI clearly runs in families, studies have had difficulty connecting the dots between genes and typical forms of the common disorder. Monaco and Fisher surveyed their huge database of children who have SLI and found a significant association with variations in *CNTNAP2* in the same region that Geschwind had pin-pointed in his autism study.

So mutant *CNTNAP2* had been implicated in SLI, autism, and brain abnormalities in children, and intact *CNTNAP2* in the fetal development of the perisylvian region. If all of this convergent evidence left any doubts that *CNTNAP2* had something to do with speech and language, the researchers, working with Geschwind, announced in November 2008 that FOXP2 bound to the gene and "dramatically" turned down its ability to express itself.[34] "We've gone from a starting point of studying a Mendelian, single gene form of a disorder to finding a cause of a more complex trait," Fisher told me. "And this is the first time we've been able to pin down a particular gene that's involved in a common form of language impairment."[35]

Helen Tager-Flusberg, a cognitive neuroscientist at Boston University who studies autism, like many other researchers I spoke with about the work, was extremely excited about the finding. "This," she said, "is opening up a whole new chapter for *FOXP2*."

—◦◦◦—

FUNCTIONAL magnetic resonance imaging offers a peek at the signals the brain sends under different conditions. So Vargha-Khadem and her colleagues recruited five members of the KE family who do not have the *FOXP2* mutation, the unaffected group, and five who do, to take part in a study that used an fMRI machine to literally see what was happening when they thought about words.

Each family member was told a noun and then asked to think of a verb without saying it. In the unaffected group, the right hemisphere showed no activity, but on the left, Broca's region lit up; in the affected group, activation occurred all over the place in both hemispheres—but not, strikingly, in Broca's area in either hemisphere. (Unaffected and affected people were compared to relatives who had the same hand preferences.)

Other experiments charted brain activation when both groups said aloud the verb that a noun triggered (generation) or simply repeated a word aloud. As expected, articulating words activated different brain regions, as did repeating a word, which does not require retrieval from memory. Again, the affected members, regardless of the task, had a pronounced underactivation in Broca's area and an overactivation in both hemispheres.

The scans showed that FOXP2, as suspected, helps construct the network in the left perisylvian region that is vital in language. But the story is more complex, as the fMRIs indicate: strong activity uniquely appears in the unaffected group far from the perisylvian region—in particular, in the basal ganglia, deep in the brain—suggesting that the circuitry for language likely courses through several regions. And the story would become more convoluted still when researchers looked at FOXP2's effects in other species.

—◦◦◦—

"Clearly, what we want to do is develop transgenic humans with the chimpanzee FOXP2 and put the human FOXP2 in transgenic chimps," Svante Pääbo told the few hundred people who had come to hear his talk at the annual meeting of the American Society of Human Genetics held in San Diego in 2007. "Obviously, that's a joke."

Pääbo, who is puckish and enjoys making his colleagues squirm a bit, was giving a presentation about how to compare different species to understand human origins, and his joke was an introduction to the mouse model of FOXP2. Wolfgang Enard in Pääbo's lab had engineered transgenic mice that had either one or two copies of the human FOXP2. "We continuously try to speak to the mice, and they won't answer," he said.

Again, Pääbo was ribbing his colleagues to raise a serious dilemma: What do you look for in mice that have the human FOXP2? Not only does the gene fail to turn them into Stuart Littles, many mice vocalizations are ultrasonic, requiring special equipment for humans to detect them. So Enard compared 323 different phenotypes in the transgenic and the "wild-type" mice to determine whether the human FOXP2 had an impact. He found only two significant differences. Humanized mice were more cautious in a new environment, avoiding the center of the cage. The humanized pups, whether they had one or two copies of the gene, also peeped more when separated from their mothers. "The two phenotypes just affect the brain and no other organs," said Pääbo.

Simon Fisher took a different approach with mice, introducing the FOXP2 mutation found in the KE family.[36] Mice born with the mutation in both copies of the gene died within a month, which in part reflects that a key function of *Foxp2* (the designation for the gene as it occurs naturally in mice) involves development of the lung. No human, incidentally, has yet been found who has two crippled versions of FOXP2. The mice they engineered to have only one mutant copy of *Foxp2*, a closer approximation to the KE family, had abnormal basal ganglia and more difficulty than wild-type mice in learning how to use a running wheel. As opposed to another lab that engineered a similar mouse, they did not find any pronounced differences in vocalizations—a reminder that animal models are, in the end, just models.[37]

Despite its limits, the mouse studies together supported the evidence from humans that FOXP2 had an impact on the brain and vocalizations and affected the formation of the basal ganglia, and that individuals had to have at least one intact copy of the gene for survival. Further insights came from an entirely different animal, the songbird.

—◦◦◦—

ACROSS the street from Geschwind's lab at UCLA, one of his collaborators, Stephanie White, has a cage tall enough for a human to walk around in that's filled with six hundred zebra finches. The males of these songbirds have a skill shared only by humans, dolphins, whales, bats, African elephants, and two other orders of birds: they can learn sounds and use them to communicate.[38] It is difficult to overstate the impact that the finding of the FOXP2 mutation in the KE family had on White's work with her birds. "It's huge," she told me in October 2007.

I watched as Michael Condro, a graduate student, shut off the lights to the bird room—they will not fly in the dark—and removed a finch. He carried the bird across the hallway and put it in a bell jar he earlier had filled with an anesthetic, halothane. The bird collapsed within seconds. Condro took the bird to his lab bench and cut its head off, placing its torso in a body bag. He then carefully manipulated a pair of surgeon's scissors to remove the brain. He superglued the tiny brain to a plate and a machine sliced it into sections. He went through all the motions in less than a minute.

Male finches rely on a region of the brain dubbed Area X, in the basal ganglia, for singing. They learn songs from their fathers, and they have two types: directed, toward females; and undirected, which they do for practice and which are unaffected by whether they are alone or with others of their species. By analyzing expression levels of *FoxP2* (yet another designation exists for birds) in different brain slices, White's lab found that undirected singers had sharp decreases in Area X. The gene, in other words, did not simply affect the embryo and development of the brain and other factors. "It could be that *FOXP2* in humans is more of a stabilizing factor in adults," White told me.

In a parallel experiment to Fisher's mouse study, a group in Germany asked what would happen to zebra finches if they lowered *FoxP2* levels. Led by Constance Scharff at the Max Planck Institute for Molecular Genetics in Berlin, the researchers took advantage of the recent discovery that certain types of RNAs could bind to specific messenger RNAs and prevent them from making a protein.[39] After designing RNAs to interfere with *FoxP2*, they injected the constructs into Area X of the birds. The interfering RNAs cut expression levels of *FoxP2* in half, mimicking the KE family. They then compared treated to untreated males and found that the reduced levels of *FoxP2* made it more difficult for the birds to accurately copy the songs from their tutors. They also found abnormal variation in song syllables that resembled dyspraxia—difficulty making proper sounds—in humans.

Scharff and White both recognized the enormous evolutionary gap between zebra finches and humans, but both stressed that this had an upside.[40] "One of the fascinating areas for me is how much of our biological foundation for language got laid down during evolution before the emergence of the capacity in humans," White said. Birds learn songs through what is known as procedural learning, which is how we learn to ride a bike or swim. We learn these things bit by bit, proceeding from one skill to the next until we can balance, peddle, and steer simultaneously. Declarative learning, in contrast, specializes in facts, some of which are arbitrary. In terms of language, grammar is procedural learning (put "ed" at the end of a verb to indicate the past tense), and our vocabulary is declarative (understanding that "cat" refers to a specific type of mammal). Intriguingly, procedural learning takes place in the basal ganglia and the frontal cortex that includes Broca's area. The temporal lobe—in particular the hippocampus—has strong ties to declarative learning.[41] Does that mean humans developed language by building on the procedural learning pathways—and the attendant basal ganglia—that had evolved as far back as birds? And how do human and chimpanzee brains compare in these regions?

—◈—

"I have a rare speech disorder, *FOXPH2*," says Stewart Young, staring into a camera attached to his computer. "I'm famous in the medical history." Young, one of the affected KE family members, posted his video, "5 Things About Me," in May 2007. Although he misnames *FOXP2*, slightly slurs words, and clearly has trouble moving his jaw, he is articulate and, unlike his cousin Laura, relatively easy to understand. He goes on to say that he has a learning disability, loves helping people, loves the Internet, and speaks British Sign Language.

Young has posted several other videos to YouTube—most begin "Hello, People"—describing his many other health problems (eczema, depression, eating disorders), explaining that he is bisexual and likes to dress in drag, showing off his piercings, telling embarrassing stories, and signing songs like the Archies' "Sugar, Sugar" and Alanis Morissette's "Ironic." He is, in turn, charming, generous, silly, and vulnerable. I e-mailed him a few months after he posted "5 Things," and he suggested we have a video chat over the Web.

Young, as in several of his YouTube videos, appeared on my screen in November 2007 wearing a bathrobe. "I learned to speak properly at eleven or twelve, but I still had lots of problems," he told me. "Many people think I'm lazy not to talk, but I always say to them, no, it's hard. I get fed up. I say to them I'm deaf because it's easier. I hate talking, to tell you the truth. But I'm living in both worlds. Hearing world and deaf world." For both selfish and selfless reasons, he appreciated working with the researchers studying the KE family's condition. "They understand us," he said. "I've had lots and lots of tests and you can get fed up, but it will help more people and us in the future."

Young, then twenty-five, told me that his speech and language skills, with years of therapy, had advanced much farther than many of his relatives', and though he said he had a low IQ, he displayed no obvious lapses in intelligence. I asked what he knew about the *FOXP2* research. "*FOXP2* can turn on and off other genes," he said. "It's a master gene. It can even turn off liver and brain, and it's quite scary in that sense." I had never heard *FOXP2* described as scary, but then I do not carry a mutated version of it.

Before we said our good-byes, Young said he did not know how his family got "the bad gene." He hoped that it might disappear from his lineage. "*FOXP2* has changed," he said, "and it might change again."

7

MIND THE GAP

NOT MANY SCIENTIFIC MEETINGS OPEN WITH A ROOMFUL OF eminent researchers pant-hooting like chimpanzees, but then "The Mind of the Chimpanzee" conference held at Chicago's Lincoln Park Zoo for three days in March 2007 marked a rare occasion in itself. For only the third time in twenty years, the zoo hosted a meeting that brought together researchers who study chimpanzees in the wild and in the laboratory. Surprisingly, it was the first meeting to focus solely on the cognitive abilities of our nearest animal relatives. "It's amazing," Jane Goodall told me during one of the breaks. She was one of some three hundred participants, including several who had followed in her footsteps at Gombe. "We're talking about things now that I couldn't talk about in the '60s. We couldn't even talk about the chimpanzee mind because chimpanzees didn't have one."

After the zoo's Elizabeth Lonsdorf, a Gombe veteran who runs an ape conservation program, kicked off the meeting by having everyone launch into a "proper chimp greeting," she introduced Kyoto University's Tetsuro Matsuzawa, one of the few researchers who studies both wild and captive chimpanzees. Since 1977, Matsuzawa and his colleagues have been conducting cognitive experiments with chimps housed at the Primate Research Institute in Inuyama, a quaint town about ninety minutes by train from Kyoto. His star research subject is a chimp named Ai, which means love in Japanese. When he showed videos of his experiments with Ai and her son Ayumu ("walk"), people resorted to human vocalizations, and gasped, ooohed, and ahhed.

Building on work he had initially reported in *Nature* seven years earlier, Matsuzawa revealed Ai using a touch-screen computer monitor to select the randomly displayed numbers zero through nine, in ascending order.[1] Each time she got the sequence right, a machine dispensed an apple slice or some other treat. Matsuzawa then ran video of Ayumu doing the same task. He began receiving number training when he was four and a half years old, and had mastered one through nine by the age of five.

More remarkably, a different computer test showed off Ayumu's spectacular short-term memory. The test began the same: the computer randomly splashed numbers across the screen. When, however, Ayumu touched, say, the number one, white blocks covered the other numbers. He then had to touch the white blocks in the correct sequence to receive his treat. On average, Ayumu saw the numbers for 650 milliseconds before the white blocks masked them. More than 80 percent of the time, Ayumu correctly sequenced five numerals. The odds of correctly guessing the sequence are 1 in 120.

In a still more difficult test, the numbers and a white circle came onto the screen. When Ayumu touched the circle, the numbers were covered within either 650, 430, or 210 milliseconds. At the shorter intervals, I could hardly see the numbers. Ayumu's success rate stayed around 80 percent, regardless of the speed. "I know," Matsuzawa said to the mesmerized gathering, which, in addition to Goodall, had attracted nearly every top chimp and bonobo researcher, including Frans de Waal, Richard Wrangham, and Christophe Boesch, as well as most of the hot up-and-comers in the field. "No one can do this."

Matsuzawa then played a clip of his graduate students failing the exercise with only four masked numbers.[2] The audience cracked up.

If anyone still needed convincing that chimpanzees have better short-term memory than highly educated humans, Matsuzawa cued up one more video of Ayumu. This time, a moment after Ayumu touched the white circle and the white boxes masked the numbers, an airplane flew overhead. Ayumu paused and looked up, waiting several seconds before returning his attention to the screen. He then touched the numbers in the correct sequence. "Ayumu gets distracted, and still he does it," said Matsuzawa, beaming.

Matsuzawa theorizes that the common ancestor to the human and chimpanzee also had a muscular short-term memory, and humans became flabbier in that skill when they acquired the ability to communicate with complex language. "There might be a sort of trade-off," he said.[3] Developing a bigger brain, in other words, came at a cost. Like a hard disk on a computer, it had to overwrite one program to make room for an upgrade, and along the way, it lost some functions. Once again, the finding demonstrates that evolution is not a climb up a ladder that moves a species ever closer to excellence. Over time, you win some, you lose some. And chimp cognition studies by Matsuzawa and others are allowing us to see with more clarity than ever before how our cognition compares. Matsuzawa's description of what he calls the Ai Project sums up the deep implications of this line of inquiry: "It's the evolutionary basis of the human mind."

—◆◇◆—

IN November 1977, one-year-old Ai arrived at the Kyoto University's Primate Research Institute to take part in what was anticipated to become Japan's first ape language research project. Tetsuro Matsuzawa, then twenty-seven, had been at the institute for a year and had done some basic experimental psychology studies with rats and monkeys, but he admittedly knew little about the field. He knew less about chimpanzees. But he was given the job of running the experiment and training the chimp.

Ai, captured in Africa's Guinea Forest and purchased from an animal trader, was at first kept in a small, windowless basement room. Years later, Matsuzawa recounted their initial interaction.

I first met her in that dimly lit basement room, with a bulb hanging from the ceiling. When I looked into this chimpanzee's eyes, she looked back into mine. This amazed me—the monkeys I had known and worked with never looked into my eyes. For them, staring straight into one's eyes carried a threat, and they would be likely to respond by opening their mouth and threatening you back or by presenting their back and assuming a submissive posture. I had simply thought that chimpanzees would be big black monkeys. This, however, was no monkey. It was something mysterious.[4]

When I visited Matsuzawa in March 2008, Ai, Ayumu, and twelve other chimps lived at the institute, and it struck me that the chimpanzee housing could not have been more different from the four-by-four-foot, bare-bulbed basement room that once held Ai.

The institute, which Matsuzawa now heads, sits on a hill in Inuyama, a quiet city that rambles along the Kiso River and is renowned for a fifteenth-century castle as well as a 1950s monkey park built as a tourist attraction. Handsome homes with traditional curved roofs line Inuyama's curvy streets. The primate facility mostly consists of drab, institutional boxes from the 1960s, but it has a stunning architectural feature that makes it stand out: an outdoor facility for the chimpanzees, surrounded by lush trees, that includes a five-story-high climbing tower. Chimps frequently scamper to the top of the tower, which looks like the sort of postapocalyptic remnant dreamed up in Hollywood, and take in the grand view of the city's homes and office buildings in the distance. Adding to the absurd beauty, ropes connect different parts of the tower, and the chimps tightrope across them and swing about as they chase each other in battle and play.

By far the more astonishing sight, however, is to watch the chimpanzees take part in the experiments. In addition to studying short-term memory, Matsuzawa and his team are analyzing whether chimpanzees can recognize individual chimps by their vocalizations and faces, the impact of hormones on cognitive performance, how chimps perceive the unity of objects and categorize, and how they communicate ideas to each other.

The chimpanzees live in large cages that divide them into two social groups. Researchers move them from the cages into the building where they conduct the studies—the "ape research annex"—via a network of catwalks. As I walked under the catwalks, the chimps moved into various laboratories off of the annex for the morning's experiments and spit on me repeatedly—the proper captive chimp greeting offered to strange humans.

The lab rooms designated for Ai and Ayumu's routines that day were each about the size of a studio apartment, split into two by Plexiglas walls that separate the humans from the chimpanzees. Following Japanese tradition, I took off my shoes, put on slippers, and took a seat with Matsuzawa and his team of researchers. The human side of each room was crowded with computer monitors, TVs, video cameras, food dishes, and machines that dispense treats to the chimps.

The Primate Research Institute in Inuyama, Japan, built this five-story tower for its chimpanzees, part of a push worldwide to improve housing for apes living in research colonies.
PHOTOGRAPH BY JON COHEN.

The animal enclosures, which looked like oversized soundproof booths from an old TV game show, had nothing in them, but slots cut into the Plexiglas allowed the chimps access to touch-screen computers.

One of the researchers pushed a button, gates clanged, and Ai and Ayumu entered the enclosures. The institute makes a point of studying mothers and their children together, following the procedures under which researchers conduct developmental experiments with human children. Ai immediately sauntered over to a computer screen and started at the number masking test. The machine was set at 210 milliseconds. "With one second, you can follow the numbers well, but at 210 milliseconds, you can't see it by eye movement," Matsuzawa told me.

Ayumu, then a randy seven-year-old, entered a separate but connected room and began doing a match-to-sample test known as the color Stroop task. Ayumu, like his mother, has learned the Japanese characters that correspond to different colors. But does he truly understand their meaning? To test this, the researchers time how long it takes for Ayumu to link the word to its color, then they miscolor the same word—showing, for example, the word "black" in the color red—and have him identify the color of the word. Matsuzawa saw that I had many different colored pens with me, and he asked to borrow three different ones: black, blue, and red. He then listed the English words for the colors in a variety of inks. He asked me to tell him, as quickly as I could, the *colors* of the words—not the words. As he expected, I slowed down when the colors did not match the words, and even stumbled. "The bottom line is it's hard to read 'red' in blue ink and say it's blue because you understand the meaning of the words," he said. He then ran the same experiment on me, but changed the words to Japanese characters, which I do not understand. I had no trouble rattling off the colors then. The bottom line of the Stroop test is that if the chimpanzees truly understand the meaning of the colors, it will take them longer to match the word "blue," when colored green, to the correct colored dot. So far, Matsuzawa told me, the preliminary data suggest they do in fact understand the meaning of the words.

Matsuzawa stressed to me that this is not ape language research. "I'm at the tail end of ape language research and the start of cognitive research," he said. "Everyone admits that apes have primitive language at the word level. No one doubts that." But the greater claims made by ape language research never persuaded Matsuzawa and most others in the field, because it is not good science. "If you do ape language research, you cannot easily repeat what you have found, but I can always repeat my findings in front of you," he said.

I watched as Ai attempted the masking test with seven numbers, and she made many mistakes. But then that is precisely the distinction Matsuzawa was making between this type of research and ape language studies. Ai succeeds with six numbers. The odds of that happening by chance are 1 in 720 tries. Instead of trumpeting the claim that Ai can do that, Matsuzawa and his team calculated her success and failure rate, compared to Ayumu at the same task. What they realized is that Ayumu and other younger chimps performed much better than their elders. It appeared that young chimps, like human children, have better eidetic memory, the ability to take a mind picture of a complicated image. Although scientists debate the existence of photographic memory in humans, the provocative idea invites further testing and may ultimately provide a finer understanding of the differences between us and them.

—◁◦▷—

MIND-BLOWING as it is to watch the chimpanzees do such experiments, I was even more struck by how Matsuzawa interacts with the animals, which reflects his deep concern for their welfare but also reveals something striking about his approach to prying inside chimps' heads.

At one point while I was watching Ayumu on the computer, I leaned up against the Plexiglas panel to take some photographs. I was not using a flash and thought I was discreet, almost invisible. Ayumu corrected me. He jumped up, stretched his arms in display, smacked the Plexiglas, and spit at me. "Stay right there, please," Matsuzawa instructed. I was completely safe, but still frightened by Ayumu's raw power. He stood just inches from my face.

As I sat frozen, trying not to react to the serious stink-eye from Ayumu, Matsuzawa slipped a jumpsuit over his clothes and a pair of thick gloves over his hands. With his racecar attire in place and what incongruously looked like a bathroom scale tucked under one arm, he headed toward an entrance to the chimp enclosure. His staff hit buzzers, and a series of metal gates groaned open, allowing him to enter the booth.

Ayumu came right over to him. "Sit," commanded Matsuzawa, in English. "Be good boy." He pointed to Ayumu and also spoke to him in Japanese. Ayumu took a seat.

Matsuzawa and Ayumu played a nonverbal imitation game, with the chimp touching his lips, patting his head, and opening his mouth in response to Matsuzawa's cues. At one point, when Ayumu realized he was not going to get a treat, he jumped up, and I was convinced he was going to bite Matsuzawa. But Matsuzawa slapped his back and took control, bulleting him with commands to sit, lie down, and even climb the wall, each of which Ayumu dutifully followed. Then they rolled around on the ground together, wrestling, until Matsuzawa, tired, just flopped into a prone position and rested. When they were done playing, Matsuzawa weighed Ayumu and checked his teeth. He then turned and did the same with Ai. He wiped the floors with paper towels to collect samples of their urine, too. "For thirty years I've been with chimpanzees in the same room, and I still have ten fingers," Matsuzawa boasted.

A few minutes later, we left the lab and headed to a stairwell balcony on the highest floor of the annex that allowed us to peer down on the massive outdoor structure. The chimpanzees spotted us immediately and began to chatter.

"Woo-ooo-woo-ooo-WOO-ooo-WOOOOO!" Matsuzawa pant-hooted.

A half dozen chimps started yelling back.

"I am sort of a member of the community," he told me. "When I pant-hoot, they have to reply because Matsuzawa is coming."

Again and again during my visit, Matsuzawa and his team told me that the chimpanzees were participants in their research. "It's not me forcing them to come to the booths," he said. "It's them willing to come to the booth. The important point is freedom. They have the freedom of choosing."

I heard the same idea when I visited the Great Ape Trust in Iowa, and the researchers there also had close ties to the individual apes. But there was something different about the Primate Research Institute, and it became obvious to me that evening as I ate fried eels with Matsuzawa. "My research target is not a single subject like Washoe, Lana, Sarah, no," he said, referring to the chimps that took part in ape language research projects in the United States. "I'm looking at three generations of a community of fourteen chimpanzees. I'm creating the community to know the social aspects of the mind."

—◦∞◦—

WHEN Tetsuro Matsuzawa was an undergraduate, he majored in philosophy, but it bored him, and he aggressively pursued his first love in his free time. "I had no interest to do anything except climbing mountains," he said. Matsuzawa climbed 120 days during each of his four years, seeking out virgin peaks over seven thousand meters high in the Himalayan and Karakoram mountain ranges. In 1973, he was part of a Japanese team that climbed Nepal's Kangchenjunga, the third-highest mountain in the world.[5]

Matsuzawa, who hung up his crampons for good in 1990 but is still fit as a climber, told me about his mountaineering days as a way of linking his career to that of the pioneering Japanese primate researcher Kinji Imanishi. Born in 1902, Imanishi founded the Academic Alpine Club of Kyoto, which Matsuzawa later joined, and traversed several glaciers in the Karakorams. In 1967, when Matsuzawa was only nineteen, Imanishi helped to establish the Primate Research Institute in Inuyama.

In an essay that Matsuzawa coauthored about Imanishi's life, he observed that his predecessor's work strove to understand the question "Human society, where did it come from?"[6] Although Imanishi studied everything from mayflies to wild horses and gorillas, he was best known for his long-term and systematic studies of wild macaque monkeys on Koshima Island, documenting their social structure, reproductive cycle, and, most notably, their "culture." Because Imanishi and his collaborators had identified individual monkeys on the island, they recognized that one of them had suddenly begun washing sweet potatoes in a freshwater stream, and that others then picked up the behavior and even modified it, washing potatoes in the sea presumably to add salt flavoring. A behavior that was "socially transmitted" thus became the definition of "culture" in animals.

Every Japanese ape researcher of note, including Matsuzawa, holds strong ties to Imanishi. "Without Imanishi, no primatology in Japan," he told me. "Imanishi is something like Louis Leakey." He is referring to the fact that Leakey jump-started Jane Goodall's observations of chimps, Dian Fossey's studies of mountain gorillas in Rwanda, and Biruté Galdikas's seminal work with wild orangutans in Borneo. Although "Leakey's Angels," with more than a little help from National Geographic, won widespread fame, few people outside of primatology circles know of Imanishi's disciples. Yet they have made a bevy of path-breaking discoveries, some of which did not initially receive the attention they deserved because of language barriers, geopolitical isolation, and cultural mores that look askance at self-promotion. In an insightful article, Frans de Waal goes so far as to suggest that Imanishi, who believed that all species were intimately connected, spearheaded a "silent invasion" of ideas that opened the field of animal culture studies and allowed researchers to anthropomorphize the animals they studied without shame.[7] "Imanishi's approach to primate behavior amounts to a paradigmatic shift that today has been adopted by all of primatology and beyond," wrote de Waal. "Imanishi's views, even though not phrased as formal theory and reaching the West only with great delay, have clearly won out."

In 1958 Imanishi moved his attention from monkeys to wild apes, visiting Tanganyika, Uganda, the Belgian Congo, Middle Congo, and Cameroon with his former student Jun'ichiro Itani. They observed gorillas and scouted for sites to study chimpanzees—two years before Goodall began her work in Gombe.[8] Imanishi and Itani in the early 1960s helped establish a research site of wild chimpanzees not far from Gombe in Tanzania's Mahale Mountains. A few years later, they sent their graduate students, Kosei Izawa and Toshisada Nishida, to Mahale, where the field scientists provisioned the chimps with sugarcane and habituated them to humans, which neither of their professors had managed to do. Nishida's close observations of the Mahale chimps led to the stunning assertion in 1968 that they did not, as Goodall and others had suggested, roam about like nomads, with the only bonding taking place between mother and child. Instead, Nishida contended, the chimps formed "unit groups" and boundaries, with individuals repeatedly breaking off and returning, creating a fission-fusion society.[9] Nishida also showed that females sometimes left for neighboring communities (the term "neighboring community" eventually replaced "unit group"), infanticide occurred, and politics roiled communities. Nishida in the early 1970s helped Takayoshi Kano survey Zaire (née Belgian Congo) for bonobos, leading Kano to establish Wamba, one of the first long-term research sites to study that species. In 1976, Yukimaru Sugiyama, who originally did fieldwork with Itani and spent time at Budongo, began a study of wild chimpanzees in Bossou, Guinea, that would uncover their complex tool use—with stones that served as a hammer and anvil to crack open oil palm nuts.[10] All told, as Matsuzawa likes to point out, counting Bossou and Mahale, Japanese researchers from Imanishi's "Kyoto School" founded and ran two of the six long-term studies of chimpanzees in the wild (the other four being Budongo, Gombe, Kibale, and Taï), as well as Wamba, which is the only site other than Lomako to conduct long-term studies of wild bonobos.[11]

In 1985, Matsuzawa published his first big finding with Ai, showing that she could identify the number of objects in front of her—a rudimentary mathematical ability—and their color.[12] The next year, he went to Africa for the first time to work with Sugiyama in Bossou. Matsuzawa eventually took over the site and works there still. In addition to building on Sugiyama's initial findings about nut-cracking, he and his students have shown how the Bossou chimps share stolen fruit with each other, only have a handedness bias for the most dangerous tool use, carefully scan roads before they cross them, and use leaves to make cushions. Captive studies have shown that yawning is contagious in about one-third of chimps, which is gauged by showing them videos of other chimps yawning; the same happens in about 50 percent of humans, and the phenomenon seems to be associated with increased in-species empathy. Infant chimpanzees, like humans, combine objects, but in contrast to human toddlers, they do not naturally stack them. Using magnetic resonance imaging, Matsuzawa and his colleagues charted how the larynx dropped in Ayumu and two other chimps born at the institute, challenging the hypothesis that this was distinct in humans and linked to the evolution of speech and language. Matsuzawa's most interesting findings, like those from the short-term memory studies, typically highlight differences between humans and chimpanzees.

From their close observations, Matsuzawa, Dora Biro at Oxford University, and others have found that the wild chimpanzees at Bossou never *teach* the complicated nut-cracking behavior. Rather, younger members *learn* by watching adults set up a stone anvil, place a nut on it, and then smash it with a second stone used as a hammer. This "master-apprenticeship" education, which de Waal has compared to the way students learn to cut sushi after years of observing master chefs, means that chimpanzee adults do not reward their young when they do something the right way or punish them when they make a mistake.[13] Mothers do tolerate their own children playing with the stones or watching them from close range, but the young learn through trial and error.[14]

The absence of evidence is not evidence of absence, and there indeed is one report of wild chimpanzees teaching. In that case, which again involved nut-cracking, Christophe Boesch detailed how one mother in the Taï National Park positioned a nut on an anvil to make it easier for her son to open it. Boesch described how another young chimp had trouble opening a nut and handed the hammer to her mother, and she then deliberately rotated the stone into a different position and showed her daughter how it was done. "Taï chimpanzee mothers show a concern about their infants' apprenticeship in tool use and, in various ways, facilitate their attempts even though at some cost to their own performance," wrote Boesch.[15]

Boesch, like Matsuzawa, has done meticulous, long-term studies of the same chimpanzee communities, so his observations carry weight. But did these actions truly come at a "cost" to the mothers, a key definition of "teaching"? Why did it happen so infrequently even in his study, which documented nearly seventy hours of mothers nut-cracking with children present? Could it be that there are behaviors that lean toward instruction but do not quite qualify as teaching?

Even more persuasive than Boesch's outlier is the fact that chimpanzees rarely, if ever, teach in Matsuzawa's observational setup at Bossou. Chimps do not have a kitchen that you can spy on to watch them use tools. They eat nuts wherever they find them and wherever they have stones handy, which makes observing the behavior a matter of luck, especially in the forest, where it is often hard to see beyond a few yards. So Matsuzawa in 1988 created an "outdoor laboratory" in Bossou at a clearing near several paths that the chimps travel on regularly. The lab has allowed his team to watch many hundreds of hours of nut-cracking at close range.

For a few months each year, the researchers place on the ground numbered stones of specific weights and dimensions and provide the chimps with piles of oil palm nuts. They hide behind a grass screen that abuts the clearing and watch the site from dusk until dawn, videotaping all chimpanzee visits. If teaching occurred, they would likely have seen it take place. And that's just when it comes to nut-cracking. The researchers also have documented the chimps using several other types of tools, including folding leafs to scoop water, fashioning a wand from a grass stem to retrieve algae from a pond, dipping for ants or honey with sticks, and using a pestle to pound the crown of an oil palm tree to extract something similar to heart of palm.[16] As Dora Biro said during her talk at the Mind of the Chimpanzee conference, "We're not seeing any examples of active teaching."

The birth of Ayumu and two other babies at the Primate Research Institute in a five-month span during 2000 gave Matsuzawa and his colleagues a unique opportunity to study another aspect of mother-infant relationships, and he stumbled upon an insight that he contends might be the root of human language.

The story begins 65 million years ago, when mammalian mothers began providing their infants with milk. About 50 million years ago, primates emerged, and their grasping hands allowed them to cling to their mothers. "Dogs, cats, cows do not cling," Matsuzawa reminded me. Monkeys evolved 40 million years ago, and they also embrace their offspring. But monkeys rarely look into each other's eyes, including mothers and infants, because what is known as a "mutual" gaze is read as a sign of hostility. Many scientists thought the mutual gaze was, in fact, uniquely human until Matsuzawa and his colleagues tested it in their three mother-infant pairs and learned the new mothers looked into the eyes of their babies twenty-two times per hour.[17] So Matsuzawa believes that 5 million years ago or so, the common ancestor to chimps and humans started to make eye-to-eye contact between mother and infant. Chimp babies, like human infants, communicate by imitation, sticking out their tongues or opening their mouths in response to an adult human, his lab has found.[18]

Humans introduced a twist in this evolutionary tale. Our babies do not cling to their mothers, who must hold their infants or they will fall. Matsuzawa suggests this difference evolved because chimpanzees have babies once every four or five years, allowing the child time to become largely self-sufficient. Humans, in contrast, take care of multiple offspring in parallel. "Chimpanzees have maybe four babies in their lives, and that makes it difficult for a species to survive," he said. "So we changed the system of rearing children, giving birth, giving birth, giving birth. And with the assistance of spouse, and grandparents, we are collaborating together to raise children."

Because human mothers separate themselves from their babies, human babies have to cry to get attention. "Not many people recognize the importance," he said. "Human babies cry in the night, but chimpanzee babies never do because mother is always there."

Add to this the fact that chimpanzee and orangutan infants, as Matsuzawa has shown, must lift contralateral limbs—a right arm and a left leg, or vice versa—to stabilize themselves when they lie on their backs. They *must* grasp. Human babies can stably lie in the supine position, allowing easy face-to-face and hand-gesturing communications. "All of these things are interconnected, and from the beginning," he said. "The underlying mechanism of communication is completely different between humans and chimpanzees because of the mother-infant relationship."

Though Matsuzawa's theory defies convincing proof, it strikes me as logical, alluring, and incredibly cool.[19] "What is the definition of humans?" he asked me. "Many people say bipedal locomotion. Decades ago, they said it's language, tools, family. No. Everything is incorrect. My understanding is the stabile supine posture, that is completely unique to humans."

Muscles, then, shaped our minds. Now that's brawn over brains.

———

ABOUT an hour after I first went chimping in the Budongo Forest in May 2008, I saw Maani, at fifty the eldest male in the community, grooming another male, Masu, then seventeen. Each chimp had one arm fully extended and holding onto a tree branch above his head. They carefully lifted the hair on each other's torso and removed bugs, dirt, dried skin, and whatever else they caught with their fingers and lips.

The chimpanzees in Gombe groom each other in the same way. In 1975, after having spent a combined seventeen hundred hours observing chimpanzees in Gombe, William McGrew and Caroline Tutin visited Mahale at the invitation of Jun'ichiro Itani. The two study sites were about one hundred miles apart as the crow flies, and only thirty miles separated the borders of the two chimp populations. On their first day at Mahale, McGrew and Tutin saw two chimpanzees clasp hands above their heads, forming an A-frame, and then groom each other—something they had never seen at Gombe. "We were dumbfounded by its elegant symmetry," McGrew later wrote.[20] Back at camp, Itani was "unimpressed" when they mentioned their discovery. "Did not all chimpanzees do this?" he asked.

Three years later, McGrew and Tutin published their finding, exploring at length the idea that the social transmission of different grooming techniques represented differences in culture.[21] Not proto-culture, preculture, subculture, or "culture" in quotes, which others had used to describe animal behavior. Simply: culture. "In 1978, Tutin and I were squashed," McGrew later explained. "Everyone *knows* that language is necessary for culture." McGrew's sarcasm under-scored how thoroughly the research community has since adopted their perspective. But it wasn't until 1999 that the tide turned, with an article in *Nature*, titled "Culture in Chimpanzees," that McGrew and Tutin wrote with the biggest names in the field: Goodall, Sugiyama, Nishida, Boesch, and Wrangham, along with Vernon Reynolds, the founder of Budongo, and Andrew Whiten, a prominent evolutionary psychologist at the University of St. Andrews in Scotland, which inher-ited stewardship of Budongo. The authors spent little time handwring-ing about the definition of culture, focusing instead on thirty-nine different behaviors they had observed at each of the long-term research sites. "We know of no comparable variation in other non-human spe-cies, although no systematic study of this kind appears to have been attempted," they wrote.[22]

In addition to unique grooming techniques, the researchers docu-mented differences in tool use, noise making, and even dancing at the start of rain. Half of the differences involved food. Interestingly, chimpanzees in Gombe did not crack open nuts, even though they had them there, and they used only objects to tickle themselves. "Fifty years ago, we knew nothing about wild chimpanzees," said Whiten. "Look at us now."

—◦◦◦—

Two of the thirty-nine cultural differences listed in the *Nature* paper involved how chimpanzees hunt for ants. Chimpanzees at many sites were observed eating ants by using a stick, either by dipping it into the ants' tunneled nests or by using it to pick the ants up as they migrate across the forest floor. Researchers had carefully measured the length of sticks and noted whether the chimps mouthed ants off of them or pulled the wand through their hands to bunch the ants and then lunch on them. Wiping versus mouthing constituted an observed cultural difference.

But a closer look at the ant's bite added an important cautionary note to supposed "cultural" differences.

Bossou has both red and black ants, and Tatyana Humle, a French researcher who worked with Matsuzawa there, knew from personal experience that black ant bites stung more. While dipping sticks herself into nests, she found that many more black ants ran up a stick, because, as she described it, they were either more aggressive or more gregarious. She decided to home in on the chimps' ant-dipping to see if there were any differences in how they handled the two types of ants. Humle found that the chimpanzees used longer sticks with the black ants, and swiping the long wand through the hand allowed the ants to be "crumpled and jumbled so that few can bite the chimpanzee before they are consumed," she wrote in a report of the observations.[23]

So if the chimpanzees at Bossou used longer sticks and a different eating technique than chimpanzees at other sites, it may well have had everything to do with ecology—and nothing to do with culture. Stick length and eating styles, in other words, could have more to do with the risk of being bitten than with some high-level cognitive function like social transmission, which at the very least would require imitation and may even suggest teaching. "Culture or not culture, that is the question?" Humle said when she presented her findings at the Mind of the Chimpanzee meeting. "Behavior is adaptive, and these are not mutually exclusive. But we need to explore how much of behavior is social learning."

—◦◦◦—

THE Yerkes National Primate Research Center has created a unique "field station" thirty miles north of Atlanta that provides researchers a unique setting to explore whether culture or "not culture" shapes behavior in its captive community. There, following two different chimpanzee communities, Victoria Horner, along with her adviser, Andrew Whiten, and the center's director, Frans de Waal, hoped to settle the cultural transmission debate once and for all.

One cool November day in 2006, I walked with Horner up a staircase that allowed us to overlook one of the corrals that house the two chimpanzee groups, which do not mix. Each corral has two thousand square feet of space shared by about a dozen chimps that entertain themselves by swinging on climbing poles, tearing up phone books, and interacting with the researchers. We brought a bucket of apples with us and tossed them down to the chimps, which fielded them expertly. A few also stuck out their tongues and gave us the raspberry. "It's an attention getter," said Horner. "The other group doesn't raspberry much."

Yerkes psychobiologist Bill Hopkins actually has published studies about the tactical use of the raspberry by the chimps, but Horner had a more ambitious agenda.[24] She wanted to see whether a different type of tool use taught to one chimp in each group would spread. The experiment centered on a "pan-pipe," a device that looks like a thin toaster. Horner placed a food treat inside the device, which a chimp could only access by using a stick to free it. Two different stick techniques would free the treat. In one, "Poke," the chimp pushed a stick through a hole in the front of the pan-pipe to move a lever that freed the food. In the second, "Lift," the chimp raised a hook on the top of the pan-pipe, which let the food slide down a chute.

After teaching a high-ranking female from one group Poke and another high-ranking female from the second group Lift, they allowed others in their communities to watch them perform the task for one week. A control group of chimps was not allowed to watch.

When given access to the pan-pipe, fifteen of sixteen chimps in each group figured out how to get food out of it, with the first successful attempts occurring within one minute for both the Poke and the Lift techniques. The Poke group only used the pushing maneuver, and, at first, the Lift group only lifted the hook. But then one chimp in the Lift group figured out how to poke and started to do both, which a few others apparently copied. The scientists then removed the pan-pipe and returned it to the groups two months later. Again, the Poke group primarily pushed and the Lift group primarily lifted. "To our knowledge, these data provide the first robust experimental demonstration of the spread and maintenance of (1) alternative traditions in any primate, and (2) alternative tool-use techniques in any

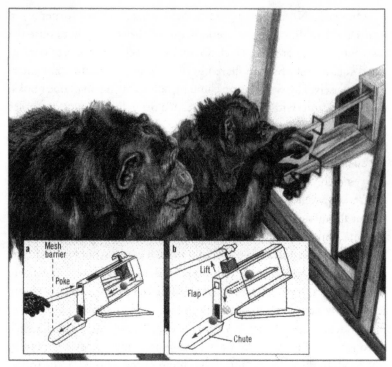

The pan-pipe device used to investigate cultural transmission of tool techniques among chimpanzees at the Yerkes National Primate Research Center. The "Poke" technique is pictured on the left; "Lift" on the right.

FROM ANDREW WHITEN, VICTORIA HORNER, AND FRANS B. M. DE WAAL, "CONFORMITY TO CULTURAL NORMS OF TOOL USE IN CHIMPANZEES," *NATURE* 437:737–40 (SEP. 29, 2005); CHIMPANZEE DRAWING BY AMY WHITEN.

non-human animal," wrote Horner, Whiten, and de Waal in a *Nature* report about their experiment.[25]

Horner told me that she did not think the chimpanzees had taught each other how to use the pan-pipe. "I think the majority of chimp learning is almost like eavesdropping," she said. Yet her findings make it unassailably clear that traditions do spread through chimpanzee communities—call it culture or what have you—much in the same way that they spread through humans.

Horner saw the limitations of chimpanzee cultures. "Chimps are very innovative, but chimpanzee culture is relatively restricted," she said. "We only have thirty-nine documented behaviors that transmit. The majority don't spread." Still, it frustrated her that many researchers, anthropologists in particular, resisted acknowledging that chimpanzees have culture—even with her unequivocal data. "In anthropological terms culture is the human niche," she said. "These things are so exclusive from the get-go. If we want to understand our place in the animal kingdom, we need to understand that the chimp/human border is so slim. Culture is just the next step. At what point are people going to give in and say, 'Yes, we are apes'? And be able to handle that? Darwin's famous quote is that it's a difference of degree, not of kind. People are just hell-bent on it being a difference of kind."

—◦◦◦—

GALLUS domesticus, the common chicken, has performed a feat similar to the pan-pipe experiment. Elephants use sticks as tools, as do woodpecker finches, New Caledonian crows, and capuchin monkeys. Egyptian vultures break open ostrich eggs with stone hammers, and sea otters open mollusks with rocks balanced on their chests. These and many other examples appeared in "The Chimpanzee Has No Clothes," a sweeping condemnation of "chimpocentrism"—the overemphasis on chimpanzees in the study of cognition—published in 2008 by C. Owen Lovejoy, a prominent anthropologist at Kent State University in Ohio known for his work on locomotion, and Ken Sayers, a Ph.D. candidate studying with him. "[We] believe that chimpanzee data have been consistently misapplied in discussions of human origins and that attempts to account for the differentiation of hominids from great apes based on a strict *Pan troglodytes* model cannot succeed," they wrote. Capuchin monkeys in particular were singled out by Lovejoy and Sayers for their chimplike intelligence. "Those damned capuchins, it would seem, may give chimpanzees their own 'identity crisis.' "[26]

In the journal's unusual format, it included extensive critiques of the article by ten researchers, and then a reply to the critiques—a whole package that ran a hefty twenty-seven pages. The most vehement criticism came from William McGrew and his collaborator and wife, Linda Marchant, whose chimp cultural studies in particular took a thrashing from Lovejoy and Sayers. "Their arguments are outdated, misinformed, advocative, and, in some cases, downright mischievous," wrote McGrew and Marchant. They asserted that the article relies on anecdotal cases and rakes over "old coals" in the "vexed and hoary issue of defining culture."

Lovejoy and Sayers predictably offered vigorous rebuttals to each rebuttal, and the academic mud fight makes for a great read, but the devil is in the big picture, not the details. "Chimpanzees, no matter how much we want them to be, are not humans," they wrote in their conclusion. "Only by drawing knowledge from all areas of zoology will we be able to view them and ourselves in a fully revealing context and address difficult questions of hominid behavioral evolution."

—◇◇◇—

AT the Wolfgang Köhler Primate Research Center in Leipzig, Germany, one morning in December 2006, I watched a chimpanzee spit water into a plastic tube to try and raise a peanut to the top so that he could retrieve it. Köhler, a father of gestalt psychology who in 1913 started an anthropoid research station in the Canary Islands, conducted many similar experiments with chimps that culminated in his 1917 book, *The Mentality of Apes*. The Köhler Primate Research Center, located within the Leipzig Zoo and a project of the Max Planck Institute for Evolutionary Anthropology, is carrying on his legacy with not just chimps but also bonobos, orangutans, gorillas— the only place in the world to house all four great apes.

I toured the center with its director, Josep Call, a psychologist from Spain who earned his Ph.D. studying chimpanzees at Yerkes. "This is a place where if you have an idea, the next day you can go test it," Call told me.

Call, with his former Ph.D. adviser Michael Tomasello—who also had moved to Köhler from Yerkes and now was director of the institute's department of developmental and comparative psychology— were just completing a massive study that would appear in *Science*, boldly proposing what they dubbed the "cultural intelligence hypothesis."[27] Call, Tomasello, and their colleagues had compared 106 chimpanzees and thirty-two orangutans living in captivity to 105 human children. Each was given a battery of tests that evaluated their "physical" and "social" cognition. Physical cognition involved issues of space, causality, and quantities, and tasks such as using a stick to retrieve a reward out of reach or remembering where a reward was in order to locate it later. Social cognition tested whether subjects could observe a problem being solved and then solve it themselves, use pointing to communicate which of several cups turned upside down held a treat, or follow a person's gaze to find a target. On average, the chimps were ten years old, the orangs were six, and the humans were two and a half. Humans and chimps performed equally well on the physical tasks, getting about two-thirds of them right, and orangutans were just slightly below them. On the social tasks, the humans got nearly 75 percent correct—about double the performance of their ape relatives. The social cognition of these young children, they concluded, was "already well down the species specific path" and showed how humans are not just social, wrote Call and Tomasello, but "ultra social." This gap in performance, they wrote, went a long way toward explaining why chimpanzee "culture" does not approach the human version of the concept.

> Some other ape species transmit some behaviors socially or culturally, but their species' typical cognition does not depend on participating in cultural interactions in the same way as it does in humans, who must (i) learn their native language in social interactions with others, (ii) acquire necessary subsistence skills by participating with experts in established cultural practices, and (iii) (in many cultures) acquire skills with written language and mathematical symbols through formal schooling. In the end, human adults will have all kinds of cognitive skills not possessed by other primates, but this outcome will be due largely to children's early emerging, specialized skills for absorbing the accumulated skillful practices and knowledge of their social group (so that a child growing up outside of any human culture would develop few distinctively human cognitive skills).

In an extension of the social intelligence hypothesis, then, the culture intelligence camp presented the bold vision that the primate brain grew because these ultra social skills pushed culture forward and made more demands on our mental faculties.

As Call stressed to me as we walked around and watched the different apes, "There's a distinction between culture and cumulative culture." For every one of the thirty-nine cultural behaviors that chimpanzees have been shown to transmit, it would be possible for a single chimpanzee to invent the action in its entirety within its lifetime. "A smart human could not invent a car or even the thing you are writing with," Call said. "Or imagine if you had to invent algebra in a lifetime and invent Arabic numerals. Without that given to you by your culture, you're not going to get there."

This idea, which Tomasello first proposed in 1999 and called the Ratchet Effect, helps to explain what makes human cognition unique. "If you raised a human baby on a desert island outside of any kind of culture, that child's cognitive abilities as an adult would be very similar to other apes," Tomasello told me. "What's really different is something in the direction of culture. All of the things we consider our highest achievements, including language, and symbolic mathematics, and social institutions like governments and universities, these are cultural products. This isn't one person's brain power. These are collective efforts. That child on the desert island wouldn't invent a language, mathematics, or a university by him- or herself."

Tomasello sat back and laughed. He began his work studying child psychology, and back then he treated children as a different species. He explained that he was not so much interested in "some evolutionary fairy tale" about how ratcheting had come about. "I'd love to know the answer, but I'm more data oriented than that," he said. "I'm more interested in the 'mechanism' question and what is it that human children bring to their learning experiences that enables them to take advantage of all this cultural stuff that other species don't bring to the table when they are developing."

Tomasello has taken much heat from the likes of Frans de Waal and Christophe Boesch—who now work in the same building—for setting this generational-transmission divide between humans and chimpanzees. "I say to them, 'There is a line, is there not?'" Tomasello explained to me, seemingly amused by the criticism. "'Isn't there something fairly different?' They will always say 'yes.' And I'll say, 'If you don't like my line, what's the alternative?' And they never come up with an alternative."

From the point of view of engaging the public, which may not believe in evolution or care one whit about the conservation of chimpanzees, Tomasello said he understands the impetus to portray chimpanzees as almost human. "But for those of us who have been trained from birth in Darwinian thinking . . . we don't need to be convinced of that," he said. Nor does he need any convincing that humans should do more to protect wild chimpanzees and treat ones in captivity more humanely. "Let's get on with it and identify the cognitive and psychological differences," he said. "It's really hard to find them." He ticked off a few examples. Chimpanzee babies don't cry. Wild chimpanzees don't point, but one-year-old babies do. We gossip, a form of social bonding, and they don't. We roll our eyes to communicate that someone odd just boarded an airplane. "We share information freely," he explained. "Apes don't do that. They groom, they have sex, and they get food."

For all his emphasis on chimpanzee and human differences, Tomasello gradually has come to the conclusion that it usually is a matter of degree and not kind. He once believed that chimpanzees did not have a Theory of Mind, but he backed off that hard-line stance after experiments with captive animals demonstrated that they could understand—to a degree—the psychological states of others.[28] He also has adopted a more nuanced view about chimpanzee cooperation: a clever study he coauthored in *Science* in 2006 concluded, "Human forms of collaboration are built on a foundation of evolutionary precursors that are present in chimpanzees and a variety of other primate species."[29]

Tomasello did not actually carry out the experiment. It involved eight chimpanzees that live on Ngamba Island in Lake Victoria in Uganda, a sanctuary established in 1998 to provide a safe home for orphaned and confiscated chimps. To study chimpanzee cooperation, Brian Hare, a Tomasello disciple also then with the Leipzig institute, and his doctoral student constructed a cage of three rooms separated by doors. By removing a wooden peg, or "key," the chimps could open the doors between the rooms.

The researchers placed trays of food on a wooden platform outside of the cage. The chimps could retrieve the food by pulling on a rope, which would bring the entire platform to the cage's edge. But there was an important limitation: the chimps had to work together to move the platform. In an ingenious apparatus designed by Satoshi Hirata, a former student of Matsuzawa's, two ends of the rope had to

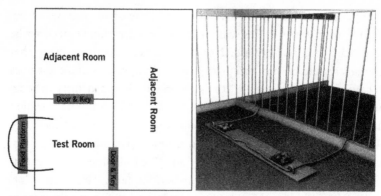

The baited food platform, designed by Satoshi Hirata, tests cooperation.
Researchers place the platform several feet away from the cage such that the
two rope ends are too far from each other for one individual to pull the
platform close enough to reach the food.

FROM ALICIA P. MELIS, BRIAN HARE, AND MICHAEL TOMASELLO, "CHIMPANZEES RECRUIT
THE BEST COLLABORATORS," *SCIENCE* 311:1297–1300 (MAR. 3, 2006).

be pulled simultaneously to move the platform toward the cage; if
only one end was pulled, it would unthread itself from the platform.
"If ever there's a Nobel Prize awarded for equipment, Satoshi Hirata
should get it," said Tomasello.

By adjusting the distance of the apparatus from the cage, the
researchers could create two different scenarios. In one, "Solo," the
platform was near the cage, which left the rope ends close enough
apart so that one chimp could hold both ends. In the other, "Cooperate," the platform was pushed away from the cage such that the rope
ends were nine feet apart, forcing the animal to find a partner to succeed. For Cooperate, both chimps had to be in the same room of the
cage, which meant that a chimp had to remove a wooden peg, "unlock-
ing" the door, to invite another one into the room. The chimps challenged with the Cooperate setup not only sought partners but also
quickly learned to discern between two potential partners based on
their experiences succeeding—or failing—to get the trays of food the
day before.

In describing the chimps' behavior, Tomasello underscored that
humans cooperate "in much deeper ways" than chimpanzees. He also
does not put much stock in the notion that they teach, noting that if

they did it regularly, Christophe Boesch and other researchers working with wild chimpanzees would have documented more instances of it taking place. And while Tomasello has done experiments showing that chimpanzees can exhibit altruistic behavior, he stressed that "humans are exceptional with respect to the breadth in which they help."[30]

Nevertheless, Tomasello has been persuaded by his experiments that several aspects of human cognition that he once thought were uniquely human exist, at least in rudimentary form, in our primate cousins. But save for the rare examples, such as Ayumu's facility with the masked numbers in the short-term memory test, we win most every cognitive competition hands down. And that includes our ability to change our minds when the experiments we conduct yield results that conflict with what we believe.

———◇———

AT the end of the Mind of the Chimpanzee conference, Jane Goodall gave a public talk, much of it about her Gombe days more than forty years earlier, at the Navy Pier Grand Ballroom. Thousands of people came to hear Goodall, as though she were some sort of spiritual guru. She had, after all, connected us to another species in a new way, showing us they not only had families and the ability to fashion and use tools, but also had feelings and, yes, minds.

As long lines of people waited for the doors to open, I sat on the bank of Lake Michigan with Brian Hare, who initiated the chimp cooperation studies at Ngamba Island. Hare was an undergraduate student under Tomasello and went on to earn a Ph.D. with Richard Wrangham, before rejoining Tomasello in his new lab in Leipzig. Hare had just conducted the first study comparing bonobo and chimpanzee cooperation, using the door/peg puzzle he had invented and the rope apparatus perfected by Hirata. But this time he had clumped the food reward in the center of the platform. When two animals successfully pulled the plank close enough to the cage for one of them to reach it, one chimpanzee monopolized 93 percent of the bounty while no bonobo ever took more than 63 percent of the food.[31] The "hippie bonobo," famous for resolving conflict with sex and being the kinder, gentler chimpanzee, had lived up to its reputation.

Humans are equidistant on the evolutionary tree from bonobos and chimps. We can no more ascribe our behavior to a chimp, in other words, than we can to a bonobo. "In the end, I think there are going to be huge differences between bonobos and chimpanzees," said Hare, who was continuing to pursue cognitive comparisons between the species at sanctuaries in Africa.

How, then, does Hare make sense of the cognitive differences between us and them? "Basically," he said, "I think apes have little bits and pieces of what we do."

—◁◌▷—

Two researchers who spent years of their lives studying cognition in chimpanzees did not attend the Mind of the Chimpanzee meeting: David Premack and Daniel Povinelli. Both had conducted extensive cognitive studies with chimpanzees, published books on the subject, and come to the conclusion that many, if not most, of the scientists in the field exaggerated the similarities between human and chimpanzee minds.

Premack was a generation older than Povinelli and had long ago stopped conducting experiments. But he continued to attack Darwin's assertion that humans were just "big-brained apes," arguing that too many of his former colleagues were "chimp huggers." He had a curmudgeonly side—he barked at me once, describing what I thought was a fascinating telephone interview the evening before as "a waste of my time, and of yours"—but I enjoyed his pluck and iconoclasm. "If you take a look at the field of primatology, which is one hundred years old, you'll only find two or three who did not confuse chimps with humans," he told me during that time-wasting interview. One of those scientists was Wolfgang Köhler. The second was Hans Kummer, Christophe Boesch's Ph.D. adviser and the world's foremost authority on the Hamadryas baboon. "And there's Premack," he said, referring to himself.

I first met Premack at the Project for Explaining the Origins of Humans, the invitation-only gatherings organized by Ajit Varki at UCSD. At the off-the-record daylong conference, which took place in a basement room of the Salk Institute, I was startled by Premack's dismissive attitude toward chimpanzee cognitive research—including that of his erstwhile student, Tetsuro Matsuzawa—but I found myself nodding in agreement, along with many of the other participants, as he laid out his arguments. Not long after the meeting, Premack published a carefully referenced scientific screed that captured many of the points he made that day.[32]

Forget the question of whether chimpanzees teach. "Teaching, the attempt to correct others, is the social side of the attempt to correct self," he wrote. "It is no coincidence that humans both practice and teach, whereas other species do neither. A species that practices but does not teach—that corrects itself but does not correct others—will probably never be found. Nor will a species of the opposite kind, one that teaches but does not practice—corrects others but not itself."

Recent claims that chimpanzees demonstrate a Theory of Mind were poppycock. "Chimpanzee mothers do not recognize that their infants lack knowledge and cannot therefore, for example, crack nuts with rocks," he chided. "Therefore, they do not teach them." He further declared that they "do not have the concept of knowledge" and said they "do not distinguish a knowing individual from an ignorant one." Humans, in contrast, understand what he called "embedded mental states," and he gave this example, especially entertaining in the pages of a staid scientific journal: "Women think that men think that they think that men think that women's orgasm is different."

Danny Povinelli had a similarly wicked wit. As head of the Cognitive Evolution Group at the University of Louisiana at Lafayette, Povinelli had studied ten captive chimpanzees, although his publications in recent years had switched from original research to synthesizing his thoughts about how the field had run off course. "My view of chimps has changed so much," he told me, explaining why he now found running experiments less interesting than he had at the start of his career. "I've personally satisfied my curiosity."

Like Premack, Povinelli assailed Darwin's contention that the "difference between the mind of the lowest man and that of the highest animal . . . great as it is, certainly is one of degree." As he and two colleagues noted in a 2008 essay, "Darwin's Mistake," just because we look alike does not mean we think alike. "The profound biological continuity between human and nonhuman animals masks an equally profound functional discontinuity between the human and nonhuman mind," they wrote. "Human animals—and no other—build fires and wheels, diagnose each other's illnesses, communicate using symbols, navigate with maps, risk their lives for ideals, collaborate with each other, explain the world in terms of hypothetical causes, punish strangers for breaking rules, imagine impossible scenarios, and teach each other how to do all of the above."[33]

Povinelli vigilantly constructs his attacks. He builds points from obtuse psychological topics that delve far beyond the Theory of Mind: relations between analogies, higher-order space, hierarchies, causality, transitivity. Yet his foundational argument is simple: chimps do not think anything like the way humans do. They perceive what they can observe, nothing else. They have no concept of "ghosts, gravity, and God," he said. Humans "reinterpret" the world, and he believes that many researchers, as they interpret chimp behavior, miss this point. "There's been a lot of crappy chimpanzee science that has gone on," he told me.

In one of his most amusing and accessible essays, published by *Dædalus*, a journal of the American Academy of Arts and Sciences—not something routinely read by primatologists, in other words—Povinelli contends that Darwin and the legions of scientists who have since promoted the "evolutionarily dubious proposition" that chimpanzees are "watered-down humans, not-quite-finished children" have caused a mess of confusion. "After several decades of being fed a diet heavy on exaggerated claims of the degree of mental continuity between humans and apes, many scientists and laypersons alike now find it difficult to confront the existence of radical differences," he wrote.[34]

Povinelli, whose interest in chimpanzee cognition stretches back to when he heard, at age fifteen, about the infamous study that said chimps could recognize themselves in a looking glass, explained that he once believed the Darwinian view that slight differences separate us from them. "But the results of over two hundred studies that we have conducted during the past fifteen years have slowly changed my mind," he wrote. "Combined with findings from other laboratories, this evidence has forced me to seriously confront the possibility that chimpanzees do not reason about inherently unobservable phenomena." If that's true, they cannot reason about what others think, believe, or feel.

He uses Matsuzawa's experiments with Ai as but one example of how science can easily fool us into granting chimpanzees mental faculties that they do not actually possess. Yes, Ai can touch numbers in a sequence and understands that four comes before five. But the evidence also suggests Ai only understands numbers as something to associate with real objects, something visible and tangible, not as "theoretical things," a strictly human conception. Ai doesn't understand that five is *greater* than four, let alone that five is one more than four—that there are relative concepts of quantity involved. Consider that human children at a young age figure out that six is *one more* than five, seven is *one more* than six, and so on. Ai—in Povinelli's estimation "the most mathematically educated of all chimpanzees"— never had that "aha" moment. Once Ai successfully connected a number to a set of objects (the number five to five circles), the researchers would introduce the next number, and it would require just as much time for her to learn it. "In other words, there appeared to be little evidence that Ai understood the symbols as anything other than associates of the object sets," Povinelli wrote.

He then brought up "zero," the definition of something that cannot be observed. Did Ai understand the concept? Although experiments showed that she could associate no objects in front of her with the number, she did not seem to know that zero was less than six, and she often confused zero and one. "However one wishes to interpret such findings, they are certainly not consistent with an understanding of the very essence of zero-ness," he concluded.

Skeptical as he has become of chimpanzees' cognitive abilities, Povinelli has made conservation a top priority in his work, and he understands why his thinking will upset people outside of the research community. In his *Dædalus* article, he appended an apologia of sorts, which rang with a poetic, earnest, and hard-earned truth.

> I have laid out the case for believing that chimpanzees can be bright, alert, intelligent, fully cognitive creatures, and yet still have minds of their own. . . . I cannot help but suspect that many of you will react to what I have said with a feeling of dismay—perhaps loss; a sense that if the possibility I have sketched here turns out to be correct, then our world will be an even lonelier place than it was before.

But for the time being, at least, I ask you to stay this thought. After all, would it really be so disappointing if our first, uncontaminated glimpse into the mind of another species revealed a world strikingly different from our own; or all that surprising if the price of admission into that world were that we check some of our most familiar ways of thinking at the door? No, to me, the idea that there may be profound psychological differences between humans and chimpanzees no longer seems unsettling. On the contrary, it's the sort of possibility that has, on at least some occasions, emboldened our species to reach out and discover new worlds with open minds and hearts.

The push to determine the amount of "humanness" in the chimpanzee has, in Povinelli's view, thrown the field into a "great freeze." His prescription for moving forward entails more than simply designing and performing better cognitive experiments. It requires taking a step backward and looking at anatomy, assessing how, specifically, the chimpanzee and human brains differ, not just in size but in their fine architecture and the genes that create it—a perspective that, incidentally, the other king of the skeptics, David Premack, wholeheartedly shares. "Remarkably, the details of the internal organization of human and great ape brain systems and structures have been largely ignored, in part because it's so difficult to study these brains, but also because most neuroscientists have frequently assumed that despite great differences in size, all mammalian brains are organized pretty much the same," lamented Povinelli in *Dædalus*. "Fortunately, even this is beginning to change."

8

HEAD TO HEAD

D O *WE HAVE CHIMP BRAINS,*" SAID TODD PREUSS, WITH MOCK incredulity.

A neuroanatomist at Yerkes, Preuss was not pondering whether humans have the brains of a chimp. He was repeating my question for effect as he gave me a tour of his lab in November 2008. He opened a refrigerator whose shelves were filled with chimpanzee brains floating in plastic tubs. "This is Pollyanna," he said pointing to one of the tubs. "This is Jenda. That's Artifee." In all, Preuss had two dozen properly preserved chimp brains in his fridges and freezers, making it one of the largest collections of its kind in the world.

He snapped on a pair of latex gloves and lifted a brain for me to inspect. The chimpanzee brain, three times smaller than its human counterpart, fit neatly into the cup of Preuss's palms. Chimp brains weigh 330 to 430 grams, while humans tip the scale at 1,250 to 1,450 grams—roughly one pound versus three pounds. I had seen many human brains in previous field trips, and to my eye, this brain simply looked like a smaller version. But that is a pervasive misconception that Preuss's research has helped lay bare.

Preuss's lab is one of a half dozen that do comparative studies of the chimpanzee brain to other species, and he is a member of an even more elite club that merges genetics, neuroscience, and anatomy research. Although his work has not received the attention that geneticists have won for isolating differences between human and chimp brains—such as the uniquely human *FOXP2*—he emphasizes that linking genotype (DNA) to phenotype (function) has vexed the field. "It's proven very, very difficult to make those connections," said Preuss, who became friends with Danny Povinelli in graduate school at Yale and for a time also worked at the University of Louisiana at Lafayette. "What do we know about the phenotypic differences between human and chimpanzee brains that can explain these genetic differences? Human brains are bigger. And that's almost all that we know. Now, there must be more differences."

Along with bipedalism and language, our bigger brain is one of the key distinguishing factors between humans and chimpanzees, and the neocortex—the outermost layer that dramatically evolved in mammals—stands out for its relative expanse. As it turns out, size alone reveals little and can even be misleading. Logically enough, brain size has a relationship to body size. The sperm whale brain, the largest of any species, weighs nearly eighteen pounds, six times more than a human's. But this tells us nothing about the intelligence of sperm whales in relationship to humans: if they were six times smarter, surely they would have let us know by now.

Comparing the brain size of a species to its total weight helps level the playing field. The sperm whale, for example, weighs an average of thirty-seven tons, which means its brain accounts for 0.02 percent of its body weight. In a human male who weighs 160 pounds, the brain accounts for nearly 1.8 percent of his weight. A male chimpanzee's brain is roughly 0.7 percent of his weight. So by this measure, the human brain has one hundred times the brain-to-body weight of the whale, and two and a half times that of the chimp. But this, again, has scant meaning: small mammals like mice have a monstrous brain-to-body ratio of 10 percent. All we can deduce from this comparison is that as the body grows in size, brain weight does not increase proportionately.

More evidence that brain size does not equal intellect comes from comparisons of modern humans to fossils of archaic humans. Researchers measure fossil skulls to determine cranial capacity and, indirectly, brain size. *Australopithecus afarensis*—which includes the superstar of fossil hominins, Lucy—had a cranial capacity similar to that of chimpanzees, and this capacity steadily grew over time in *Homo habilis* and then *Homo erectus*. But our closest known archaic relatives, the Neandertals, had a larger cranium than us. And they left no evidence they were smarter. They had larger bodies, though, which likely explains their bigger brains.

Head-to-head comparisons of modern humans further underscores that size does not equal acumen: people with celebrated mental musclepower do not necessarily have particularly large brains. Consider Albert Einstein, no slouch in the smarts department. Einstein's brain only weighed in at 1,230 grams, below the average of 1,400 grams for adult men.[1]

In 1973, the psychologist Harry Jerison proposed a more sophisticated algorithm to compare the intelligence of different species.[2] He called it the "encephalization quotient," EQ being analogous to IQ. After factoring in brain/body relationships, EQ calculates how much the brain's size differs from expectations. In the case of mammals, Jerison used the brain/body ratio in cats and assumed that brain size in other species should increase at about two-thirds the rate of the body size. So the cat, by definition, has an EQ of 1. In this scheme, humans do beautifully. We have an average EQ of 7.5. Chimpanzees are one-third of that. Bottlenose dolphins have a stunning 5.3, and the rat a meager 0.4. Chimps may not like the idea that dolphins are more intelligent, but the EQ has a certain logic to it that beats the brain-to-body ratio by itself.

Seductive as the EQ is, critics soon pointed out its weaknesses, too. White-fronted capuchin monkeys have an EQ of 4.8—more than twice that of chimps. Capuchins clearly have a lot going on upstairs, but the notion that they are twice as smart as chimps seems far-fetched. Chihuahuas did better than German shepherds and other large breeds, and, as one paper quipped, "few would claim that Chihuahuas represent the pinnacle of canine intelligence."[3]

Brain Weight, EQ, and Neurons in Mammals

Animal	Brain Weight (in grams)	EQ	Number of Cortical Neurons (in millions)
Human	1,250–1,450	7.4–7.8	11,500
Bottlenose dolphin	1,350	5.3	5,800
White-fronted capuchin monkey	57	4.8	610
Capuchin monkey	26–80	2.4–4.8	
Squirrel monkey	23	2.3	480
Chimpanzee	330–430	2.2–2.5	6,200
Rhesus monkey	88	2.1	480
Gibbon	88–105	1.9–2.7	
Whale	2,600–9,000	1.8	
Old World monkey	41–122	1.7–2.7	
Marmoset	7	1.7	
Fox	53	1.6	
Gorilla	430–570	1.5–1.8	4,300
African elephant	4,200	1.3	11,000
Walrus	1,130	1.2	
Camel	762	1.2	
Dog	64	1.2	160
Squirrel	7	1.1	
Cat	25	1.0	300
Horse	510	0.9	1,200
Sheep	140	0.8	
Lion	260	0.6	
Ox	490	0.5	
Mouse	0.3	0.5	4
Rabbit	11	0.4	
Rat	2	0.4	15
Hedgehog	3.3	0.3	24
Opossum	7.6	0.2	27

Source: Gerhard Roth and Ursula Dicke, "Evolution of the Brain and Intelligence," *Trends in Cognitive Sciences* 9:250–57 (May 2005).

—⟨∞⟩—

RESEARCHERS have attempted to compare the brains of species with several other metrics. Because large-brained animals have much more cortical folding, or gyrification, than smaller ones—mice have relatively smooth-looking brains, not cauliflowers like ours—brain anatomists have created a gyrification index.[4] They also have estimated the number of cortical neurons, compared the size of specific brain regions, and even analyzed the ratio of one type of brain cell to another. (Chet Sherwood a biological anthropologist at George Washington University, went so far as to show an increase in glia-neuron ratios in humans, a fine-honed detail that I especially like.[5]) But none of these parameters yield entirely convincing gauges of cognitive abilities. In the end, the attempt to understand the functional differences between a human and a chimpanzee brain requires more sensitive tools than a scale and a calculator.

Fortunately for Todd Preuss, he entered the field just as improved machines for soft-tissue imaging and DNA sequencing emerged, offering unprecedented windows into chimp and human brains. But Preuss's first major finding about a human/chimpanzee brain difference relied on a decades-old technique, which showed that the most important hindrance to clarifying these questions had less to do with technology than with mind-set. Neuroscientists for too many years had primarily relied on animals that are cheaper and more plentiful—and easier to experiment on—such as the mouse, rat, and rhesus monkey, to dig into an understanding of how the human brain works, Preuss said, repeating a complaint made by many others, including the Nobel Prize–winning codiscoverer of DNA's double helix, Francis Crick, in 1993.[6] This has led to many confusions about what makes the human brain special. "People don't think hard about these things because it's the model that drives the science," complained Preuss.

Valid concerns about resources and ethics have kept studies of the chimp brain on the sidelines. Few captive chimps exist in research settings, and they cost far more to house and care for than any animal routinely used by neuroscientists. The research climate also no longer would allow, say, the removal of a chimp's frontal lobes or cutting into a chimp's skull to insert electrodes—both of which took place in the middle of the twentieth century, when experimenters had no special concern for the welfare of great apes.[7] So rodents and monkeys, which routinely have their brains physically probed and are sometimes even "sacrificed" for the sake of neuroscience, became lab favorites. Still, Preuss and a handful of others recently have shown that they can study the neurology of chimps without causing the animals harm, by harvesting their brains at autopsy or subjecting them to minimally invasive procedures such as blood tests or neuroimaging. Their discoveries make it plain that when it comes to brain differences, we have a lot to learn.

—◦◦◦—

IN the language of evolutionary biologists, "human specializations" are the things that distinguish us from other species. In 2006, Preuss listed every known human specialization of our brain. The tally weighed in at two succinct paragraphs. "What slim pickings!" he reckoned. "It seems extraordinary that neuroscience has so little to offer on a matter so fundamental as what it is about our brains that makes us human."[8] Researchers still debate whether humans have uniquely asymmetrical brains. And it was not until 1997 and the use of MRIs to compare ape and human brains that a commonly held notion about a supposed key difference was dispelled: the relative size of the frontal cortex.

The frontal cortex is what allows us to paint and compose, to plan, speak, remember, to control muscles. As Katerina Semendeferi and her colleagues at the University of Iowa showed, the notion that humans, especially smart and creative ones, have large frontal lobes dates all the way back to the Greeks. But when Semendeferi was a graduate student in the late 1980s, she looked for scientific studies that proved this. "I assumed these things had been investigated and looked at extensively, given how much has been written about the topic and how interesting it is," Semendeferi told me. But she could not find the evidence. "I was shocked."

Semendeferi, who after earning her Ph.D. moved to the University of California, San Diego, met with me in her office, which has spare decorations save for a model of an orangutan brain. She laughed frequently as she recounted her improbable route to becoming one of the world's experts on comparative studies of ape and human brains. A native of Greece, Semendeferi began her graduate work as a paleontologist studying human evolution, but then switched to ape brains when she realized how little scientists had uncovered about their anatomy. She began her project by writing letters to about one hundred zoos around the United States, asking them for the autopsied brains of chimpanzees, bonobos, gorillas, and orangutans. "Nobody, nobody was asking for the brains of these animals anywhere in the States," she said.

As her collection grew, she followed the classical methodology of cutting the ape brains into thin slices, staining them, and then viewing these slides with an old-fashioned magnifying projector to map the distribution of different types of cells. But she had difficulty navigating through the various brain regions. "There was no atlas, nothing there to help me orient these brains, and they're very convoluted," she explained.

Then in 1993 Semendeferi learned that Hanna Damasio ran a lab nearby that studied MRI scans of living humans, and she thought that 3-D, X-ray-like pictures of the soft tissue of ape brains might help her anchor the slices and better visualize how all the pieces fit together. The university devoted the expensive MRI to exams of living humans, however, not the brains of dead apes, so the project began on the sly. "We snuck in at night with a gorilla brain in a bucket, and that's how the whole MRI thing started," she recalled. "We had to take many precautions, treating that bucket as if it were from another planet."

Soon, Semendeferi and Damasio had scans from many different ape brains, and they decided to compare them to MRIs from the brains of living and dead humans, mainly to confirm the established "fact" that humans had larger frontal lobes. "We did not believe our results initially," she said.

Semendeferi found that humans do have larger frontal lobes than other apes, but when she moved to the next step and calculated the volume of the lobe relative to the entire hemisphere that it came from, the distinction seemed much less impressive. It turns out that the human frontal lobe makes up 36.7 percent of a hemisphere's total volume. The chimpanzee's frontal lobe is 35.9 percent of the hemisphere. In other words, there is no difference. A comparison of surface area to the entire hemisphere, another measure of size, found that the chimpanzee had a slightly *higher* value than the human.[9]

Monkeys had a relative volume of 28.1 percent, so Semendeferi believes that the great apes did evolve a larger frontal lobe. But that

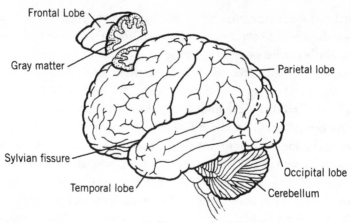

Frontal Lobe

Gray matter

Parietal lobe

Sylvian fissure

Occipital lobe

Temporal lobe

Cerebellum

Map of the Brain
The cerebral cortex, the outermost layer of the brain, has gray matter comprised of neurons that communicate with each other via their axons, which create the white matter below. The cortex divides into four lobes. Broca's area (language production) is in the frontal lobe and it connects to Wernicke's area (language comprehension) through the Sylvian fissure.
ADAPTED FROM ADRIENNE L. ZIHLMAN, *THE HUMAN EVOLUTION COLORING BOOK*, ILLUSTRATIONS BY CARLA J. SIMMONS (NEW YORK: HARPERRESOURCE, 2000).

applies to all hominoids, not just humans. She later repeated the experiment with living great apes, including bonobos, and confirmed her results. Semendeferi then extended the analysis to every other major brain region and again found no statistically significant differences—other than the cerebellum, the ball at the base of the brain that coordinates movements. It was relatively *smaller* in humans compared to the whole brain.[10]

Although the size of the brain, or even its major components, tells us next to nothing about what makes human cognition so extraordinary, Semendeferi stresses that size may well help expose differences between species. "A human brain and a great ape brain are not simply enlarged versions of each other," she said. "It's not like you can photocopy a slice, enlarge it, and then you just get the same thing two or three times magnified. That's not how it works."

Semendeferi noted that size does correlate with function in more defined brain regions. The cortex, the brain's so-called gray matter, depends on the white matter beneath it. White matter mainly consists of axons, which are wrapped in protective sheaths of myelin and send signals from one place to another. More white matter typically means an increase in the number of axons and in their width, which work together to speed the conductivity of signals. An MRI study, which Hanna Damasio again helped conduct, found that deaf people had less white matter in a ridge in the brain—called Heschl's gyrus—that processes incoming sounds.[11]

With that in mind, Semendeferi measured the ratio of white matter to gray cortex in other gyri. Humans had more white matter in the gyri of their temporal lobes. Chimpanzees and bonobos had unusually high levels of white matter compared to orangutans and gorillas, even though the larger apes have much larger brains.

When Semendeferi examined other well-defined, discrete brain regions she again found intriguing volumetric differences between species. In the prefrontal cortex, humans had an enlarged Area 10, which plays a role in learning rules and retrieving memories. The frontal lobes of orangutans—the most solitary great ape—had unusually small orbital sectors, which studies have linked to learning social rules. In comparison to bonobos, chimpanzees had larger dorsal sectors in their frontal lobes, a region critical to motor skills. One section of the human amygdala, a part of the more primitive limbic system that lies under the neocortex, is larger; this may help explain our hypersociability, as this region of the brain is involved in facial and vocal recognition, reactions to peer pressure, independence, and evaluating whether we should trust another person. Orangutans, interestingly, had the smallest amygdalas.[12] A comparison of Broca's area, the region in the frontal lobe associated with language production, found that humans had larger "minicolumns" than the other great apes.[13] These vertical stacks of nerve cells act like parallel processors and are, as my *Science* colleague Michael Balter elegantly described them, the "basic modular unit of neural information processing, one that can respond to many simultaneous stimuli at once."[14] It takes little imagination to see how language would benefit from a minicolumn advantage.

Semendeferi believes scientists must compare many more samples than she and others have analyzed to date, because a great deal of variation occurs between individuals within the same species. She also notes that the most important anatomical differences between humans and other species likely have to do with systems—how different areas of the brain connect—and not just the size of the isolated parts. "We have not really identified and measured any neural system across species," she told me.

Undoubtedly, a big-picture perspective, drawing on neural architecture all the way down to the genetic level, will go a long way toward revealing what makes the human brain special. But so will an even closer analysis of the brain's architecture, as Todd Preuss has shown.

—◦◦◦—

JUST as two homes with the same interior space—with the same volume—can have a profoundly different look and feel, the cells that build the brain can have entirely different staircases, plumbing, wirings, and windows. Preuss made his name in the field in 1999 with a paper that reached precisely this conclusion.

Preuss closely analyzed what is known as the primary visual cortex, an area in the occipital lobe at the back of the head. The primary visual cortex, which affects perceptions of space, motion, and luminosity, has a diverse variety of nerve cell bodies and their dendritic and axonal extensions, which separate into different layers. For more than a century, neuroscientists have used stains to highlight the cellular structure of brain slices. While staining slices of brain from four monkey species, orangutans, chimps, and humans, Preuss stumbled upon distinct differences in one sheet of the primary visual cortex called V1. Preuss realized the V1 architecture might explain why other studies had found humans to be more sensitive to a contrast in luminosity than macaques. Studies of the brains of people with developmental dyslexia also had found unusual pathology in the very area of V1 that seems to have more structural changes—probably because of more complex cellular connections—in humans.

This might seem like a minor detail, something that would hold little interest outside of a neuroscience conference. But as Preuss and his colleagues asserted when they published their findings, this was "the first documented feature of brain organization, not obviously related to differences in brain size, that distinguishes humans from apes."[15]

The same year Preuss announced his finding, Patrick Hof, a neuroscientist at the Mount Sinai School of Medicine in New York City, described a rare cell, the so-called spindle cell, in the brains of both humans and apes, but not in other primates. The small community of researchers that conducts comparative studies of human and chimpanzee brains rarely holds meetings to discuss findings, but at one such gathering at the National Academy of Sciences branch in Irvine, California, in November 2006, Preuss and Hof gave back-to-back talks.[16] Hof explained that he had stumbled onto spindle cells while studying Alzheimer's disease. The corkscrew-shaped cells, first described in 1881 and observed in a chimp brain as early as 1927, had received relatively little attention until Hof and his colleagues analyzed the brains of three humans with Alzheimer's in the early 1990s.[17] (Two of Hof's coworkers, incidentally, had wonderful names for people studying the chimpanzee/human differences: Esther Nimchinsky—which echoes Nim Chimpsky, the infamous research subject that overturned the ape language field—and John Allman.)

In the Alzheimer's brains they examined, the researchers found that the population of spindle cells, also called Von Economo neurons, or VENs, had decreased by about 60 percent. An exhaustive study later showed that twenty-three species of primates and thirty lower species did not have VENs in their brains, but bonobos, chimpanzees, orangutans, and gorillas all did.[18] The density of VENs, in elegant phylogenetic order, progressively declines from bonobos and chimps to gorillas and finally orangutans. And the cells are relatively larger in humans than in any other species, which suggests their axons may reach into more parts of the brain. "The bigger they get, the more human you are," Hof said at the "New Comparative Biology of Human Nature" meeting.

But it is the function of VENs that gives them their real cachet. They reside in a deep recess of the brain, the anterior cingulate cortex that sits at the border of the neocortex and the limbic system. A lesion in the anterior cingulate cortex can lead to mutism in humans, and it evolved in parallel with Wernicke's area, which is critical to understanding language. That led Hof to believe that it might play a role in the development of complex communications in both apes and humans. Imaging studies of live humans also indicate that the anterior cingulate cortex works overtime in people who are more socially aware and have more self-control.

Allman in 2006 offered more evidence that VENs have strong ties to social skills. Frontotemporal dementia affects as many humans under the age of sixty-five as Alzheimer's disease. A primary difference between the two conditions is that people with Alzheimer's "show intact social graces," as Allman and his fellow researchers put it, until near the end. Frontotemporal dementia, in contrast, leads people to lose empathy, become hypersexual, and even stop talking. "They can see someone get in a car accident and be in pain, and they don't feel anything," Hof told me.

Allman and his team looked at VENs in the postmortem brains of seven people who had frontotemporal dementia, five who had Alzheimer's, and seven controls who had no known neurological problems. They found that people with frontotemporal dementia had a 74 percent drop in VENs from the controls. (Alzheimer's brains, it turned out, did not show a significant loss of VENs, conflicting with Hof's original finding. He told me this likely reflects the fact that the first brains he analyzed were severely diseased, and Allman's follow-up studies included everything from mild to severe cases.) So VENs, once again, stuck out as a brain component that helped make us and other great apes extraordinarily social creatures.

Hof closed his talk at the meeting with something of a bombshell: he and Allman had just found VENs in humpback, fin, and sperm whales. "Here we have a very, very fascinating case of convergent evolution," he said, explaining that the cells must have evolved independently in the large cetaceans and primates millions of years ago. He noted that whales also exhibit self-awareness, form coalitions, cooperate, transmit cultural traditions, and use tools. In 2009, Hof, Allman, and Sherwood reported that they also had found VENs in the brains of elephants and dolphins, two other remarkably social species. "VENs represent a possible obligatory neuronal adaptation in very large brains, permitting fast information processing and transfer along highly specific projections and that evolved in relation to emerging social behaviors in select groups of mammals," they concluded.[19]

—◦◦◦—

TODD Preuss had moved beyond simply staining slices of cortex to hunting for microscopic differences between human and chimpanzee brains, tying together findings from genomics with the more old-fashioned techniques. "Houston, we have a problem," he said, by means of illuminating the limitations of using only brand-new genetic technology to understand the brain. Preuss was speaking to colleagues at the Irvine colloquium held by the National Academy of Sciences. "We have now a lot of genomic information about comparing humans to chimpanzees and other known primates and we have hundreds of differences. But we don't have anything like that number of phenotypic differences to match these genetic differences. So we've got a lot of genetic variation with basically nothing to explain it. Where are all the phenotypes?" Preuss's pioneering work, like Semendeferi's MRI studies, has tried to push the field into the modern era.

Working with a team that included Daniel Geschwind at UCLA, Preuss compared the brains obtained from five humans and four chimpanzees to see if they could find "expression" differences, which assesses the degree to which genes are turned on in each species. Basically, when a gene is turned on, its DNA produces messenger RNA, which then tells a cell to make the gene product (usually a protein). By measuring levels of messenger RNA, researchers can then quantify which genes are turned on and to what degree. In the study, they found eighty-three genes that were turned on higher in human brains.[20]

One of these genes came from a family called the thrombospondins, which codes for proteins that, among other things, lead to the formation of synapses, the ends of neurons that allow them to communicate with each other. "If you grow neurons in a dish, they don't make many synapses," Preuss told me. But add the right thrombospondins, as another research group first described around the time Preuss discovered the genetic difference, and boom, synapses flourish.[21]

All too often, gene expression studies find something interesting like this but do not go on to examine what the gene actually does in the body. "It's real easy to spin a story about what these genes do or don't do," said Preuss.

For Preuss, genomics provides a guide for conducting comparative histological studies, the analysis of those slices of human and chimpanzee brain tissue that helps to pinpoint their components. The proteins that genes express, such as the thrombospondins, can appear in different types of cells and in different quantities. So the same protein may have different functions in different places. Identifying that a gene is overly expressed in a human brain, then, only offers a starting point. It is the fingerprint at the murder. Finding the suspect, establishing motive, and proving guilt is the end game. "The frustration is that the genomics is very high throughput and gives you hundreds of differences, some of which are probably real and some of which are probably not," Preuss said. "Once you get away from RNA and into proteins and histology, the work becomes orders of magnitude more labor intensive and orders of magnitude slower. We can survey hundreds or thousands of genes with a microarray of the RNA, and the histologically based studies of these are pretty much one gene at a time. They go very slowly."

In 2007, four years after Preuss reported that human brains had higher levels of thrombospondins, he and his team published a study on the distribution of thrombospondins in the brains of humans, chimpanzees, and macaque monkeys. The main difference they found was that humans had a higher density of one of the thrombospondins in a frontal cortex region where synapses form. "This is the first study to relate changes in gene-expression levels in human evolution to changes in the localization of proteins within the brain," they declared.[22]

Humans, they concluded, have "greater plasticity" in the cortex, which could, in turn, allow them to reorganize a network of neurons based on what they experience. Thrombospondins may help explain why humans are unusually adept at learning and remembering.

—◦◦◦—

LIKE cartographers during the Age of Exploration, researchers who study the genetic differences between the brains of humans and chimpanzees keep revising their vision of what this world looks like; other than the contours of the continents, they have drafted few reliable landmarks. Yes, proteins coded for by the thrombospondin and *FOXP2* genes distinguish the human from the chimp brain, but they remain rarities. For the most part, researchers are at the stage of discovering forests, not trees.

UCLA's Daniel Geschwind devised a smart strategy to determine which trees in the forest of brain genes lead to human specializations, and it relies on comparing chimpanzees to humans. Just as Geschwind used subtle differences between people like left handedness and dyslexia to make sense of normal brain architecture and its relationship to language, he measures faint but detectable differences in gene networks (the forests) between the brains of chimps and humans to find specific gene interactions (the trees) that look unique to us. He first takes slices from the same region of the brain in each species and maps the genes that are expressed at the same time. His maps resemble airline routes, with hubs representing major cities of genes and spokes connecting to more remote genetic locales. Genes that are coexpressed with lots of other genes sit at the hubs. "A gene's position in a network has huge implications," he said.

Like Semendeferi, Geschwind and his team analyzed gene expression in specific slices of cortex dissected from chimpanzees and humans to identify which of the four thousand genes in the brain are coexpressed. They discovered that 17.4 percent of the connections in the cortex are unique to us. Separating the hubs and spokes that are uniquely human, Geschwind isolates an intriguing network of genes, but the role they play in the construction or function of our brains remains as fuzzy to him as the interior of the New World was to sixteenth-century adventurers. Still, he has created maps that he and others can use and refine, which eventually should lead to the land of milk and honey: the specific genetic factors that make our brains unique.

Another approach to unraveling the specific genes that create the human brain relies on the molecular clock used to calculate rates of evolution. When DNA mutations accumulate more quickly than predicted by the molecular clock, it suggests that they somehow benefit the organism (as opposed to a slower accumulation of changes that have no effect). In 2004, Bruce Lahn at the University of Chicago identified genes involved with growth of the brain that had gone through such positive selection.

Primary microcephaly is a disease that leaves humans with a chimp-sized brain. The brain is a shrunk version of the normal one, with the same basic architecture, although people with the disease have severe mental impairments. Building on research that identified mutated genes responsible for the abnormality, Lahn found that two of them,

Unique Neural Hubs and Spokes

The left panel shows the coexpressed genes in a slice of human cortex. This is the airline route for that country in the brain, and it is impossibly complex. The right panel shows the human cortex—but removes the hubs and spokes that also appear in the chimp cortex. These are the "uniquely human" connections, including two major hubs.

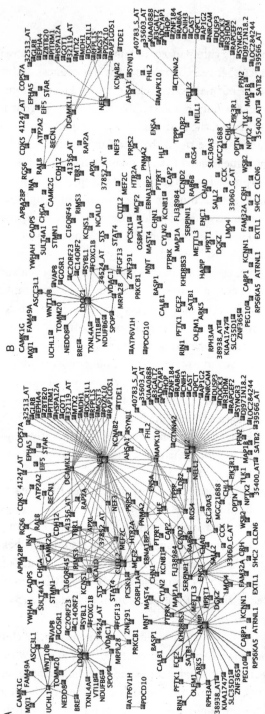

FROM MICHAEL C. OLDHAM, STEVEN HORVATH, AND DANIEL H. GESCHWIND, "CONSERVATION AND EVOLUTION OF GENE COEXPRESSION NETWORKS IN HUMAN AND CHIMPANZEE BRAINS," *PROCEEDINGS OF THE NATIONAL ACADEMY OF SCIENCES* 103:17973–78 (NOV. 21, 2006).

ASPM and *MCPH1*, likely went through positive selection in humans. Two other genes associated with the condition, *CDK5RAP2* and *CENPJ*, showed evidence of being positively selected in primates as compared to rodents. All four of these genes help to regulate the embryonic growth of the cells that go on to make neurons.[23]

Then in 2006, researchers reported the discovery of *MGC8902*, forty-nine copies of which appear in humans and ten in chimpanzees— far more than the standard two copies in the genome. The gene itself oddly contains a section that repeats itself, *DUF1220*, which codes for a protein with an unknown function that appears in high levels in neurons.[24]

Finally, the supposed "junk" DNA in the genome, the portion that does not harbor genes that code for proteins, contains sections that regulate gene expression and appear to have shaped our brains. A sophisticated computer program designed by Katherine Pollard at the University of California, Santa Cruz, allowed her to query the genomes of several species for the most rapidly evolving regions. The stretch she dubbed human accelerated region 1, or HAR1, had more changes than any other between humans and chimpanzees, and it is expressed in neurons found in the neocortex of developing human embryos. Pollard and her colleagues also found it in adult brains.

The publication of the HAR1 findings yielded newspaper headlines blaring nonsense. "Revealed: The Gene That Gave Us Bigger Brains," said one. "Scientists ID Genes That Make Humans Smarter Than Chimps," announced another.[25] The names of these putative brain makers resemble license plates, and to my mind, that is roughly how much information they contain. They can lead us to a car and maybe a driver. But in and of themselves, they do not tell us anything about the traffic pattern on a big city's freeways, which, more or less, resembles the frenetic but organized way that signals fly around the brain and allow us to compute an algebraic equation, compose a song, or cry at a movie. Then again, as genome scientists repeatedly point out, these are early days.

—◦◦◦—

JIM Rilling sat in front of a large computer monitor in his lab and pulled up a spectacular image of a chimpanzee's brain. Rilling, an

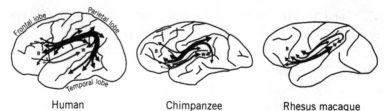

Human   Chimpanzee   Rhesus macaque

Using diffusion tensor imaging, Jim Rilling has mapped the synaptic pathways that explain why only humans have complex language. These diagrams show the essential role of the arcuate fasciculus, which connects Broca's area to Wernicke's, in humans versus chimpanzees and rhesus monkeys. The arcuate is much more elaborate in humans and reaches past Wernicke's area and into the lower temporal lobe.

ADAPTED FROM JAMES K. RILLING, MATTHEW F. GLASSER, TODD M. PREUSS, ET AL., "THE EVOLUTION OF THE ARCUATE FASCICULUS REVEALED WITH COMPARATIVE DTI," *Nature Neuroscience* 11:426–28 (APR. 2008).

anthropologist who works at Emory University in Atlanta and collaborates closely with Todd Preuss, had created the image by putting the postmortem brain of a chimp into an MRI scanner and using a relatively new technique that tracks the diffusion of water molecules through tissue. Called diffusion tensor imaging, or DTI, it has allowed Rilling to delineate tracts of white matter, the axons, that link different parts of the brain. Different shades represent different tracts. "The biggest and most important differences in neurobiology between humans and other species are going to come down to how the brain is wired," Rilling told me. So he, too, is trying to make sense of the networks.

The Sylvian fissure designates the area that ranges from Broca's area (language production) in the frontal lobe back to Wernicke's area (language comprehension), which sits at the border of the temporal and parietal lobes. The bundle of white matter fibers that surrounds the Sylvian fissure sends messages between Broca's and Wernicke's. Rilling homed in on this tract, called the arcuate fasciculus, in macaques, chimpanzees, and humans. In human brains, DTI shows that the arcuate fasciculus runs from Broca's to Wernicke's and then beyond into the lower temporal lobe. Chimpanzees, in contrast, have an arcuate but it typically does not reach past Wernicke's homolog and occupies less space. Macaques have a smaller arcuate still.[26]

Rilling pulled up the DTI from a human and pointed to that

deeper region in the temporal lobe that abuts Wernicke's area. "This part of the brain is involved in lexical semantic processing—processing word meaning," he said. "So we think what the tract is doing is taking this information about word meanings and sending it up to the frontal lobe so that when you're having a conversation and you're deciding what you want to say to someone to express what you're thinking, you use the pathway to reach back into the temporal lobe and retrieve the words that go with the meanings you're trying to convey."

Chimpanzees, in essence, do not have the wiring to do this. But they do have a lot of the same wiring. Preuss, who was looking at the images with us, stressed that it is a quintessential difference of kind, not type. "Chimps don't have the same connections to these regions, but we have pretty good reason to think they have homologs to Broca's area and Wernicke's area," he said. "They must be doing something." No, chimpanzees do not sit around and argue about the subtle shades of meaning to a word, nor do they share our obsession with finding nuanced words to label everything we see (the arcuate fasciculus), feel (empathy), and think (a Theory of Mind). But they do communicate in relatively sophisticated ways compared to many other species, and these detailed wiring diagrams of the arcuate fasciculus give researchers a new way to compare and contrast *systems*—not just the size of a region, its architecture, or its cells.

Rilling has used another modern imaging technique, positron emission tomography, better known as the PET scan, to explore other aspects of the chimpanzee brain. PET scans can produce images of the brain at work by injecting into the blood a tracer such as radioactively labeled glucose. When they are active, neurons gobble up this sugar, allowing the PET scan to reveal specific areas of the brain that turn on under a given condition.

When humans rest, their minds wander. They talk to themselves, plan their vacations, think about last night's dinner, or visualize hitting a baseball over a fence. Brain imaging studies of humans show that when they move from a "resting state" to a task that demands their attention, some regions deactivate, suggesting that even taking a break requires work from some neurons.

Chimpanzees, unlike many other primates, can recognize themselves in the mirror, and there is some evidence that they are capable of understanding the intents of others. This led Rilling to wonder whether they experience a resting brain state of introspection similar to humans'. He tested the idea in eight adult humans and five adult chimps. After receiving the radioactive glucose, the humans sat quietly in a private room and the chimps rested in their cages for about an hour, the time it roughly takes for the brain to soak up the sugar molecules. All the study subjects then received PET scans.

Both chimpanzees and humans displayed loads of activity in the medial prefrontal cortex.[27] Rilling explained to me that in human studies, researchers have linked this region to a Theory of Mind. "It raises the possibility that if the same areas are involved in chimps, they can be doing something like mentalizing or mental self-projection just like humans do," he said.

Lisa Parr, a Yerkes psychologist who collaborated with Rilling on the resting state study, did a similar PET scan comparison of face recognition in chimpanzees and humans. "Chimps are really good face processors," she told me when I visited her lab in November 2006. Parr has become a world authority on the chimp's ability to recognize faces, a key facet of communication and social cognition. "People really like to talk about Theory of Mind, but I think that's too high of a process," she said. You also have to design complicated experiments to observe Theory of Mind, and even then, it's trying to observe a concept. Facial recognition, though lower on the scale of cognitive abilities, has the virtue of being witnessed easily.

Parr designed the PET scan studies around a test that gave chimps a reward if they could determine which two of three chimpanzee faces shown on a computer screen were the same. A second, similar test replaced the faces with objects. Studies with humans indicate that they use different parts of the brain to identify faces and objects. Parr wanted to see how chimps compared.

After drinking Kool-Aid laced with radioactive glucose, the five chimps worked on either the face- or the object-matching tasks. As with humans, the face recognition task lit up a region in the temporal lobe known as the fusiform gyrus, while the object task primarily activated a different region.[28] "Everything we've been able to discover so far says they do things the same way," she said.

Such sophisticated imaging techniques can, without causing harm, probe activity in each hemisphere, lobe, sulcus, gyrus, neuron, and axon that until now has remained invisible. As they track the contents, connections, and functions of every part of the brain, scientists are clarifying, like never before, how humans are almost chimpanzees, but why, in the end, they are not.

—◦◦—

ONE of the most pronounced differences between the human and chimpanzee brains has nothing to do per se with comparisons between size, specific connections, or types of cells. It has to do with development, and the start of brain growth and its completion. At birth, the human brain is only 25 percent of the adult brain, and the chimpanzee's is about 40 percent of what it will become. But at one year old, the chimpanzee brain has completed 80 percent of its growth. Humans will not reach 95 percent of their ultimate brain size until age ten. The late evolutionary biologist Stephen Jay Gould, in his seminal book *Ontogeny and Phylogeny*, wrote that this relative retardation in human brain growth was "a factor of paramount importance in human evolution."[29]

Jean-Jacques Hublin, an anthropologist who directs the department of human evolution at the Max Planck Institute for Evolutionary Anthropology, believes the prolonged growth of the brain may be particularly critical during childhood. At birth, most of the neurons that we will have for life are there, but the white matter continues to grow. "After birth, it is mostly the development of the wiring of the brain," Hublin told me when we met in December 2006 in his Leipzig office, which is filled with art depicting Neandertals, his main scientific specialty. "All the cables are connecting all the cortical areas, one to the other, and in this critical period there are a number of connections which are reoriented or deleted or created. So humans have this amazing feature in which most of the wiring of the brain takes place when they are interacting with other humans and the environment. It gives this incredible plasticity to the human brain, and it gives us a better chance in all the education processes that take place for many years after birth."

Hublin came to this controversial area because of his belief that archaic humans, based on their fossils, completed their skull growth earlier than do modern humans.[30] If Hublin is right—and he certainly faces challenges from many colleagues who think he is wrong—then our ancestors were more chimplike in their social structures, and maybe their intellectual abilities.[31]

Hublin pointed to studies recently completed in children with autism that back his theory. Brains of autistic children, who can have severe social problems, mature at an accelerated pace. At birth, autistic children have a normal head circumference. But at one to two years of age, head circumference becomes abnormally large. Between two to four years, their brains are 10 percent larger than normal. At three to five years, children with autism have 15 percent heavier brains.[32]

Eric Courchesne at the University of California, San Diego, who conducted the studies, has theorized that the failure to develop long-distance white matter connections leads to the "major physical and psychological social functions in autism." This neatly fits with Hublin's own thinking. "The longer time you have to build up all your connections, the more complex they can be, and the more you have a chance to model them as a result of social interactions," Hublin told me. "With these kids, especially in terms of interactions with other individuals, they don't have very sophisticated ways to build up social interactions."

Any attempt to address the evolution of the human brain and the speed of its growth ultimately requires a more holistic approach than the decapitated view researchers often present. A woman's pelvis, because humans are bipedal, is narrower than a chimp's, and women cannot safely give birth to babies with excessively large heads. We also rear our children for longer periods than chimpanzees, which is supported by the way in which men and women partner and extended family members chip in. And the brain, which accounts for only 2 percent of our body weight, has a voracious appetite, demanding 20 percent of the energy supply, which has reconfigured our digestive tract and our diets.

The body and mind connection, then, with complete disregard for spirituality, runs deep.

THREE

BODIES

Man still bears in his bodily frame the indelible stamp of his lowly origin.

—CHARLES DARWIN,
Descent of Man, 1871

9

WALK THIS WAY

ANIEL LIEBERMAN HAS SKULLS COMING OUT OF HIS EARS.
A professor of biological anthropology at Harvard University, Lieberman had packed his office with so many skulls that on my visit in February 2008 it reminded me of a shrine at the Killing Fields in Cambodia. But these were not victims of genocide, and the collection included everything from modern humans to Neandertals, *Homo erectus*, chimpanzees, monkeys, hyraxes, and even sheep. Lieberman, who dubs himself an experimental musculoskeletal biologist and paleoanthropologist, specializes in "how and why the human body looks the way it does." Which means he compares anatomical studies of skulls and many other bones and muscles.

Lieberman kept one skull out of view, hidden in a box.

"Do you want to see Toumaï?" he teasingly asked, well knowing what I would answer. Lieberman did not actually have the skull of Toumaï, formally known as *Sahelanthropus tchadensis*. The oldest hominin fossil yet found, Toumaï dates back perhaps 7 million years. But he had been given a cast made from the cranium, jaw, and teeth that the team led by Michel Brunet had found in Chad in 2001. Lieberman took it out and held it in his palms as though it were as precious as a crown jewel. "There's no way it wasn't bipedal, and yet it's unbelievably chimpy," he said.

I asked him how he could make this distinction.

The incisors and the brain size were chimp, he said. The ridge of the brows, the canine and cheek teeth, and the "slightly pulled out face," all had the markings of the human lineage. "But the big change is back here," he said, turning the face away from me. "The whole back of the skull has been rotated down."

Lieberman showed me a nearby chimp skull. "See that nuchal plane, and look at the nuchal plane in Toumaï," he said, gesturing to a prominent hump on the back of the skulls. "That's forty degrees different. That's how you know it's a biped." What could be seen of the foramen magnum, the hole at the base of the skull through which the spinal cord enters, also had a shape more similar to hominins than to chimpanzees. "It's basically a bipedal chimp," Lieberman concluded.

Although chimpanzees can walk upright for short periods, "habitual" or "committed" bipedalism requires different mechanics, and it marks the first critical anatomical distinction between the great apes and our more human ancestors. Hypotheses abound about how and why bipedalism evolved, and while some have only a few oddball proponents and others have prominent scholars behind them, none by themselves offer a thoroughly convincing explanation. Some are downright screwball. "Speculations on the origins of bipedalism are often fascinating exhibitions of ingenuity—expressing, above all, that it is a theater for intellectual daring," wrote Jonathan Kingdon in *Lowly Origin*, the first book devoted entirely to the subject.[1] A zoologist and artist based at Oxford University, Kingdon ends his "saga of origins" with a refrain commonly invoked by many who have thought deeply about bipedalism: it "remains a tentative and ongoing tale."

The primate morphologist Russell Tuttle of the University of Chicago provided the simplest and most entertaining distillation of the competing bipedalism theories. Tuttle, who incidentally came up with the term "knuckle-walking" to describe the way chimps move across the ground on all fours, coined comical names for the various two-legged camps.[2] "Peek-a-Boo" contends that we humans stood up when the African rainforests gave way to savannas of open grasslands, and we needed to see farther to find food and hunt. Similarly, "Tagalong" says that we joined herds of animals that crossed savannas to scavenge food, and we switched to two legs to walk long distances and also carry our babies efficiently. "Hot to Trot" argues that standing upright allowed us to cool our bodies more efficiently in the hotter open savanna. "Trench Coat" hilariously suggests that males stood up to show their penises to females, and "Swingers Go First" says using arms to brachiate through the trees reoriented our skulls in favor of bipedalism. "Schlepp," from the Yiddish word for "lug," posits that bipedalism freed the hands to carry food and offspring to a home, and "Hit 'Em Where It Hurts" extends this theory to bearing weapons and other tools. "Two Feet Are Better Than Four" looks at the reduced energy required to move as a biped. "Wet and Wild" claims we went through an aquatic phase that helped us learn to stand, whereas "All Wet" mocks the idea.

Many of these hypotheses are pocked with problems. The savanna-based proposals are specious given that early hominin fossils have been found in what were wooded areas.[3] The natural implication of Trench Coat would be that males walked before females, and no evidence exists to back that. If Swingers was valid, chimps, bonobos, and gorillas—all knuckle-walkers, and all evolutionarily younger than orangutans—should brachiate or habitually walk, and they do neither.

I asked Randall Susman, who has studied ape and human locomotion for decades—and pioneered field research of bonobos in what was then Zaire—whether there was any new compelling evidence in this age-old debate. "I have no idea why people became bipedal," said Susman, a professor at the State University of New York in Stony-brook. As for the various hypotheses, he said, "take your pick." In August 2009, Susman personally favored what he called the "Multiplex Theory," which holds that several of these forces combined to evolve the bipedal human. "Bipedalism only happened one time in all of mammalian evolution," he reasoned. "One animal, namely us and our progenitors, became bipedal. If it was a single cause, wouldn't you suspect it would have happened more than once?"

When scientists do publish new evidence supporting one theory or another, they easily garner headlines and inevitably rile colleagues—who are at once unconvinced and appalled by the supposed revelations about how humans acquired one of their most distinctive features. For example, after observing wild orangutans in Sumatra for a year, Susannah Thorpe from the University of Birmingham in England observed that they walked upright on thin branches, using their hands to stabilize their bodies.[4] The Sumatran orangutan is the only great ape that mainly lives in trees, and Thorpe charted 2,811 "locomotor bouts," meticulously documenting the diameter of the branches the orangs walked on. Thorpe, working with the anatomist Robin Crompton from the University of Liverpool, speculated in a 2007 report in *Science* that human bipedalism evolved when our ancestors similarly stood on weak branches to forage for fruit, which in Darwinian terms granted them a selective advantage. This suggested that humans had an intermediary stage between knuckle-walking and standing tall, unassisted.

For the theory to stick, they suggested the common ancestor to all extant apes lived in trees just as the orangutan still did, but as the forest canopy became fragmented and gave way to savannas, the African apes were forced to develop a broader locomotor repertoire. Gorillas and chimps thus also became expert vertical climbers, which requires hindlimb and forelimb motion similar to knuckle-walking. When chimps walk upright, they use a bent-knee, bent-hip posture that does not resemble the straight-limbed bipedalism of orangutans in trees or of humans. Humans, rather than evolving the unique ability to walk upright, merely retained an adaptation that existed in other ape ancestors. If true, hominins simply remained "Upwardly Mobile," while other apes did not. This radical notion removes bipedalism as the original dividing line between us and them.[5]

Many other researchers guffawed at Thorpe and Crompton's Upwardly Mobile thesis. Fossils of early humans showed evidence of knuckle-walking. No other ape has the cranial shape found in humans that is linked to the orientation of the bipedal spine. Hand-assisted arboreal bipedalism also dramatically differs from human walking. And orangs do not look particularly humanlike when they walk on thin branches. "Orangutans are awkward bipeds, using partly abducted hips, with their knees facing laterally, and walking on the sides of their feet," wrote a group of anthropologists in a stinging critique published by *Science*.[6] "In these respects, chimpanzees and bonobos appear to be better bipeds."

Thorpe and Crompton punched back. They even made curious reference to Poko, a chimpanzee raised in a narrow parrot cage that mainly walked bipedally yet still did not maintain the humanlike posture of an orangutan. But at the end of the day, the hypothesis that hand-assisted arboreal bipedalism evolved into human walking was just a stroll around another intellectual cul de sac. Dan Lieberman's harsh reaction reflected the way many in the field viewed the argument. "Very unimpressive," he told me. "What does the orangutan have to do with human evolution? So what? The fact that orangs sometimes stand up to feed in a tree doesn't explain why we're bipedal."

In fact, Lieberman and other critics of Thorpe and Crompton were shifting their support to a resurgent viewpoint, Two Feet Are Better Than Four. The idea had fallen out of favor in 1973, when researchers trained two chimpanzees to run on a treadmill quadrupedally and bipedally.[7] Masks strapped onto their faces measured the amount of oxygen they used when they ran at different speeds. The chimps sucked up the same amount of oxygen—an indirect measure of energy used—regardless of whether they were on all fours or standing up, and the same was true for two capuchin monkeys included in the study. Bipedalism, the researchers concluded, offered no energy advantage. But three anthropologists from different universities revived this explanation in 2007 with a similar study that compared five chimps to four humans.[8]

In the original experiment, the two chimpanzees were young, with an average weight of 17.5 kilograms, about 38.5 pounds. The anthropology trio suspected this biased the findings, and their chimps ranged in age from six to thirty-three years old, weighing between 33.9 and 82.3 kilograms, about 75 to 180 pounds. They found that the cost of quadrupedal and bipedal walking differed between individual chimpanzees, and human walking was a startling 77 percent less expensive. When they further analyzed the biomechanics, they showed that chimpanzees, because they have shorter legs, hit the ground harder with each step and had to take more steps to travel the same distance. "Knuckle-walking is very, very costly," Lieberman asserted. "It's a compromise that only evolution can dream up between an arboreal, tree-climbing and -hanging athlete—if you've ever watched chimps in a tree, they're phenomenal—that can also quadrupedally locomote."

Why, then, the switch to bipedalism? "It enables you to travel longer distances, presumably between patches of disparate resources, at less cost," Lieberman said. But if you're a biped, you can no longer gallop, he noted, which makes it more difficult to move quickly and ambush prey.

Roughly 1.8 million years ago, *Homo erectus* emerged, and these creatures not only could walk, but also possessed a skill that further separates humans from chimpanzees and all other apes: they could no longer gallop, but they could run for long distances.

—✧—

THE study of endurance running has helped Lieberman make his mark in the field of human evolution, and the day I visited, I watched him teach his undergraduate class on the subject. The class was held in the august Harvard Hall, and, with a crew videotaping the packed room so students could later browse it at their leisure, the setting was theatrical and even grand. It was snowing outside, and many entered with coats so fluffy and furry they resembled arctic explorers, a lovely and absurd contrast to the day's focus on human foragers and hunter-gatherers in sub-Saharan Africa.

"Any of you like to run? Like to jog?" Lieberman asked rhetorically. "Any of you who can run six miles, you can run down any mammal. Even I, a middle-aged Harvard professor, I have run down jackrabbits. You have to make them gallop. As soon as an animal gallops, it cannot pant." Endurance running, he explained, also helps hunter-gatherers scavenge: when they see vultures circling in the distance, they can race to the site and beat other carnivores to the carcass. And you don't even need weapons like spears, which the evidence suggests only began to be used by human ancestors about four hundred thousand years ago.[9]

In 2004, Lieberman teamed up with the morphologist Dennis Bramble of the University of Utah to pull together a sweeping review of the evidence about endurance running and human evolution.[10] "The bizarre thing about humans is not that we can run, but that we can run so long for endurance distances," Lieberman told me. "That's what has to be explained. The ability to run a marathon is not something that's just freakish for some human being. It's an intrinsic ability of human beings. . . . We have all of this anatomy that evolved from running."

Lieberman and Bramble's provocative argument ties human endurance running to a wide range of anatomical features that we do not share with chimpanzees. For starters, we have relatively huge rear ends. "You don't use your gluteus maximus for walking," said Lieberman. "It's very easy to see. Walk around the room and hold your butt and nothing happens. You start running, and it starts crunching up." Indeed, Lieberman, Bramble, and their colleagues formally proved this by placing electrodes on the gluteus maximi of volunteers who ran and walked on a treadmill, demonstrating with electromyography that the muscle contracted little unless a person was running. The gluteus maximus, they concluded, plays a crucial role in stabilization during running.

Tendons attach muscles to bones. Long tendons in the legs of humans also serve as springs, cutting the metabolic cost of running in half. Chimpanzees and gorillas have tiny tendons.

When it comes to bones, humans have much longer legs than arms, which means our stride during running is typically twice as long as that of a chimp, six rather than three feet. To reduce the skeletal stress of running, we spread the force over a larger area, leaving us with larger surface areas on many joints including the top of the femur, the knee, and the sacroiliac that links the spine to the pelvis. Relatively short toes lower the mechanical and metabolic costs of running, and possibly reduce the risk of injury.

Lieberman allows that "an incredibly tiny" number of other species can run for many miles over several hours, including horses, dogs, wildebeests, and hyenas. But humans have one other unique feature that makes them exceptional endurance runners. "We're the only ones that can do it in the heat of the day," he said. And our extraordinary thermoregulatory feats derive from another aspect of human biology that differs markedly from chimpanzees: our skin.

SKIN is the human body's largest organ, and, when peeled off an adult, averages twenty square feet and weighs nine pounds. Nina Jablonski, the head of anthropology at Pennsylvania State University, runs through these details in her book *Skin*, which repeatedly emphasizes that humans take for granted its importance.[11] And one of the skin's most critical functions is to cool us down.

Human skin contains eccrine glands that secrete sweat, which, in turn, evaporates and cools the body. Dogs, in contrast, rely mainly on panting to cool down, and larger mammals like cows use a jumble of arteries in their brains to keep that critical organ from overheating.

"Chimpanzees are fairly good sweaters," Jablonski told me in September 2009. But they only have about 20 percent of the number of eccrine sweat glands that are found in humans. "Chimpanzees can engage in short bursts of activity, but they cannot run a marathon. After a sporadic burst, they have to pause and literally cool down."

Not only do we humans have more eccrine sweat glands to prevent our blood from boiling; we have very little hair for an ape. Yes, we have patches of hair on the head, at the crotch, under our arms, and typically, peach fuzz on our limbs and torso. But the hairiest humans never have the fur coat that every other ape wears, and this, again, helps us keep our cool. "Hair inhibits the cooling process," explained Jablonski. "Once the hair becomes saturated with sweat, the rate of evaporation drops, and it's very hard for the skin to be cooled. It's essentially covered by a wet blanket."

Jablonski suspects chimpanzees and humans millions of years ago lived in shady terrain, and we shed our hair after we started spending more and more time in the scorching hot savannas. Put that together with increased activity, like endurance running, and too much hair becomes a liability.

Humans, famously called the "naked ape" in the title of the zoologist Desmond Morris's 1967 book, had a new skin problem when they lost most of their hair.[12] Archaic humans, like chimpanzees today, presumably had pale skin under their fur. When they shed their coat, that skin could easily sunburn, ultimately leading to cancers. Too much ultraviolet radiation also fries the B vitamin folate, which cells need to properly divide and copy themselves, and, when in short supply, leads to birth defects such as spina bifida. As an adaption, human skin began to produce more melanin, a pigment that works as a natural sunblock. So humans developed black skin.

Evolution involves trade-offs, and one of the downsides of black skin is that ultraviolet rays also synthesize vitamin D, which helps keep bones strong. In tropical regions, black-skinned people receive ample vitamin D, but as humans moved away from the equator, they had to adapt once more, leading to lighter skin. In an elegant study that uses NASA satellite data to calculate UV radiation at different parts of the Earth, Jablonski and George Chaplin in 2000 conclusively showed for the first time that latitude determined skin color.[13]

So as we humans evolved away from the common ancestor with chimpanzees, we developed more sweat glands, lost most of our hair, and grew more flexible in our skin color. These changes dramatically altered our thermoregulation and allowed us to run for long distances. On the downside, they increased our susceptibility to skin cancer, birth defects, acne, bone abnormalities, and racism. Those are significant differences for two species that are so closely related.

A comparison of the human and chimpanzee genomes underscores that some of our major differences are in fact skin deep. Genes can be physically linked to each other, and inherited as groups, which means they travel together through evolutionary time. The epidermal differentiation complex, a few dozen genes involved with many aspects of forming the outer layer of skin, is one of these clusters. When the chimp and human genomes were first compared side by side, the epidermal differentiation complex stood out as having the most variation of any cluster. In other words, it had changed more rapidly than any other group of genes found in both species.

Some of these changes involve keratin, a protein produced by keratinocytes, the main type of cell in the epidermis. "There are possibly two dozen different forms of keratin that are unique to humans and different from those in chimpanzees and other nonhuman primates," Jablonski noted. In addition to helping our naked skin deal with abrasions, specialized keratin is found in cells that line human sweat glands, which may help us deal with our high flow of perspiration. "Studies of these skin differences is still in its infancy," she said.

There are many mysteries about the skin of our ape cousins, too. When I searched the scientific literature for "chimpanzee and skin," I located articles dating back to 1934, but most focused on the sexual swelling of females or on diseases, and I found only a handful that looked at the subject more deeply. One showed that captive chimpanzees exposed to little sunlight had, as would be expected, vitamin D deficiencies, and another, by Jablonski, made an evolutionary argument about human skin color. It was striking that none explored how bonobos came to have black faces, while chimpanzee faces range from black to pale and freckled. Jablonski suggested to me that it again has to do with ultraviolet exposure and how close to the equator the animals live. But I have seen both light- and dark-faced wild chimps living in the same community in Uganda, and several chimp specialists have told me that they cannot identify the different subspecies by looks. I also have seen recently captured wild bonobos in Congo and ones living in zoos in San Diego and Leipzig; the ones born and raised in captivity far from the equator may have slightly lighter faces, but they are still mostly black. When I mentioned this, Jablonski posited that bonobos may have more sensitive melanin-producing cells that spit out more pigment when they are exposed to ultraviolet rays, which seems plausible enough. But it still does not address how such a change came to be, and why the variation among chimpanzees exists.

—◁◇▷—

In addition to skin, Jablonski studies the evolution of lower primates, and in October 2004 she stumbled on an astonishing fossil. She had traveled to Nairobi after the discovery of monkey fossils in the Tugen Hills of Kenya, and while sifting through the collection she came across a molar that looked more ape than monkey. She then spotted an incisor that also appeared ape-ish to her. A comparison to teeth at the Kenyan National Museum revealed that they matched chimpanzee teeth. No one had ever found a chimpanzee fossil.

The teeth had been uncovered in digs led by Sally McBrearty, an anthropologist at the University of Connecticut who also suspected they had not come from monkeys. In further excavations at the site, McBrearty's team discovered two more chimp teeth.[14] The researchers dated the teeth to about half a million years old, and while the teeth did not by themselves reveal anything profound about chimpanzee evolution, their location indicated that chimpanzees once lived in the East African Rift Valley—farther east than anyone had imagined—at the same time as *Homo erectus*.

The absence of a chimpanzee fossil record in part indicates that small chimp populations existed in the semi-arid areas that were inhabited by archaic humans, who are the main quarry of archaeologists and paleontologists. Unlike archaic humans, chimps also do not live in caves, which are dry, somewhat protected from scavengers, and contain dirt that may cover and safeguard bones over the long period necessary to form a fossil. In the denser and wetter rainforests that chimps favor, the acidic soil and abundance of predators mean that dust returns to dust more quickly, so much so that the scientists who study living, wild chimpanzees can only gather the bones if they collect them shortly after death.

Given the enormous attention that human fossils receive in attempts to unravel our history, the absence of chimpanzee fossils represents a colossal gap in the evolutionary record, making it especially difficult to understand how chimpanzees evolved when they split off from our common ancestor. But Dan Lieberman and many in the field believe the changes in chimps must have been far fewer than in humans; simply look at the similarities with the much more ancient gorillas, they say, which also are knuckle-walkers, have protruding jaws, and feature prominent canines. Others, including the paleoanthropologist Tim White at the University of California, Berkeley, argue that the common ancestor to humans and chimpanzees was a unique ape, making anatomical comparisons of modern humans and extant chimps or bonobos—which also have evolved substantially from the common ancestor—more of a curiosity than an explanation of how we evolved. White's interpretation gained momentum in October 2009 when he and his colleagues provided a truckload of new data: 110 fossils of a hominin that lived 4.4 million years ago in what today is Ethiopia. The species, called *Ardipithecus ramidus*, did not knuckle-walk or climb vertically and had relatively small canines; in addition, the females and males were about the same size. White and his team spent fifteen years assembling their fossils into a partial skeleton, reconstructing crushed bones with the help of CT scanners, and identifying fossil pieces from at least thirty-six individuals. The fossils, according to the researchers, indicate "that the last common ancestors of humans and African apes were not chimpanzeelike and that both hominins and extant African apes are each highly specialized, but through very different evolutionary pathways."[15] An exhaustive analysis of the modern-day desert surrounding the trove exposed that 4.4 million years ago it had been a woodland, filled with monkeys, kudus, and elephants. The fossils all but killed the theory that bipedalism evolved in human ancestors to help them navigate the grassy savannas.

White's team has assembled some evidence that *Ardipithecus ramidus* could be an ancestor of *Australopithicus*, but it also may well be an extinct lineage, with no connection to modern humans.

Still, even if that were true, it is the oldest intact hominin—the much more ancient Toumaï, after all, is just a skull, and the rock star of fossils, the *Australopithicus afarensis* skeleton of Lucy, is only 3.2 million years old. In that sense *Ardipithecus* provides a clearer picture of what the common ancestor looked like and how it behaved than anything previously described. Its skull housed a brain that was roughly the same size as that in a chimp, but it had a much smaller face. Its wrists, elbows, feet, and hands suggest that *Ardipithecus* clambered about in trees by putting weight on the hands, a locomotive tactic called arboreal palmigrady. The small canines in males in particular have far-reaching implications to some paleoanthropologists. Male chimps have large incisors, primarily to fight with other males. The smaller, humanlike canines in *Ardipithecus* males, reasoned the anthropologist C. Owen Lovejoy, who was part of White's team, implies that they did not fight for females in estrus, as chimpanzees do. The lack of dimorphism between the sexes also suggests that males were not chosen by mates for their size, and that, combined with the little canines, Lovejoy said, means males were more cooperative in child-rearing and even maintained pair-bonding in a form somewhat like monogamy. "The markedly primitive *Ar. ramidus* indicates that no modern ape is a realistic proxy for characterizing early hominin evolution—whether social or locomotor," the team wrote.

Paleoanthropologists, molecular biologists, anthropologists, and other "ologists" surely will argue about the meaning of *Ardipithecus* for years to come. What behavioral inferences do the bones really reveal? Does its chimp-sized brain mean it faced the same cognitive limitations as chimps? Is the species even a hominin? What most interests me is that *Ardipithecus* spotlights the differences between humans and chimpanzees and sounds a cautionary note. Maybe chimpanzees are not, as some prominent evolutionary biologists have argued, "time machines" that reveal our ancestors. And intently as we try to learn about humans by studying the bones, blood, and behavior of chimpanzees, the data inevitably have more to say about their history than our own.

—◦∞◦—

"IT is unclear whether the common ancestor was more similar to chimpanzees or bonobos, or equally different from both." That disclaimer appeared on a nascent Internet project called the Museum of Comparative Anthropogeny, or MOCA, part of the Center for Academic Research and Training in Anthropogeny (CARTA) at the University of California at San Diego. MOCA's mission: to catalog every known human-specific difference from great apes. By the end of 2009, it had identified twenty-four domains for investigation, which ranged from culture and genetics to general life history and pathology. In addition to the usual suspects, the differences listed include construction of shelters, gift giving, verbal joking, milk composition, cleanliness, storage of food, psychic tears, fatness at birth, mind-altering drug use, and perception of ethnic differences. In all, there were more than five hundred entries. Anatomy and biomechanics alone accounted for forty-two uniquely human traits.

MOCA is also providing researchers around the world with virtual access to one of the largest collections of chimpanzee skeletons ever assembled. The skeletons came to CARTA via Jo Fritz, the head of the Primate Foundation of Arizona. When chimps at the foundation died, their carcasses were carefully fed to dermestid beetles, scavengers that dine on the skin and muscles until the bones look as though they've been through a dishwasher. The bones then were placed in acid-free yellow boxes, with one individual per box. When Fritz was forced to close up shop, she decided to donate her bone collection. "I am so excited that that's where they are, and that they will be used," she told me. "And they don't call them numbers. They call them the names that we called them."

The collection boasts fifty-one chimpanzee skeletons, most complete, and when, in July 2009, I first visited the transformed UCSD lab that served as their new home, Andy Froehle, a graduate student in biological anthropology, was preparing the bones for computed tomography scans. Froehle had to encase each bone from each individual in bubble wrap, which he then labeled. The plastic wrapping did not show up on the CT scans and allowed technicians to safely handle the precious material. There are roughly 208 bones per individual. Eventually, all of them will be posted on the MOCA Web site so that scientists anywhere can view the 3-D scans of the bones.

As part of the preparation process, Froehle had spread bones across most every table and countertop in the room. One table had two chimpanzee arms and hands, each of which had about thirty bones. Splayed out on the surface, each bone in its proper place, they appeared even more enormous than when they are on a living animal, yet one detail jumped out for just the opposite reason: chimp thumbs are tiny. Elsewhere, Froehle had arranged a set of ribs, and the ordered, U-shaped bones resembled a wave on an oscilloscope as it tracks a sound moving from soft to loud to soft again. A pelvis, split in half, incongruously sat next to a skull that had the trademark thick brows and pulled-out face of a chimp. Some sixty human skullcaps, arranged in neat rows like a platoon, covered another countertop. The room could have been set up for a still life art class taught by Georgia O'Keeffe. This aesthetic was accented by the art hanging on the walls, also donated by Fritz, including a giant photo of freckle-faced Geronimo, now one of the skeletons, and a poster of a 1925 Italian advertisement for Anisetta Evangelisti that shows a chimp chugging from a bottle of the liqueur that it apparently has filched from a wooden case under its feet. "We're very weird, for apes," Froehle said.

Pascal Gagneux, the associate director of CARTA, and Alyssa Crittenden joined Froehle to lead me on a tour of the bones. Gagneux, one of the few laboratory scientists who has studied wild chimpanzees, is a walking encyclopedia of chimpanzee/human differences. Crittenden, a postdoctoral student in biological anthropology, not only knew chimp bones but had for a year lived in Tanzania and studied the hunter-gatherer Hadza community. Froehle was the master puzzle maker who knew each bone by name, where it fit in, and the function it played. I asked them which forces they thought most powerfully shaped the anatomical differences between the human and chimpanzee, and they agreed on three: diet, locomotion, and reproduction. Or as Gagneux joked, "Eating, dancing, and sex." They also underscored that the three are braided together. Both species move about to hunt and gather food, and humans became bipedal—whether you ascribe to Peek-a-Boo, Tagalong, Schlepp, or Two Feet Are Better Than Four—to do this more efficiently in their environment. Bipedalism affected how women birth, as well as the size of the newborn's head and its brain development.

As but one specific example of the interconnectedness of anatomical evolutionary forces, they pointed to the top of the head, where some male chimpanzees have a pronounced ridge that runs the length of the skull like a Mohawk haircut. This sagittal crest, which does not exist in modern humans, is where the jaw muscles attach. "The argument is we don't need the same muscles because of the shift in our diet," explained Crittenden. In particular, when we began to mechanically process food—pounding meat, for example—and, later, cook, it reduced the need for large masticatory muscles. Chimp males also need the strong masticatory muscles to battle each other for status, territory, and access to females. Archaic humans like *Australopithecus* had a sagittal crest, too. In 2004, researchers reported that a gene that helps form the masticatory muscle, *MYH16*, has a mutation in modern humans that inactivates it.[16] Every other nonhuman primate they evaluated had an intact *MYH16*.

The researchers dated the emergence of the mutation by using a molecular clock approach to compare the human version to chimpanzees, orangutans, macaques, and dogs—each of which has an active form of *MYH16*. The mutation dates back 2.4 million years. This matches the fossil record, which shows large masticatory muscles for *Australopithecus afarensis* (which emerged 3.6 to 2.9 million years ago) that had shrunk by the time of *Homo erectus* (which arrived maybe 1.8 million years ago). The smaller masticatory muscle, the investigators suggest, in turn removed constraints on cranial capacity and is directly linked to the expansion of our brains.[17]

Gagneux moved on to the delicate skull of a two-month-old chimpanzee, Kiki, and pointed out that the cranial plates already had moved together. "This is very delayed in humans," he noted. Human infants have soft spots on their heads because the plates, or fontanels, have to fold over each other during the birthing process for the head to come out of the narrow entrance of the pelvis, the inlet, a bipedalism by-product. With a larger birth canal, chimps have no such obstruction in their head size. The chimp skull also is thicker, or more robust. "If you have to do an autopsy on a dead neonate chimp, you need a hacksaw," said Gagneux, quoting the venerable UCSD pathologist Kurt Benirschke.

In the mouth, chimpanzees and humans have the same "dental formula": thirty-two teeth at adulthood, with two incisors, one canine, two premolars, and three molars on each side, top, and bottom. In addition to the famously large canines, the last molars—the wisdom teeth that erupt at adulthood—get much more space in chimpanzees than in humans because chimps have an elongated face. "In humans, before modern dentistry, eruptions and the impaction of wisdom teeth followed by septicemia was a major cause of death," Gagneux said. "There's a huge cost to having shortened the face. You can kill yourself by trying to grow the tooth that defines you as an adult."

Knuckle-walking chimpanzee limbs differ markedly from human limbs. The arms, in particular, are extremely long. By body weight, chimps are 16 percent arms and 24 percent legs. Humans, in contrast, are 32 percent legs and 8 percent upper limbs.[18] Chimp hands are downright gigantic, save for the thumbs, and the fingers curve to help them grip. The bones do not show it, but the massive muscles in chimp arms have led to the oft-repeated assertion that they are five to eight times more powerful than humans, an idea that the media gives ample play each time a pet chimp attacks a human and rips the person's face to shreds. That "fact" comes from studies published in the 1920s that were refuted two decades later by more careful experiments. Chimps, it turns out, indeed are stronger, but they can only pull about 25 percent more pounds relative to their body weight than humans.[19]

Because they are knuckle-walkers, the chimp scapula, or shoulder blade, is much larger than that of humans and attaches to the rib cage at a different angle. Not only does the torso have an extra rib on each side—thirteen, rather than the twelve average in humans—but it twists more. Although some large men and pregnant women have barrel chests, that description applies to all chimps. "They are quadrupeds, and the musculature is totally different because it has to hold everything in, so it's a totally different rib cage because the happy inside guts are right above the ground," explained Crittenden.

The barrel ribs sit above a longer and narrower chimp pelvis. The shorter and broader human pelvis takes the shape of a bowl, flaring out from the waist along the iliac blades to make room for the large gluteus muscles we need for bipedalism and endurance running. Gagneux tapped the area on the back of the ilium where the rounded gluteus maximus on humans attaches. "Chimps have no butts," said Gagneux. "Zero. You actually see the gut passage inflating when they go."

Froehle found a stillborn chimp head and placed it in an adult

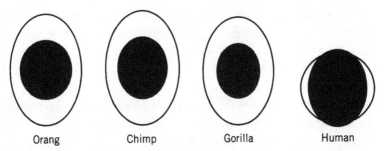

A drawing of newborn head circumferences pushing through, from left to right, the pelvises of an orangutan (Pongo), a chimp (Pan), a gorilla, and a human (Homo).

FROM KAREN ROSENBERG AND WENDA TREVATHAN, "BIRTH, OBSTETRICS AND HUMAN EVOLUTION," *BJOG: AN INTERNATIONAL JOURNAL OF OBSTETRICS AND GYNAECOLOGY* 109:1199–1206 (NOV. 2002).

female's pelvic inlet. Daylight peeked from around the entire circumference. "Human infants go through more twists and turns in the birth canal," he said. "It's pretty amazing because in order for the baby not to be damaged in humans, the entire process has to be massively slower," Gagneux said. "Chimp birth is very fast, and it doesn't look traumatic at all. There are no obstetric complications, and no one assists them."

Gagneux, who specializes in reproduction, noted that chimpanzees have a tiny bone in the penis, the baculum, that is absent in humans. In the *Selfish Gene*, the evolutionary biologist Richard Dawkins suggests that human males lost this bone in something akin to the Trench Coat theory of bipedalism. The baculum helps maintain an erection, which makes it useful especially as males age or develop mental problems, diabetes, or neurological complications. But for females selecting a mate, the baculum hides problems. "It is not implausible that, with natural selection refining their diagnostic skills, females could glean all sorts of clues about a male's health, and the robustness of his ability to deal with stress, from the tone and bearing of his penis," writes Dawkins. "But a bone would get in the way!"[20]

Bipedalism has thoroughly altered the femur, which in a human stretches toward the middle of the body from the hip joint to the knee. The chimpanzee femur, in contrast, takes a straighter line from hip to knee. Crittenden pulled out the wonderful *Human Evolution Coloring Book*, written by the physical anthropologist Adrienne Zihlman, that shows side-by-side drawings of the pelvis and lower limbs of the chimpanzee, *Australopithecus*, and modern human. Most intriguingly, the angle of the human femur means that we place our weight on the outside of the knee joint—a key feature of bipedalism—whereas chimps put their weight on the inside.

Beneath the femur lies the tibia, aka the shin bone, and it recently has attracted interest because of how it fits into the ankle joint. As the anthropologist Jeremy DeSilva found after studying videotapes of wild chimpanzees in Kibale, when their toes point upward, the ankle flexes forty-five degrees, aiding them in vertical tree climbing. Human ankles, in contrast, only flex about twenty degrees, at most, when walking.[21]

The chimpanzee foot distinctively has an opposable big toe, which helps it grab while climbing. Interestingly, *Ardipithecus* also has an abductable hallux, as this toe is known, but *Australopithecus* does not. The hallux on *Australopithecus* and on all hominins down to modern humans points forward and lines up with the other toes. The human big toe also contains two bones, absent in chimps, that are embedded in the tendon and lift the toe upward, helping to stabilize the bipedal foot. And the human heel bone, the calcaneus, is blockier because we use the heel to "step off" when we walk, shifting our weight from one heel to the ball of the foot and the big toe. "It's really precarious because there's a point where the weight of your entire body is actually on that," Crittenden said. "You transition from one foot to the next when you're walking. So it's a real balancing act, which is why kids look like they're drunk when they're learning to walk."

From head to toe, chimpanzee and human anatomy differs markedly. Researchers have worked diligently to make sense of these differences, especially when it comes to locomotion, doing extensive comparisons of chimpanzees, modern humans, and ancient hominins. But one critical locomotive difference, as far as I can tell, has never been addressed in a scientific study: Why can't chimpanzees swim?

—◦◦◦—

LONG before Rick Swope rescued the drowning Joe-Joe, the Detroit Zoo had given careful thought to the relationship that chimpanzees had to water. The new island exhibit Joe-Joe lived in was built with a state-of-the-art design that aimed, with great fanfare, to herald in a new era of chimpanzee attractions at zoos. "Very simply, this will be the best chimpanzee exhibit in the world," promised the zoo's director, Steve Graham, at a press conference on October 28, 1986. "Nowhere is there a chimp exhibit so capable of simulating a full range of behaviors, and as biologists have shown, chimps more closely resemble man than any other primate."

The exhibit, then projected to cost six million dollars, would display chimps freely roaming around the spacious island during the warmer months. The chimps could climb concrete structures and use sticks to probe artificial termite mounds for food, allowing visitors to observe the astounding tool use that Jane Goodall had first documented two decades earlier. Male chimpanzees could drum on custom-made logs as part of their power displays, just as they do in the wild. "The first chimp that walks in there is going to think he died and went to heaven," said Graham.[22]

Until 1982, when Graham took over, the Detroit Zoo housed chimpanzees in cages and displayed them in a popular show. For fifty years, zoo-goers had thrilled to the circuslike attraction, which at first featured Jo Mendi, an actual star on Broadway and in movies. Purchased by the zoo's director with his own money because the Depression-strapped Detroit Zoological Society balked at the expense, Mendi brought in ten cents a head and ended up adding handsomely to the enterprise. Dressed in Bermuda shorts, shoes, collared shirt, and a tie, Mendi would delight the crowds by lacing his shoes, pouring tea, riding around the stage on a bicycle, walking a tightrope, and dancing.[23] Because zookeepers could not safely manage older chimps, Mendi's fame was fleeting, but for several decades he was replaced by fresh-faced recruits that went on to ride ponies and even motorcycles. Graham was at the vanguard of a movement that thought captive chimpanzees should live and act like chimpanzees. "We have to get as far away as possible from chimps in tutus and chimps riding motorcycles," he said. "Chimps are not imperfect human beings. They are perfect chimps."[24]

As part of his ambitious efforts, Graham sought help from Richard Wrangham, then a professor at the University of Michigan. Trained in zoology at Oxford and Cambridge universities, Wrangham had done his Ph.D. thesis on chimp behavior in Gombe. Wrangham recommended one of his graduate students, Susan McDonald, for the Detroit Zoo job. "Richard called and said, 'We've got this fantastic opportunity to do something in Detroit that's never been done before,'" recalled McDonald more than twenty years later. She jumped at the chance.

McDonald's job required far more than helping the zoo design the exhibit. The zoo wanted to create a new colony of chimpanzees, mixing Joe-Joe and another male, Chuck, that it previously had displayed in its show with adult females and infants purchased elsewhere. So it sent McDonald around the world to visit zoos and their administrators, learning everything she could about exhibit design and the behavioral issues that surface when introducing new chimps to each other. McDonald spent one month at both the Taronga Zoo in Sydney, Australia, and the Arnhem Zoo in the Netherlands, which would serve as the main models for Detroit. Each of these had celebrated exhibits that featured chimpanzees cruising around more than half an acre of open space. These two zoos also had a key feature that the Detroit Zoo wanted to replicate: moats. Because chimpanzees could not swim, the water created a barrier between the humans and the animals. Instead of viewing chimps behind bars, zoo visitors could peer across what looked like a river and see the animals in a faux natural habitat.

Three chimpanzees had drowned in the Arnhem moat. The first drowning occurred in April 1972, less than a year after the exhibit opened, when the zoo attempted to introduce three new chimps to an established colony. The old-timers harassed the newcomers so severely that two of them ran into the moat and one drowned. The second drowning occurred ten years later, when a mother fell into the moat and dropped her one-year-old, which died. The final accident took place in 1984 and involved a chimp that earlier had crossed the moat when it was covered with ice; he drowned when he tried to jump across, again after power battles with others in the group.[25]

As McDonald learned from her colleagues at Arnhem, the Dutch zoo tried to prevent drownings by allowing new chimpanzees to explore a small moat that had netting at the bottom—pulleys attached to a tractor could then haul out any animal that fell in—before they put them in the enclosure with the large moat. They also decided not to allow chimps near the moat if it was even partially frozen over.

On May 5, 1988, the Detroit Zoo broke ground for the Chimps of Harambee exhibit, which would eventually cost $8 million and stretch over a prodigious 3.7 acres. The event, complete with a high school band, featured none other than Goodall herself.

Five months later, the nine chimpanzees that would join Joe-Joe and Chuck began to arrive. Joe-Joe and Chuck, each seventeen, would live in a new colony with the six adult females: Beauty, Bubbles, Peggy, Sarah, Trixi, and Tanya. The group also would include five-year-old Abby and the two-year-olds Akati and Suzy. The chimps came from zoos in Dallas, Knoxville, Cleveland, Chicago, and Germany. All of the older ones, including Joe-Joe and Chuck, were born in the wild, and one, Tanya, for a time had worked in a circus. The two youngsters, Akati and Suzy, had been hand-reared by humans, as often happens in zoos when mothers reject their newborns.

McDonald, who was hired to be the manager of the colony, knew from her studies of other zoos that introducing chimps to each other was a dangerous business. At a zoo she visited in New Zealand, an infant was killed after being introduced to an adult female and male. Arnhem lost an adult from injuries sustained during an introduction. That zoo, with limited success, went so far as to dose chimpanzees with Librium during the initial meet-and-greet period.[26] "We were very nervous," said McDonald.

Adult chimpanzees first met each other one on one, with Plexiglas and then mesh barriers between them. Keepers kept fire hoses and extinguishers at the ready to break up any fights. To protect the orphans, they formed a "nursery group" with Beauty and Bubbles as the adoptive mothers. Because neither of the older females had had their own babies, McDonald and her team showed them documentaries about Gombe chimps that included mothering behavior. Adult males only met females in estrus to reduce aggression. "The most nerve-wracking day was when both males came in with both infants in the room," said McDonald. Nothing happened, and soon the males even played with the babies. "It was phenomenal," she said. Problems certainly surfaced: over the first six months they lived together, Chuck badly wounded Joe-Joe four times, including one attack that required the amputation of digits. But all in all, McDonald and her team were confident the chimpanzees were ready to be displayed to the public when the exhibit formally opened on October 19, 1989.

The Chimps of Harambee first restricted the animals to the indoor enclosure that would house them during the frigid Detroit winter, but the outdoor space would be what set the attraction apart. In addition to the termite mounds and logs for drumming, chimps could climb fallen tree trunks (anchored to the ground with steel cables) and concrete structures that reached nearly forty feet into the air. Living trees, surrounded by electrical netting to prevent chimps from stripping the bark, added to the natural appearance of their island. Eight view spots would allow the public to watch the chimps from as close as twenty feet. The zoo imagined it would one day have a barge on the "river" that would offer even more intimate vistas.

In keeping with an expect-the-unexpected philosophy, the zoo took several extra safety precautions in its cutting-edge exhibit. The island on which the chimps lived was split into two partitions by an indoor enclosure for the chimpanzees. The larger south side was surrounded by a water-filled moat that measured at least twenty-one feet across, five feet wider than even the most skilled chimp supposedly could jump. If chimpanzees did fall in, they could hoist themselves out by grabbing a cable made of woven stainless steel that was attached about a foot from the moat's edge on the enclosure side of the exhibit. The moat, which dropped to just over six feet deep at its lowest point, had a gentle slope on the enclosure side so that a chimp could wade into the water and easily return to dry land. The north side of the island featured a dry moat more than twenty feet deep, with a sheer wall on the public side. Worried that a chimp might figure out how to scale the wall, the zoo hired veteran mountain climbers to assess the "escape potential." The mountain climbers also attempted to ascend the exterior walls of the indoor chimp housing, but they could not reach the roof.

Shortly after the chimpanzees first entered the outdoor enclosure on April 16, 1990, one of the females, Trixi, waded into the water but stopped when it reached her knees. None of the other chimps seemed interested.[27] Six days later, tragedy first struck.

Two keepers, Karen Hock and Gloria Copeland, were picking up trash on the public side of the moat on that Sunday morning. Tanya, a twenty-one-year-old female, ambled up a small hill across from them. Tanya had become "enculturated" to humans when she worked in the circus, and she was well liked by the keepers. The other chimpanzees also quickly warmed to her dominant but benevolent demeanor, and she was an immediate leader in the new community. "She just started running," remembered Karen (Hock) Stadele, who at the time was in her early twenties and had just completed her undergraduate studies at the University of Michigan. "It looked like her intentions were to jump over the moat. We saw some people out eating ice cream. Our thought was maybe she wanted to have some ice cream. She was used to those sorts of treats since she was a young-ster." Maybe the human-friendly Tanya wanted to visit with Gloria and Karen, another possibility they later considered. Or perhaps her exposure to water as a young circus chimp made her less fearful of the moat.

Whatever Tanya's motivations or intentions were, she did not make the jump to the other side and splashed into the moat. "She looked liked she breathed water in pretty quickly," said Stadele, an intellectual property lawyer who remained shaken by the memory nearly twenty years later. "I do remember her face. She had no idea what to do. She went up and down a few times. She flailed. She didn't know what was going on. She couldn't help herself at all." Maybe Tanya could have grabbed the underwater steel cable near the bank to rescue herself, but as Stadele asked, "How were they to know that was there?"

The keepers had received express orders not to dive into the moat to rescue a drowning chimp, as they could be dragged underwater or mauled by the powerful animals. Stadele, who had run marathons and lifted weights, had mixed emotions about obeying the orders, but she did. Following their training, she radioed for help, using a special code so as not to alert anyone who might inadvertently have picked up their channel. "Gloria was so panicked she couldn't even radio, she could not speak," said Stadele. "I remembered the codes and every-thing. I couldn't believe it." The two women then ran and retrieved a Styrofoam flotation device hanging on a nearby gate.

By the time they returned, other zoo workers had fetched Tanya from the moat, and a veterinarian was on the scene, attempting to give her CPR. But it was too late. "We were all distraught—crying, sad," said Stadele. Keepers, she said, had worried about the water from the start, but they had been given no input into the exhibit's design. "It never feels good to be on the I-told-you-so side of anything," she said. "I think the risk was taken because it was felt, yeah, it might very well happen, but it's worth it."

McDonald said Graham, the zoo's director, came up with the idea for a moat, and when she raised the possibility of slowly introducing chimpanzees to the water, as the Arnhem Zoo did as a precaution, it did not take. "Everyone was just hoping they wouldn't go near the water," McDonald said.

The next month passed without a serious incident, but on June 29, Bubbles scaled an exterior wall of the indoor enclosure that the mountain climbers had deemed escape-proof. "When Bubbles got out, we were not well versed in what to do," said Stadele. "I remember looking down the hallway and there were men with loaded guns." Bubbles jumped on the roof of one building and over to a parking lot used by the staff, flipping a keeper who tried to subdue her. From there, she headed to the zoo grounds, climbed a tree, and was finally shot with a tranquilizer and carried back to the exhibit.

In the aftermath, the zoo coated the building's exterior walls with slippery goop, strung electrical wire along the roof's perimeter, and made Stadele and the other keepers take shifts patrolling the roof with fire extinguishers they presumably would spray on any escapees. Stadele, already wary of what she thought were dangerous locking mechanisms on the indoor cages, was incredulous at what she perceived as the lack of concern for keeper safety. "I don't quite know how I would have reacted if a chimp started climbing up and I was up there alone," she said.

One month after Bubbles's breakout, Joe-Joe went into the moat. "Thank God I was not at work that day," said Stadele. "Call me selfish, but I could not have handled another stress like that."

Shortly after Rick Swope rescued Joe-Joe, Karen Stadele, outraged about yet another accident, quit her dream job. The Detroit Zoo closed the south section of the exhibit with the wet moat, and, at the recommendation of an advisory committee, installed an electric fence on the exhibit side of the moat and netting under the water. Today, the zoo no longer has water in the moat. "It was one of the things I wanted to change as soon as I got here," Ron Kagan, director of the zoo since 1993, told me. The drained moat, however, offered chimpanzees an easy escape route, so the zoo had to dig it several feet deeper to make a higher wall, a complex and expensive project. "It took us a long time to do it, but I'm very glad we did," said Kagan, who called the water-filled moat "lethal containment." No other exhibit at the zoo, he emphasized, has water unless the species *can* swim. "You don't have containment that could cause death for anything else. It really wouldn't make sense."

The Detroit Zoo's experience with the moat illustrates that even the humans who know the most about chimpanzees have a shaky understanding of how they relate to water. While many avoid it, as Goodall noted all the way back in her first *National Geographic* article, Joe-Joe's fear of Chuck overrode any fear he had of water, and Tanya drowned in pursuit of an ice cream cone or some companionship. So chimps' apprehension around water is far from absolute, and it does nothing to explain why no chimpanzee has ever been observed freestyling, backstroking, butterflying, or even dog-paddling.

———⚬⚬⚬———

I have asked most every chimpanzee researcher I have met whether they have seen a chimpanzee swim or read a witness's account of it. None had. I have also asked if they have studied the question or even given the idea serious thought, and few had. This is not a novelty question. Chimpanzees can learn to skateboard, ice skate, and ride motorcycles. None of those skills would have helped them survive in the wild. But swimming would allow chimps to cross the rivers that typically bound their natural habitats. They could escape pathogens, find new foods, and mate with strangers. All of these theoretically could have made them more fit, and surely would have had an enormous effect on the evolution of their immune systems, their metabolism, and their reproductive capabilities. Yet no one studies this.

The rote answer that researchers offer, frequently with the dismissive tone that smart people use when they are guessing, is that chimpanzees do not have as much body fat as humans. But Michael Phelps, one of the fastest human swimmers of all time, did not have much body fat when he broke thirty-seven world records and won fourteen Olympic gold medals. Clearly, not much body fat is required to swim.

A peculiar study from Cameroon assessed the role of body fat and buoyancy, in part in an attempt to understand why blacks performed so much worse than whites in international swimming events. (The Olympics did not have a black swimmer even qualify for an event until 1984.) I am not insinuating that blacks are a proxy for chimpanzees or any such nonsense, nor were the authors. It was simply a way to compare two groups of people that had distinct morphologies. The researchers matched thirteen whites to thirteen blacks living in Cameroon by age, weight, and stature, and found that whites were more buoyant. A closer analysis discovered that the whites had more body fat in their upper trunks.[28] Body fat, if you believe this study, may have an important link to swimming performance. Yet black-skinned humans of course can and do swim.

Others contend that the chimpanzee's center of gravity is different, as is their muscle mass. "It may be weight distribution and trim control," Frank Fish told me in October 2006. This delightfully named zoologist at West Chester University in Pennsylvania specializes in mammalian swimming. He said the heavy front part of the body might force a chimp to face down in the water. "I don't think anyone would do the actual experiment to demonstrate this," said Fish, adding that a virtual experiment with a physical or computer model might be able to address the question.

Maybe the answer has nothing to do with their anatomy. "Chimps rarely live near a body of water in which they *could* swim," said Nina Jablonski when I put the question to her. "And rivers and other bodies of water throughout Africa are thick with crocodiles and have been for millions of years. So there are all sorts of disincentives to swimming."

Elaine Morgan, a Scottish journalist and writer for several television series, has for more than forty years pushed the idea that humans have a special relationship with water that led them to bipedalism. For Morgan, who has written six books and countless articles on her

"Aquatic Ape" theory—called All Wet by its many skeptics—swimming is tangential to her main argument, which is that geologic shifts flooded the terrain inhabited by early hominins, and they began wading and standing erect out of necessity. Because water reduces the forces of gravity, these hominins supposedly could have learned to walk without putting undue stress on their spines. Along the way, like many other mammals that enter the water, they learned to swim and lost their hair.

Morgan points to the proboscis monkey, which lives in mangrove swamps in Borneo, as a primate that has exhibited this behavior. "More often than not when it climbs down a tree, what it encounters is water," she wrote. "If the water is deep the monkeys swim. They are excellent swimmers and can swim for miles, and may be seen diving into the water from the tree tops."[29] (Several monkey species in fact can swim.)

Morgan did not hatch the Aquatic Ape idea. In 1960, Alister Hardy, then a marine biologist at Oxford University, wrote an article for *New Scientist* magazine, "Was Man More Aquatic in the Past?" Hardy notably did not publish the theory in a journal, complete with data from his own observations and references to the work of others. Rather, he put it forward as a "speculation" that might "at first seem far-fetched." Man, he proposed, may have been "cradled in the sea," an idea he stumbled upon some thirty years earlier when he read a passage that pondered why the subcutaneous fat and hair of humans and chimpanzees differed so dramatically. Maybe humans had evolved like whales, seals, and penguins, he thought. "He would have to raise his head out of the water to feed; with his hands full of spoil he could do so better standing than floating," Hardy mused. "It seems to me likely that Man learnt to stand erect first in the water and then, as his balance improved, he found he became better equipped for standing up on the shore when he came out, and indeed also for running. He would naturally have to return to the beach to sleep and to get water to drink; actually I imagine him to have spent at least half his time on the land."[30]

Morgan built upon Hardy's fanciful imaginations of the Aquatic Ape with scientific arguments, but no compelling evidence has surfaced in four decades to support her claims. Entire conferences and

Web sites have blasted the theory point by point. In *Skin*, Jablonski allows that the theory has "had the greatest popular appeal" of any when it comes to trying to explain how humans lost their hair. Ultimately, though, Jablonski raises the Aquatic Ape to dismiss it, which she does in six terse paragraphs.

Robert Yerkes believed chimpanzees did not like water, but recognized, accurately, that the aversion is not absolute, and one Morgan acolyte published a dissertation as a graduate student that documented a dozen instances of chimpanzees and other great apes wading.[31] In 2008, *National Geographic* photographers and videographers working in Senegal with the researcher Jill Pruetz documented chimpanzees that live on a savanna entering a natural pool and crouching until the water reached their waists.[32] Most impressive of all, Terry Wolf, the wildlife director at Lion Country Safari in West Palm Beach, Florida, described several instances to me of chimpanzees there crossing the waterway between the five islands that hold them.

Wolf, who had worked at the facility for thirty-eight years, told me that he has never seen any of the two dozen chimpanzees there swim in the six-foot-deep canals that separate the islands. But he has seen two of them cross the nine-foot span from island bank to island bank. Bashful, thirty-three years old in 2009, had made the trip at least four times, Wolf told me. "He had a couple different methods," said Wolf. "He was very crafty in his younger days—he doesn't try it anymore."

In one technique, Bashful found ridges on the bottom of the canals and balanced himself on them to walk across. "He's tall and could reach down five and half feet," said Wolf. "He literally laid his head back so he was facing the sky, and you could see him feeling the bottom across. If he reached a center point that was deeper, he'd launch himself with floating objects like logs or planks to help him in this endeavor. One time he did it front of Jane Goodall, and she was tickled."

On other occasions, Bashful would bounce off the ridges to keep his head above water. Wolf also saw him use the water lettuce, a floating weed, to help himself across. "It has a lot of buoyancy but also has a lot of dense matter you can't get through," Wolf said. "Once he hit water lettuce, he was thrashing his arms overhead like a swimmer would do with this windmill technique. I was very worried about it and wondering whether I was going to be getting wet. But he was hitting water lettuce and staying afloat. He wasn't making much headway, and this went on at least one and a half minutes, which is an extremely long time when you're worried about someone drowning. He finally made it to other side and then collapsed in exhaustion. I'd never seen a chimp that exhausted. It was a big relief for him to get back on land. I've never seen him try it again."

Another chimp, Barbara, escaped from her island several times. The first time she made a crossing, Barbara was actually chasing after Wolf. "Barbara was very motivated," he said. "She hit bottom, held her breath, crawled, and found herself on other side. She said to herself, I did that once I can do it again." Eventually, Lion County Safari had to move Barbara to the Jacksonville Zoo.

Like Rick Swope, Wolf has had to rescue a chimp from the water. "It literally is like pulling a sack of concrete off the bottom of the canal," he said. "It's like you have weights pulling you down. I had to roll her up on the island. Luckily, I gave her mouth-to-mouth and CPR, and she spit in my face." But Wolf also has lost chimps to drowning. "I don't like to go public about it, but it's true," he told me. "These chimps don't all catch on like Barbara." And truth be told, he would be happier if the enclosure did not have water. "I didn't invent this system," he said. "I inherited it."

—◦◦◦—

MAYBE the reason humans swim and chimpanzees do not is because swimming is learned, and chimps do not teach. Indeed, contrary to the popular perception that human children will naturally swim if thrown into water, drownings are the second leading cause of injury-related death of children between ages one and fourteen years of age in the United States, according to the Centers for Disease Control and Prevention.[33] In children fifteen and older, most near drownings and deaths occur in natural settings like oceans and rivers, indicating they knew how to swim but faced challenging conditions such as riptides, waves, freezing cold water, or eddies. But in the four-and-under group, about half of all deaths and more than 80 percent of nonfatal injuries took place in swimming pools. This age bracket also had more fatalities than any other, accounting for 30 percent of all deaths among one- to four-year-olds.[34]

Further underscoring the importance of teaching children to swim, a study of eighty-eight unintentional drownings in the United States compared each child to matched controls and found that formal swimming lessons for children between ages one and four reduced the risk of drowning by 88 percent.[35] The study was targeted at changing the swimming guidance prescribed by the American Academy of Pediatrics, which explicitly does not recommend lessons for these young children because of concerns they will lose their natural fear of water.

But humans over the past century have taught chimps many "unnatural" skills, which is why they have become staples in circus acts and movies. Why has no human ever taught a chimp to swim?

In 2007, the *Los Angeles Times* Sunday magazine published a cover story about a supposedly seventy-five-year-old chimpanzee named Cheeta that starred in Tarzan movies and had retired to a house in Palm Springs. What caught my eye is that Cheeta lived with his supposed grandson, Jeeter, and this younger ape "loves the water, even swims," the article said. The article was written in the voice of Cheeta, not the actual author, which immediately raised my eyebrows high.[36] I e-mailed the owner of the chimps, Dan Westfall, to verify the claim and see if I could come witness this myself. His reply: "Chimp cannot swim. Jeeter like to go in and hold to the side. Dan & Cheeta." Westfall's me-Tarzan-you-Jane syntax delighted me, especially since Cheeta, again, apparently was the one speaking. R. D. Rosen wrote an article for the *Washington Post Magazine* the next year, "Lie of the Jungle," that exposed Cheeta as a fraud, amassing evidence that the chimp was nowhere near seventy-five nor a former Tarzan sidekick.[37]

Jo Fritz in point of fact gave swimming lessons to a chimpanzee. In the 1970s, Fritz was contacted by a couple in Phoenix who owned a pet chimpanzee, Casey, then about four years old. The couple had a swimming pool in their yard and feared Casey, like a human toddler, might fall in and drown. So they turned to Fritz for help. "We really knew nothing except there were stories out there that they didn't swim," remembered Fritz. "He was out running around with the family all the time and splashing at the edge of the water, and they were determined to teach him to swim."

Fritz spent two ninety-minute sessions with the couple working with Casey in the swimming pool. They placed him on an inflatable mattress and attempted to teach him how to dog-paddle. "He wasn't terrorized of the water," said Fritz. "He would laugh and splash people. But each time we would gently push the float out from underneath him, he'd just go down. There was never any conception that went through his mind that he could hold his head above water and dog paddle or that he could even try."

To Fritz, this was all the evidence she needed that a chimp could not learn to swim. "He was at the age to learn if one was going to learn," she said. "Now hear me: I taught a five-year-old chimp to knit. If you can teach a five-year-old to knit, to take the stitch off the needle and put it on another needle, don't you think if their life was threatened by drowning they'd at least dog paddle? Casey didn't. And it has nothing to do with intelligence."

One of the most interesting theories I heard about why chimpanzees do not swim came from Dan Lieberman. Although Lieberman specializes in anatomy and locomotion, he did not point to differences in chimp and human body mass, fat ratios, muscle structure, arm length, or anything else biological. Rather, he raised a fundamental evolutionary principle and wondered why it is that *humans* swim. "Natural selection drives creatures toward better performance," explained Lieberman. "Humans must be among the really, really poor swimmers out there. We *can* swim. There are people who enjoy doing it. But nobody would argue that we're good at it."

As Lieberman sees it, there is no reason to think that natural selection operated on human swimming abilities. In essence, the ability to swim may be something of an accident, what evolutionary biologists call an "exaptation." "It's just like we have an ability to type or play basketball or play soccer," he said. "It's a feature we can do as a result of selection for something else. And frankly, it's not much of an exaptation, because we're not very good at it."

Mediocre as humans may be at swimming, the skill can save our lives, and, as Rick Swope demonstrated that day at the Detroit Zoo, we can use it to save others, including chimpanzees that happen to fall into moats. But Lieberman made an important point that I think people often forget when comparing how the bodies of humans and chimpanzees function. Just because differences exist, it does not mean they evolved. These exaptations—a term Stephen Jay Gould helped coin—force us to accept that not every distinction has a deeper meaning. In a bawdy essay, "Male Nipples and Clitoral Ripples," Gould assailed scientists who came up with fanciful theories to explain the utility of his two title characters. Sure, the relatively large female breast in humans—chimps only swell when they are lactating—is an adaptation. But the male nipple does not exist because men once breastfed, he argued. And the clitoral orgasm, contrary to the pronouncements of some theorists, is not an adaptation to give women sexual pleasure so that they will pair bond. This, in Gould's words, is "a flawed view of life." In reality, the embryo is neither male nor female at first, and the male breast is a smaller replica of the female one and the clitoris is a smaller penis. "I am now convinced that many structures (including male nipples and clitoral orgasm) have no adaptive 'why,' " Gould wrote.[38]

Any discussion of tits and clits—a title Gould considered for his essay—inevitably leads to Sigmund Freud, who intoned that "anatomy is destiny."[39] Freud was interested in whether a person was born male or female, but I have no interest in arguing the truism on that level. I am intrigued by the broader implications. Whether driven by an adaptation, an exaptation, or some other force, anatomy does chart our destiny. From the foramen magnum to the opposable big toes, the epidermal differentiation complex to *MYH16*, subcutaneous fat to the pelvic inlet—all of these matter when trying to understand the separate evolutionary roads that humans and chimpanzees walked or knuckle-dragged after diverging from a common ancestor. Sometimes, though, to riff on the Freudian riff, a nipple is just a nipple.

10

CARNAL KNOWLEDGE

PASCAL GAGNEUX NEEDS FRESH CHIMPANZEE SPERM TO DO HIS research, and he has gone to great lengths to obtain this precious commodity, including enlisting me to help him fetch a sample during my February 2007 visit to the Primate Foundation of Arizona.

Gagneux, who in addition to his scientific aptitude is a splendid artist and craftsman, has spent many hours fashioning devices to coax sperm from chimpanzees. He began by sculpting a silicone version of a female chimp rear end, complete with an estrus-swollen vagina that looked as lifelike as any of the human versions peddled in sex shops. But the male chimpanzees at the Primate Foundation that were recruited to help with the project did not see it that way, and the masturbator sat unused on a counter in a supply room. "It's a nice chimp butt, but I thought it was a bonobo butt when I first saw it," Jim Murphy, the foundation's colony manager, admitted to me. "Maybe that's why they don't like it."

Gagneux's next attempt relied more on medical science than art. He modified a piece of PVC pipe to create a modified version of what's known as a Penrose drain, which physicians use in wound patients to help remove pus and other liquid discharge. For the chimps, the wall of the pipe had been jerry-rigged with an add-on compartment that holds warm water, and latex coated with K-Y jelly lined the interior. Rachel Borman, who had worked at the foundation for ten years as an animal handler, was given the job of selecting a sperm donor and encouraging him to produce a sample.

Borman first "gowned up" to protect her clothes. The lucky donor today was a sixteen-year-old named Shahee. Borman asked me not to follow her into the space that held the caged chimps, as the presence of a stranger would break the mood. So I peered through the glass portal in a door. "I'm just going to go in there with these other guys to make him jealous because he's not erect," Borman told me as she entered the chimp space. She did a quick pass by Shahee's rivals and returned to the supply room for the modified Penrose drain.

Love tunnel in one hand and a training clicker in the other, Borman walked toward Shahee. After a few clicks, he stuck his erect penis through the bars. Borman held up the PVC pipe, and said, "Good boy! Good boy!" She then gave him an M&M, and walked back to the lab space. "He did it," Borman said proudly.

Borman cracked open the tube. Lying on the tan-colored latex was a huge chunk of spunk. I say chunk because most of it had coagulated into what is known as a plug, about one quarter of which usually melts in the warm vaginal vault. Using a popsicle stick, Borman transferred the ejaculate into three vials. "It's fun for the chimps to do this because it's enhancement for them," Borman explained, as she completed her odd task. "They love it." She added a buffer solution to the vials and capped them.

After a few minutes, Borman went back in to retrieve a second sample from another chimp, but she returned empty-handed. "He's not in the mood today," she said. "He started but he wouldn't finish."

Borman carefully packaged the vials and handed them to me. My job was to shuttle them to San Diego when I flew home that evening, and then drive them to Gagneux's lab at UCSD so that he could study them while the sperm were still alive. As we exited the chimp enclosure, we passed Shahee. He spit on me.

On the way to the airport, I realized that the chimp sperm created something of a dilemma. I had the vials in my daypack, the only bag I had brought for my short trip to Arizona. If I wanted to carry the bag with me on the flight, I would have to pass it through security, and surely the screeners would question the liquid in my vials. What would I say? It was hair conditioner? Packed in laboratory vials? If I told the truth, would they think I was a modern Il'ya Ivanovich Ivanov, the Russian scientist who tried to breed a humanzee? But if I checked my small daypack as luggage, would that raise suspicions that I was a drug smuggler or some such, leading to a search and a humiliating outing as a chimp cum courier?

I gambled that the security checkpoint was a higher risk, and I checked my daypack at the ticket counter. My bet paid off. Before I knew it, I was back in San Diego, sperm in hand, making a late-night rendezvous at UCSD with Gagneux, one of the world's only chimp sperm interrogators.

—◦◦◦—

A giant sperm sculpture hangs from the ceiling of Pascal Gagneux's lab. He based his model on a 160,000-times magnification of a primate sperm—chimp and human look nearly identical—and used PVC pipe as a backbone, papier mâché for the head, and a nylon fabric pillow casing for the tail. People tend to think of sperm as cylindrical, but they are actually wide and narrow, Gagneux told me. "When they move around they resemble a surfboard tumbling around in the waves," he said.

Although chimp and human sperm look the same, they have important differences, the most pronounced being volume. There is great variation between individuals, but the amount of ejaculate is roughly the same between the two species—just over three milliliters, about half a teaspoon. Chimpanzee sperm count, however, is an order of magnitude higher than the sperm count in humans. Healthy humans have 20 million sperm per milliliter or more, with studies involving hundreds of men finding an average of about 66 million per milliliter.[1] Chimpanzee sperm counts in five males at Yerkes ranged from 214 million per milliliter to 6.3 billion, with an average of 2.5 billion.[2]

Logically enough, higher sperm counts require larger testicles, and indeed a study published in 1981 found that a pair of chimpanzee testicles weighed about 118 grams (about four ounces), roughly three times more than humans'. The research, led by Roger Short of the MRC Reproductive Biology Unit of Edinburgh in Scotland, went on to compare thirty-three primate species.[3] As it turns out, gorillas and orangutans have the smallest testicles among the great apes. Especially striking, however, are the ratios of body weight to combined testicle weight. Chimp testicles, at the upper end of the scale, make up 0.27 percent of their body weight; this ratio drops to 0.06 percent in humans and 0.02 percent in gorillas. Short proposed that the dramatic differences evolved because of mating systems.

In gorillas, one large male typically dominates a harem of females. Orangutans have a similar mating system, though smaller males do achieve some reproductive success with females by, in essence, raping them. With chimpanzees, and, to a lesser degree humans, females are much more promiscuous and have many male partners. Sperm from one individual, then, must compete with sperm from other males to fertilize an egg. The premise of sperm competition—which an insect researcher first proposed in 1970—contends that the gorillas and orangs produce small amounts of sperm because a little bit is enough to do the job.[4]

Gagneux believes that a chimp with a longer penis may also win out in the sperm competition, because it might be able to displace the plug left by the ejaculate of a male who mated with the same female earlier. "For chimps, bigger truly may be better," he told me. Researchers at the International Center for Medical Research in Franceville, Gabon, pulled out the rulers and studied this possibility.[5]

The size of the chimpanzee penis may come as a surprise, in part due to the popularity of *The Third Chimpanzee*, by Jared Diamond, who asserted that the human male had a "relatively enormous, attention-getting penis, which is larger than that of any other primate."[6] Diamond, a physiologist, ornithologist, and evolutionary biologist at the University of California, Los Angeles, who won a Pulitzer Prize for his book *Guns, Germs and Steel*, went on to jest that modern science had a glaring failure, as it had yet to form what he called "an adequate Theory of Penis Length." Diamond's average human had a five-inch erect penis, the chimp a mere three inches, and the gorilla and orang just barely longer than an inch.

Gagneux, who once worked at the San Diego Zoo, told me that an erect orangutan's penis is the size of a human thumb, a gorilla's a human pinky—so I suspect Diamond was in the ballpark on those two. But the Gabon study, conducted by Alan Dixson and Nicholas Mundy, shows that Diamond badly underestimated the chimpanzee and furthered men's proclivity to have delusions of grandeur about their own endowment.

Erect chimpanzee penises, "induced (manually) by one of the authors," averaged 14.4 centimeters, or about five and one half inches. The penis length of eleven chimps ranged from ten to eighteen centimeters—just over seven inches. Dixson and Mundy noted that Robert Yerkes and James Elder had in 1936 measured the lengths of four erect male chimps and recorded a similar range: fifteen to twenty centimeters. "The chimpanzee penis is narrower than that of man (2 to 4 centimeters wide at its base and less at the tip according to Yerkes and Elder) but is probably longer, when erect, than that of many human males," they concluded.

The researchers could not formally prove that a longer penis displaces the plug left by a shorter one, but they made a compelling circumstantial case—and illustrated the extraordinary steps some scientists will take to investigate an unanswered question. First, during routine exams, they anesthetized females and used a rod to measure the depths of their vaginas. "The rod (sterilized and coated with surgical jelly) was inserted to maximum depth, with the horizontal plate pressed firmly against the sexual skin surrounding the vaginal introitus," their report notes. Most males had a long enough erect penis to reach beyond the opening of the cervix when vaginas were not swollen from estrus, thrusting the Theory of Penis Length and copulatory plugs toward the trash bin. But when they measured females during estrus—which is when chimps copulate the most and are most fertile—the swelling increased the vaginal depth by as much as 50 percent. The exceptionally well-hung Nestor, who had the gigantic eighteen-centimeter erection, it turned out, theoretically could deposit his plug at the cervical opening of all but two of the fifteen estrus-swollen females measured. This gave him an appreciable advantage; Valliant, the next in line at 16.9 centimeters, only made it to the cervical opening in eight of the females, and the success rate dropped steadily among the other males.

The findings challenged the assertion that a long penis provided an evolutionary advantage because it worked as a magnet for the opposite sex, à la the plumage on a male peacock. (Diamond, to his credit, scoffs at anthropologists who make this argument about humans, maintaining that the "display" advantage of a long penis is to increase status with other *men*.) Rather, in chimps at least, a long penis worked in concert with a huge number of sperm per ejaculate to increase a male's chances of literally spreading his seed.

Gagneux's own interest in chimpanzee sperm takes an even closer look at their competitive edge, measuring motility, speed, the immune responses they trigger, and the sugars that adorn their surfaces and allow them to bind to cells in the walls of the uterus or the fallopian tube.[7] He then compares them to human sperm, which he hopes might clarify fundamental riddles about infertility.

He prepared some of the sperm I flew in for examination, diluting a sample and placing it on a microscope slide. The microscope was connected to a computer screen, so I could watch in real time. The sperm did not so much resemble surfboards tumbling in the waves as bugs flittering about on the top of a pond. "Wow, look at that," said Gagneux. "It's pretty sweet, huh? There's nowhere near that many in humans."

Gagneux's lab space was adjacent to that of his collaborator Ajit Varki, who helped uncover the functioning of the sugars, known as sialic acids, on cell surfaces. The sialic acids that adorn the surface of sperm have become the central focus of Gagneux's work, too. Humans, as Varki discovered, lost the ability to make Neu5Gc, and Gagneux suspected that Neu5Gc also played a role in fertilization. He hypothesized that Neu5Gc helped female chimpanzees, in a process called "cryptic female choice," select the most compatible, high-quality sperm. The sugar acted like the fuzzy part of Velcro and attached to barbs formed by sugar-binding proteins on the surface of the cells in the uterus or fallopian tubes. Neu5Gc, as Gagneux imagined it, might "sweet talk" the female reproductive system.

Gagneux's Neu5Gc ideas had a critical implication for human fertility. Although we cannot synthesize Neu5Gc, we ingest the sugar when we eat meat and dairy products, and it, in turn, can then be incorporated into our cells. Does Neu5Gc coat the surface of human sperm? Is it found more readily on the sperm of men who eat lots of animal by-products? Do women then have an immune response to the extremely foreign Neu5Gc that selects *against* the survival of the sperm? "It could be that men who eat loads of meat pass a threshold and become infertile," suggested Gagneux.

I left Gagneux shortly after midnight, and he was still cranking away on the fresh chimp and human sperm samples he had the good fortune to collect during the day. He phoned me the next afternoon, elated. "I found some Gc on three of the four sperm samples from humans!" he said.

Scientific knowledge grows in much the same way as a plant. You need time-lapsed footage to see the branching of a new limb or the budding of a flower in action. But some days, you can observe something that you did not notice the day before, be it a leaf unfurling or a root bursting through the soil. The evidence of Gc on human sperm was one of those days, and it made him feel confident that if he kept watering this idea, some day he would actually harvest fruit.[8]

—◇◇◇—

IF you randomly snipped sentence fragments from a few copies of *War and Peace*, put them in a pile, and then looked for words at the end of each fragment that overlapped with words inside another fragment, you could reconstruct the novel. But it would be a difficult exercise, to say the least, and it would be more confusing still if Tolstoy used the exact same stretch of words in different passages. Making matters more complicated still, imagine if some of these repeated stretches were palindromes that read the same in either direction ("do geese see god") or if the paper could be read from either side and some words were mirrors of each other ("mood" and "boom"). That is the dilemma researchers faced when, as part of the chimpanzee genome project, they cut up the Y chromosome and tried to unscramble the sequence: it is littered with long, repeated stretches of nucleotides.

To circumvent this problem, a team led by David Page, the director of MIT's Whitehead Institute, took the bacterial artificial chromosomes that each had large chunks of Clint the chimpanzee's Y—paragraphs rather than sentences or words—and mapped how they fit together. The researchers then sequenced the DNA, and as they reported in 2010, the properly spelled out As, Cs, Ts, and Gs contained a startling message.[9] In Page's words, when Clint's Y was compared to the human Y, the two were revealed to be "horrendously different."[10] Those horrendous differences mostly relate to the genetics of sperm production.

To understand the differences, turn back the clock to 300 million years ago, when birds emerged and the sex chromosomes first splintered off from the other autosomes. At first, the proto-X and proto-Y could "recombine" with each other and swap DNA during meiosis. But then chunks of DNA in the proto-Y flipped over, or inverted, and the Y no longer could match up with the X and share genetic information except at its ends. The Y chromosome then pulled in new pieces from other autosomes, including the genes that make sperm. It also recombined with itself, creating extra copies of many genes.

Typically, pairs of chromosomes are twins, and if a gene is damaged on one, its twin can do the equivalent of a kidney donation and transplant the functioning DNA. But most of Y has nothing to swap with on X, so many of the ancestral genes degenerated over time and disappeared, making Y a runt of the genome, a mere 60 million bases in humans out of the 3.2 billion total in humans. As for functioning genes, scientists have identified only seventy-eight on human Y, as compared to one thousand genes on X. In chimps, Y is smaller still with half as many bases and only thirty-seven functioning genes. Dogma suggested that the Y, in both species, slowly was crumbling away.

Page's group focused on the "male-specific" region of Y, which means they ignored the parts that are identical to X and compared just over 20 million bases in each species. When they could match up sequences, they were about 98.3 percent similar, which resembles other chromosomes. But fully one-third of the chimp and human chromosomes carried sequences unique to each species. Page and his colleagues also found that chimps had lost many ancestral genes and humans had acquired a few new ones. In the final tally, which compares the number of unique genes (not the number of genes themselves, which is biased by extra or fewer copies of the same ones), chimp Y has only two-thirds of the gene repertoire of the Y in humans. To put a finer point on it, human and chimp genes differ by 1 percent in every chromosome except for Y, in which the level of difference seen in males, 33 percent, is about the same as what you see when comparing the autosomes of humans and chickens. Yes, chickens. Page suggests that a major driving force behind the differences is sperm competition.

GIVEN the sexual habits of chimpanzees, sperm competition relies on more than penis size and copulatory plugs, the number of spermatozoa per ejaculate, the motility and speed of the sperm, the immune responses to it, and the Y repertoire. Male chimpanzees, and the closely related bonobos, may copulate several times each day. So scientists have attempted to factor in the amount of sperm they can produce, their so-called sperm reserve, and their ability to repeatedly ejaculate. This has led to some bizarre experiments with chimpanzees.

A team of scientists, once again based at the primate research facility in Gabon, developed some unusual techniques in their hunt for answers. First, the researchers trained six chimps for semen collection. "This was accomplished by attracting them to the bars of their enclosures with orange juice and grasping and stimulating the penis (usually already erect) with a gloved and lubricated hand," the team explained.[11] The approach, they said, yielded much higher volumes of ejaculates than electrostimulation, which was used with other groups. (The details of those electrostimulation experiments, incidentally, make masturbating a chimp by hand seem much less odd: one group anesthetized chimps with ketamine and then inserted a rectal probe, three centimeters in diameter and eighteen centimeters long, that delivered "stimulation" in five-volt increments up to fifteen volts.)[12] The "hands-on" researchers attempted hourly masturbation from the chimps and succeeded in coaxing up to six ejaculations from the same individual.[13] In fact, the total amount of ejaculate *increased* after successive masturbations. Although the sperm count in the liquefied fraction of the plug decreased, it was not a statistically significant drop and the average totaled 369 million per ejaculate—six times higher than what is seen in an average man. "This large reserve capacity . . . may have advantages in the multi-male breeding system of chimpanzees," the researchers cautiously concluded. *May* have advantages?

Chimps, then, hold a huge "sperm reserve." Humans, in contrast, have fewer than two ejaculates in their reserve.

How, human males in particular might ask, does a chimp achieve an erection and ejaculate so many times in succession? One intrepid group of researchers, led by the Freudian psychologist William Lemmon and his graduate student Mel Allen, theorized that vaginal contractions may help the penis along. This, Lemmon and Allen suggested in an unparalleled moment of scientific creativity, led to the evolution of the female orgasm.[14]

Lemmon, who died in 1986, created the Institute for Primate Studies on his farm in Norman, Oklahoma. The institute at one time had forty chimpanzees, which lived in buildings adjacent to his farmhouse, and young researchers who worked there included Sue Savage-Rumbaugh and Roger Fouts, both of whom would gain fame for their ape language studies. The chimps Lemmon bred or housed included the sign-language stars Nim Chimpsky and Washoe. Lucy, another signing chimp owned by the institute, achieved a measure of notoriety herself after a biography about her revealed she used to masturbate while flipping through *Playgirl*, sometimes using a vacuum cleaner hose for stimulation.[15] "Lemmon had allowed, even encouraged, a sexually charged atmosphere to flourish around his chimps," wrote Elizabeth Hess in her elegiac *Nim Chimpsky: The Chimp Who Would Be Human*. "It was all grist for his Freudian mill."[16]

To find supporting evidence for their ejaculation thesis, Lemmon decided to manipulate female chimpanzees to orgasm. Allen, the grad student, performed the manual stimulation of the circumclitoral areas and vaginas. "Most of these females permitted stimulation to continue to sexual arousal," wrote Allen and Lemmon. "One of them allowed stimulation to continue to orgasm on ten separate occasions." They carefully recorded the duration of perivaginal contractions, secretions, hyperventilation, spasms, facial expressions, and the number of digital thrusts required to trigger contractions (20.3, for the record). They took special interest in one chimp that was lactating and therefore never became swollen from estrus. "On one occasion, when both clitoral and vaginal stimulation were being concurrently provided, this female reached back to grasp the thrusting hand of the experimenter and tried to force it more deeply into her vagina," they noted.

Anticipating criticism from colleagues who earlier had asserted that female chimpanzees do not have orgasms, Allen and Lemmon remarked that this erroneous conclusion may be linked to one behavior that seems singular to climaxing women. "Of considerable interest is the fact that relatively few blatant emotional responses (e.g., vocalizations) associated with orgasm in women were manifested by the chimpanzee, even when intense vaginal contractions were being palpated," they wrote. "In fact, a female colleague who observed several experimental sessions found it incredible that this female chimpanzee was experiencing anything even approaching the emotion characteristic of orgasm in women."

With their stimulating data set presented and parsed, Allen and Lemmon leaped into an oh-so-authoritative explanation of how the female orgasm evolved. "It is our hypothesis that the orgasmic response in the anthropoid female (Infraorder: *Caturrhzni*), or perhaps in mammals generally, has evolved for the purpose of stimulating the orgasmic response in the male." To bolster their supposition, they used the sophist strategy of raising potential shortcomings and blithely dismissing them. The assertion that the female orgasm merely represented a "byproduct of selection"—Stephen Jay Gould's argument about "tits and clits"? Pshaw. As for the fact that women routinely become pregnant without having had an orgasm, maybe climaxing is just one more component of "female choice," a way for her to increase the chances of fertilization if she really likes that particular male.

But the pièce de résistance is Allen and Lemmon's contention that the evolutionary roots of the female orgasm are obscured because human males in modern times do not have enough intercourse. Men, they note, on average copulate about twelve times a month. Our ancestors, they reasoned, had a chimpier existence, making it "likely that men copulate and experience orgasm by means of coitus less often today than was typical during the vast majority of hominid existence." Now factor in that a man ejaculates more quickly if he has not had an orgasm for several days. "A female coital partner cannot be expected to obtain orgasm by means of copulation with a male having a lower threshold to orgasm than herself," they declared. "Therefore, if men today copulated more often, their ejaculatory thresholds would rise to a more species-typical level and orgasm would result from appropriate stimulation rather than from inappropriate sources, i.e., from vaginal contractions instead of from vaginal corrugation or constriction."

Not surprisingly, no one ever replicated Allen and Lemmon's groundbreaking study. I suspect their arguments had some impact, though: they caused hyperventilation, facial contortions, and spasms in humans of both sexes—brought on by the blatant emotional response known as uncontrollable laughter.

—◦◦◦—

I live about thirty miles from the San Diego Zoo, which hosts one of the few communities of captive bonobos anywhere in the world. I visit the bonobo exhibit frequently, and more often than not, I see males humping males, males and females engaging in "missionary position" coitus, or two females genito-genital rubbing their crotches. Sex, in the words of the primatologist Frans de Waal, who began studies of the San Diego Zoo group in 1983, "is the key to the social life of the bonobo."[17] Scientifically named *Pan paniscus* because they at first were wrongly thought to be "pygmy" chimps, they more accurately should have been named *Pan satyrus*, de Waal jokes, to reflect their kinship with the libidinous satyr of Greek mythology. (In a fitting historical note, the first live "chimpanzee" taken to Europe around 1641 may have been a bonobo and was called a satyr.)[18]

De Waal has, more than any other single researcher, helped popularize the notion that bonobos are the hippie chimp, the sexy ape, the make-love-not-war species. Without question, he has carefully documented that captive bonobos have different sexual habits and are less aggressive than either captive or wild chimps, *Pan troglodytes*. As his revolutionary data from observing bonobos at the San Diego Zoo show, males prefer to mount females in the missionary position, more sex occurred between two females G-G rubbing than in any other combination, and juvenile males mouth-kissed and had oral sex with each other. Females also regularly had sex with males even when they were not fully swollen in estrus, and evidence suggests they also stay swollen for a longer period than chimps.

De Waal's bonobo findings call into question attempts to connect the dots between the sexual behavior of chimps and the sexual behavior of humans. Bonobos and chimpanzees, after all, are equidistant to humans on the evolutionary tree. So an alpha male chimp that dominates the females in a group is just as related to us as the bisexual bonobo female who dominates males in hers. "In the same way that paleontologists prefer their fossil finds to belong to a human ancestor rather than to an extinct side-branch, experts on ape behavior sometimes claim that their subjects are the only or best model for the last common *Pan*-human ancestor," de Waal and his fellow researchers wrote in an article with the teasing title "The Other 'Closest Living Relative.'" "Although bonobos are occasionally proposed for this distinction, chimpanzee models have long reigned supreme in this arena."[19]

De Waal argues that chimp behavior also plays to human biases about why we do what we do. "Recent discoveries of social systems in bonobos . . . which are based upon sex rather than warfare, fruit sharing rather than hunting and meat sharing, life-long maternal influence, and female rather than male dominance have met with a mixed reception," de Waal acknowledges. "While they have delighted some, they have obviously disturbed others." He believes the bisexual, horny bonobo makes many puritanical Americans uncomfortable—he had to first publish explicit photographs of bonobo sex in a European magazine. "Anyone interested in the reconstruction of our evolutionary past will need to face the implications of having a sexy, female-centered close relative," he declares.[20]

In ape research circles, de Waal's findings have not so much disturbed his colleagues, who roundly admire his careful and methodical work. But several have questioned whether what he observed accurately captures bonobos in general. "What Frans has seen is certainly not wrong," the bonobo researcher Gottfried Hohmann told me, sitting in his office at the Max Planck Institute in Leipzig. "He knows how bonobos in captivity behave, and particularly in San Diego." Hohmann since 1989 has done field research of wild bonobos at Lomako and more recently has established a new site at LuiKotal. His group also studies captive bonobos at different facilities, including the Leipzig Zoo. "If you focus on one captive group, you're likely going to get a black and white shot," he said. "If you look at several, you get gray. And ideally you'd get information from the field, but de Waal's information is very hard to obtain." Hohmann further pointed out that the San Diego Zoo bonobos de Waal studied were related to each other—and most were juveniles. Hohmann emphasized that captive animals do not have the same sexual behavior as those in the wild for the most obvious of reasons: they are confined and cared for. "In zoos, they get fed all the time, and they can't escape," he said. "They have to do *something*. And much of their sex is conflict resolution."

These arguments initially were crafted by Craig Stanford, who "in a fit of devil's advocacy" gave a talk at a primate meeting in Brazil in 1996 challenging de Waal's portrayal of the sexy bonobo and other supposed differences from chimpanzees. Stanford, an anthropologist at the University of Southern California who studied wild chimpanzee hunting and meat-eating behavior, furthered the revisionist depiction of the bonobo as extremely similar to the chimp in his 2001 book, *Significant Others*. "Sexy apes versus brutal ones represents an appealing dichotomy—our evolutionary options laid out in simpler terms," wrote Stanford. "It may be high time, however, to throw some cold water on the lurid descriptions of bonobo sexuality."[21]

Stanford and Hohmann's criticisms did not change Frans de Waal's views. An amiable, impassioned man who speaks English with a lilting Dutch-tinged accent, de Waal had gotten somewhat hot under the collar when in 2007 the *New Yorker* published a lengthy review of bonobo science that featured Hohmann—with a main theme the debunking of de Waal—and moved the debate from the cloisters of the academy to the dinner party table and the editorial pages of the *Wall Street Journal*. The *New Yorker* article colorfully captured "the pop image of the bonobo—equal parts dolphin, Dalai Lama, and Warren Beatty."[22] "I didn't like the article," said de Waal, tartly.

The author of the piece, Ian Parker, thoughtfully dissected the "appeal of de Waal's vision" of the sexy, hippie chimp. "Where, at the end of the twentieth century, could an optimist turn for reassurance about the foundations of human nature?" asked Parker. "The sixties were over. Goodall's chimpanzees had gone to war." He asked de Waal for his rebuttals to the criticisms launched by Hohmann and others. So what, de Waal said, if he only observed captive bonobos? Captive chimpanzees did not G-G rub, and wild bonobos do. Craig Stanford countered that wild chimps had similar amounts of sex. De Waal remarked that this detail was only true if you ignored homosexual sex in bonobos. And Takayoshi Kano, who headed the longest-running bonobo field site, at Wamba in Congo, and who had observed more wild interactions than anyone else, supported the assertion that females dominated and the species was more peaceful than chimps. But some of the strongest assaults—including Hohmann's assertion that G-G rubbing may not even be related to sex—went unchallenged by de Waal in Parker's account.

De Waal parried with a riposte in the newsletter of the Skeptics Society. Titled "Bonobos, Left and Right: Primate Politics Heats Up Again as Liberals and Conservatives Spindoctor Science," the article chided Parker for traveling all the way to Congo to watch a few bonobos "quietly sitting in the trees, eating nuts" and, more sharply, for promoting the idea that de Waal's depiction of the bonobo—and, by extension, the common ancestor with humans—was wish fulfillment, a fairy tale, a politically palatable myth designed for liberals. In one particularly bare-knuckled passage that made it clear de Waal was no prim academic in a bow tie, he sarcastically attacked Hohmann for challenging that G-G rubbing didn't qualify as sex. "Fortunately, a United States court settled this monumental issue in the Paula Jones case against President Bill Clinton," wrote de Waal. "It clarified that the term 'sex' includes any deliberate contact with the genitalia, anus, groin, breast, inner thigh, or buttocks. In short, when bonobos contact each other with their genitals (and squeal and show other signs of apparent orgasm), any sex therapist will tell you that they are 'doing it.' "[23]

We do not need bonobos to justify our lustiness, homosexuality, or the dominance of women in some relationships, or to clarify for us the definition of the word "sex." Similarly, we cannot rightly use alpha male chimpanzees to explain away the brutish cad as nothing more than biology at work. We are all of these things, and all of them are shaped by both nature and nurture. For example, in *Demonic Males*, the Harvard primatologist Richard Wrangham and the writer Dale Peterson explore the link between human violence and the behavior of male apes, and they urge people to gauge carefully how much they read into the very connections the book makes. They use the bonobo as something of an antidote to the chimp, arguing that humans and chimpanzees are not tangled together in an inexorable "evolutionary dance of violence." And they give the example of rape by orangutan males—which is common and has even involved a documented rape of a woman working as a cook at a research camp—to strongly caution against viewing ape behavior as "an excuse for evil," noting that "even if animal parallels tell us about ourselves, they justify nothing."[24]

A lot can change in a million years, which is roughly when chimpanzees and bonobos diverged from each other and seemingly went on their separate sexual ways. Given that humans shared an ancestor with these apes at least 5 million years ago, many more evolutionary forces have separated us from both chimps and bonobos. These, ultimately, are the differences that hold the most interest. Men do not produce billions of sperm in an ejaculate or need a long penis to push away a copulatory plug. Is that linked to the fact that we are more monogamous, and that men, in general, know which children they have sired and help raise them in nuclear families? Is this tied to the way men and women share food with their sexual partners and their children? Does our inability to synthesize our own Neu5Gc sugar help explain fertility problems that many couples face? Only humans have both males and females that choose to be exclusively homosexual. If humans do have a homosexual gene that is connected to exclusively homosexual behavior, as some researchers have reported, is it absent in apes—or more expressed in bisexual bonobos?[25] And, perhaps most consequential of all in an evolutionary sense, why don't women have visible genital swellings, and how does that affect our ability to reproduce?

—◦◦◦—

MY wife, Shannon, and I are the same age, and at thirty-one, we had our first child. She conceived the first month we tried to get pregnant, and everything proceeded without a hitch. The only fertility issue we ever discussed was how *not* to get pregnant.

When we decided to have a second child four years later and threw away the birth control, we were stunned that everything did not immediately click into place. Then Shannon had a miscarriage. Our pillow talk soon included discussions about induced ovulation, hysterosalpingograms, human chorionic gonadotropin, intrauterine insemination, and dilation and curettage. After five years and three more miscarriages, we quit, content that we had a lovely, healthy daughter. Then lightning struck: Shannon, at forty-two, became pregnant and carried to term. Then lighting struck again, and at forty-four, she gave birth to our third child.

We had received so much misleading and inaccurate information from "experts" during our miscarriage odyssey that I decided to write a book about the science of pregnancy loss, *Coming to Term*.[26] One of the most alarming studies I came across assessed miscarriage rates in 221 healthy young women, with no history of fertility problems, whose urine was checked daily for spikes in human chorionic gonadotropin, hCG, the levels of which dramatically increase when the fertilized egg implants itself in the uterine lining. The hCG test is the basis for home pregnancy pee sticks. They detected a miscarriage rate of 31 percent. Given that some loss occurs after conception and before implantation, many leading researchers estimate that the actual miscarriage rate is closer to half of all conceptions. And rates increase as women near menopause.

That is a huge amount of loss. While half of it is due to chromosomal abnormalities—nonviable conceptions—I was baffled that humans would make so many mistakes. When I looked for evidence of miscarriage in other species, I was startled to discover that it was rare, even in chimpanzees. I asked Kurt Benirschke, a pioneer of studies into chromosomes and reproduction as well as the founder of the San Diego Zoo's Center for Reproduction of Endangered Species, why he thought humans miscarried so much more frequently than other great apes. "We don't screw at the right time," he said.

Benirschke's reasoning is that chimpanzees mate most frequently when female genitals are visibly swollen, indicating that they are in estrus and ovulating. They may also get olfactory cues about ovulation, too. Humans in contrast have no obvious signs of ovulation, other than cervical mucus increasing and becoming more stretchy, which most men never notice. So humans, this theory holds, frequently fertilize eggs that are "overripe" and prone to chromosomal problems that lead to miscarriage. Studies dating back to 1922 in other species, ranging from frogs and fish to rabbits and sheep, have demonstrated that overripe eggs lead to what scientists gently call "reproductive wastage."[27]

Adding support to Benirschke's ideas, Tobias Deschner, who set up the Max Planck Institute's endocrinology lab, made a remarkable observation of female chimpanzees in heat. For his Ph.D. dissertation, Deschner investigated the effects of sexual swellings in the chimpanzees that Christophe Boesch had habituated in the Taï National Park in Côte d'Ivoire. "From an evolutionary point of view, you'd definitely say this swelling is such a big thing and seems to be costly, there should be some advantage to it," Deschner told me. Indeed, when you see a fully swollen female—in "maximal tumescence"—she appears to be lugging around a large water balloon below her rear end.

Given the costs of finding the right time for conception, it could be that miscarriage is *the* strongest force of selection in human evolution.

—◦∞◦—

JUST as proposing an origin of bipedalism has become something of a cottage industry, researchers have floated many theories to explain why Old World primates, a group that includes apes and monkeys, display swellings. In tribute to Russell Tuttle's categorization of the various ideas for how humans came to walk on two legs, I have nicknamed the mainstay primate swelling theories as "Who's Your Daddy?" "I'm All Yours," "Heat Shield," "Fight for Me," and "Well Hung."

Who's Your Daddy? suggests that female swelling serves to confuse paternity by extending the mating period. A chimpanzee cycle lasts about thirty-six days, and she is maximally swollen for only about ten of those, according to several studies. Because males cannot tell precisely when the receptive female is ovulating, they will be more likely to invest in raising any offspring or at least not to harm the female or the infant (and infanticide does occur among chimps).[28] Diametrically opposed, I'm All Yours argues that swelling confirms paternity by advertising that a female is about to ovulate, another way to increase the odds of paternal investment. Or swelling may be a Heat Shield, a way for females to travel safely to another community to mate—strange females often are attacked if they are not swollen— and diversify the gene pool.

Males also battle over swollen females, so maybe Fight for Me is a way to select the fittest mate. Swelling, then, would benefit males, too. Sure, they can have sex with abandon; they can frequently ejaculate huge numbers of sperm, and females do not expect much in terms of help to raise the offspring. "But the problem is when several males and several females are in a group, sex can be very costly to males: they might end up in an escalated fight," Deschner explained. "So you should only do it if it's worth doing it. And maybe you should only go for a female that's worth it." Well Hung, a variation of this, is the phenomenon described earlier in which only the male with the longer penis can displace the copulatory plug of rivals during maximal tumescence, when the depth of the vagina substantially increases to give him a meaningful advantage.[29]

Deschner's study focused on two new additions to the Theory of Swelling sweepstakes, Trust Me and Don't Rush.[30] Trust Me posits that if many males and many females are in a community, it is a dreamy world for the males: females compete for their affection, and the males become choosy. A female with plump genitalia is showing off her fitness, much in the way that in some cultures, according to anthropologists anyway, large rear ends in women are thought to indicate to men that they have the best hips for giving birth, or that large breasts signal that a woman will be able to feed their children well.

Don't Rush suggests even more straightforward marketing: that as the swelling increases, it indicates when a female is about to ovulate and is more fertile. Successful fertilization typically occurs from about three days before to the day an egg is released from the ovary. A female thus could mate with many males, netting the benefits of disguising paternity, yet she also stands the highest likelihood of being impregnated by the dominant male that can monopolize her during her most fecund days.

Deschner did something that all too rarely happens in the contentious Theory of Swelling debate: he compared the different predictions made by Trust Me and Don't Rush with the data he collected in the field. Specifically, he cataloged the size of swellings at different time points, the timing of ovulation, and the behavior of males. Working with Boesch and others, Deschner followed more than fifty chimps in the Taï National Park for various times between 1998 and 2001. He collected urine from twelve females over forty-two menstrual cycles and then analyzed the hormones that rise during ovulation. Videos documented the size of a female's swelling. To observe consorts between males and females, Deschner and his team tracked the group from dawn to dusk.

Swellings gradually grew by up to 20 percent as females approached ovulation and peaked, but they remained relatively large for four days after ovulation. Alpha males had about four times as many copulations on the day before ovulation until one day after; by three days post ovulation, they lost interest. Interestingly, they also did not prefer the females with the largest swellings, instead seeking partners who experienced the most increase compared to their nonestrus size. So they did not simply rely on absolute swelling size, but rather took advantage of the peak in an individual. Alpha males also intervened more frequently when another male tried to mate during the ovulatory window, and the lowest-ranking males more often chose the least optimal times, conceptionwise, during the female cycle to mate.

Based on these observations, Deschner determined that swelling size, by itself, does not provide a good indicator of a female's fertility—casting doubt on the Trust Me theory. "It's the change over time in one individual that carries the information," Deschner told me. The data nicely fit Don't Rush, with males paying close attention to a graded signal. And there was one extra bonus for the females: a dash of Who's Your Daddy? The alpha males ultimately had not developed a system to perfectly time copulations to ovulations, as their interest continued for a few days after peak fertility. The lower-ranking males had many opportunities to mate at times outside that window, further obscuring the paternity of offspring.

Deschner makes no claims about the relevance of the study to understanding human reproduction. "I'm not an anthropologist," he stressed. "I do not come from the direction that says, okay, humans are the things that interest me, and I study chimpanzees because they are our closest relatives and then make big implications about humans."

I, too, am not an anthropologist. But data speak, and these are offering a few loud messages about us. Clearly, these chimpanzees did mate at the optimal times, and sperm from the fittest males had the best opportunities to fertilize eggs. Although the actual miscarriage rate in chimpanzees remains unclear, Deschner's findings strongly support the idea that, unlike humans, chimps make a point of screwing at the right time with, theoretically, the best sperm available.

The role of swelling in chimp mating highlights the fact that humans have evolved to conceal a woman's ovulations and peak fertility. To be sure, some research suggests that women do not conceal ovulations—subtle changes in the face, the breasts, and body odor are evoked. But consider a study that took urine samples from thirty-six women and asked them to predict their own ovulation.[31] Across the eighty-seven cycles between them, the women's guesses were inaccurate 57.5 percent of the time—and these women, unlike, say, *Australopithecus* females, had calendars and knowledge of their menstrual cycles. The researchers unequivocally concluded that "this data supports the notion that ovulation is concealed to the vast majority of women." And having written a book on miscarriage and interviewed more than one hundred women about the details of their sex lives, I am thoroughly convinced that, as limited as it may be, women have a far keener sense of what is happening in their reproductive organs than their male partners do.

Why, then, do women have concealed ovulations? Maybe they remove barriers to males mating frequently. This, in turn, could reduce conflicts between males in a community, and also increase monogamy. Maybe they were a by-product of the upright stance of bipedalism, an exaptation along the lines of swimming. Maybe, following the Who's Your Daddy? argument, they protect women and their infants by confusing paternity. Or maybe the common ancestor to humans and chimpanzees—and bonobos—had concealed ovulations and the fact is that swelling evolved later in our modern ape cousins.[32]

There currently is no satisfying way to sort out a definitive Theory of Concealed Ovulation. To do so will require more data. And scientists including Pascal Gagneux, Frans de Waal, Gottfried Hohmann, Richard Wrangham, Christophe Boesch, and Tobias Deschner, who actually are testing hypotheses, with both wild and captive chimpanzees, will start to provide the answers to two of the deepest questions modern humans have ever asked: How did I come to be, and what will I leave behind?

11

IT'S A CHIMP'S LIFE

THE EVENING BEFORE I FIRST SAW A CHIMPANZEE IN THE WILD in May 2008, the research team I was going chimping with at the Kibale National Park in western Uganda teased me that I was about to lose my virginity. I did not sleep well that night. In part, I had the virgin-on-a-honeymoon jitters, a mix of thrill and fear over something I had imagined and wanted to do for many years that was about to become a reality. Then there were the midnight monkeys, the black and white colobus that were endlessly screeching and racing across the roof of my cabin, and, at times, seemingly trying to break through the walls or bust open a window. I had a mosquito net over my bed, and when I turned on my flashlight to make a visit to the bathroom, I noticed that a diverse insectaria, several of whose members appeared to have been abusing steroids, had taken residence on the mesh. Adding to my insomnia, I had to rise at 5:30 A.M. and was worried that my alarm would fail and I would wake after everyone had left for what promised to be a rigorous hike through the thirty-two square kilometers (about twelve square miles) of rainforest that was home to some fifty chimps in the Kanyawara community.

I did not lose any sleep over the possibility that a chimpanzee might attack me. Gombe had problems, especially with Frodo, an alpha male that had grown into a bully and particularly liked to harass Jane Goodall, dragging her around and stomping on her. More horrifying still, in 2002 Frodo ripped a baby from the back of a woman who lived in a village that abutted Gombe and who had been walking through the park. Frodo partially ate the baby girl and then left her remains in a tree. But Frodo's behavior was deemed aberrant for chimpanzees habituated to humans. Goodall believed the way in which she had initially provisioned the Gombe chimps, by feeding them bananas to make it easier to observe them, had created unusual amounts of tension and aggression.[1] Kibale, in contrast, had never provisioned chimpanzees. Yet a Kibale chimp had also snatched a three-month-old baby from a mother who had set the boy in the shade while she harvested her potatoes. "Saddam," as the locals named him, had been shot and killed for the offense.[2]

I did ask many questions over dinner about what to do if a chimp charged me, and I was assured it would not happen unless I did something that would indicate a threat, say, by intentionally placing myself between a mother and her child. The researchers make a point of keeping their distance from the chimpanzees, both to reduce the chance of harming the animals by spreading disease and to respect that they are visitors to their home.

I made it to the 6 A.M. breakfast on time, and thirty minutes later our hike began. I soon was focused on two species that I had not expected to think about this day: ants and elephants. The half dozen men in the team mostly followed each other single file on skinny trails that wound up and down the forested hills, but fallen logs and streams sometimes forced us to scamper through bushes that were lousy with small black ants. Each brush with a bush led to mounds of ants clumped on my neck. I quickly scraped off the ant balls, but sentries inevitably would make the trip down my shirt and start nibbling on my back and destinations farther south. Paco Bertolani, the primatologist leading the hike, seemed entertained by my discomfort, and the ants were no more of a nuisance to him than a little rain in the damp forest.

Originally from Rome, Bertolani had studied wild chimps with Christophe Boesch in Côte d'Ivoire and with Jill Pruetz in Senegal, where he helped habituate a community that lived on the savanna in Fongoli, and documented how they sharpened sticks to stab bush babies in the cavities of trees—the first evidence of chimps making tools for hunting.[3] Bertolani was only now working toward his doctorate, as he had taken many journeys away from chimps, including five years as a computer programmer, a few years in the Amazon working with freshwater turtles, and a stint on the Mentawai Islands, during which he went native and framed his face with a thin-lined tattoo (his forehead has what looks like a bird's body over his nose, and its long, curved wings arc around his eyes). But chimping kept calling him back.

Bertolani knew more about chimpanzee behavior than many a bona fide Ph.D., and he loved his work, which at Kibale focused on using GPS devices to precisely map the routes used by individual animals to venture between feeding and nesting sites. "I just like to stay in the forest, silent there with these chimps," he explained to me. "It's amazing staying with these animals and trying to understand what they think. And you can see that 5 million years ago, we were similar to this creature. Everyone asks where do we come from, right? I don't have any problem. I'm sure we don't come from God; I don't have any creationist or religious beliefs. We are animals like chimps. So where do I come from? I try to answer the question by coming into the forest and not looking somewhere else."

When Bertolani first pointed to trees felled by forest elephants and earth that their hulking feet had trampled, I was more surprised than alarmed that they were in our midst. But when we heard the elephants clomping and crushing their way through the dense forest below, and the Ugandan field assistants with us began shooting worried looks at one another and switched from English to whisper in their native language, I realized these were nothing like the gentle giants I had seen balancing on platforms in circuses, trumpeting in countless zoos, dressed up in downtown Bangkok to dazzle tourists, and ambling down a New Delhi highway with a turbaned swami aboard. These forest elephants, I quickly learned, sometimes intentionally rush humans and can cause serious harm, even death. Bertolani said not to worry. If they charge you, just run in zigzags and you'll escape, he said. Anyway, the elephants usually paid him no mind. "Maybe they know that I like them," he said. One of the field assistants laughed. "They stay away from you because of your tattoos," he said.

After about an hour of hiking, we turned a corner and arrived at our first destination, an enormous fig tree, *Ficus capensis*, that Bertolani knew was in full fruit and thought would similarly be full of chimps. He was right. First I saw Tongo, a twenty-eight-year-old female with her three children. Two were juveniles, Lanjo and Tuber, and one an infant, Tsunami. Makoku, twenty-six, was the only older male around. We sat for a few hours and watched them languorously pluck figs, which hung in bunches like grapes. Quinto and Wilma, two sixteen-year-old females that had immigrated to the group a few years earlier, when they became sexually mature, ambled in and joined the breakfast gathering.

I was struck by the silence. It was nearly ten in the morning before I heard a chimp vocalize, a subordinate pant-grunt from Wilma to Makoku when he approached her part of the tree. The quiet reflected the fact that there were no other adult males there to pull rank with Makoku, and also that wild chimpanzees often have little need to communicate, at least with their voices. I felt like I was in a monastery watching a group of monks do their business, which the researchers documented in great detail, including charting every movement and using long sticks with plastic bags on the end to collect urine samples that fell toward the ground.

But I was even more struck by something else that was missing: old chimps.

Bertolani and I spent the entire afternoon following Makoku. He fissioned off by himself, eating the leaves of various plants—the researchers call them terrestrial herbaceous vegetation—and minding his own business. Makoku called it a day a good hour before dark, interestingly using the same night nest he had made high in a tree the night before. I felt like I was watching an old man, because of both the wear on his face, and his lumbering, solitary behavior. But I was almost twice his age.

Chimpanzees in the wild have a life expectancy of a mere thirteen years, a fact that only became clear in 2001 when Richard Wrangham, the Harvard primatologist who heads the Kibale research program, collaboratively collated long-term data from the forest with information gathered by Boesch's group in Côte d'Ivoire, Goodall's team at Gombe, Toshisada Nishida's project in nearby Mahale, and Yukimaru Sugiyama in Bossou, Guinea.[4] In her earlier writings, Goodall designated Gombe chimps as "old" when they reached thirty-three.[5] The oldest living wild chimp on record is Auntie Rose of Kibale, estimated to be sixty-four when she disappeared in July 2007.[6]

Humans on average live to be sixty-seven. The range varies dramatically based on where a person lives, with a baby born in Swaziland between 2005 and 2010 having a life expectancy of 39.6 years and one born at the same time in Japan enjoying the prospect of 82.6 years.[7] Still, wild chimpanzees and humans have starkly different life spans, which is further highlighted by the oldest documented person, France's Jeanne Calment, who was born in 1875 and died in 1997 at a staggering 122 years old.[8]

A bramble of factors account for these differences. New research contends that nothing compares to a grandmother's love, we truly are what we eat, and genes are destiny. Other root causes factor prominently in the tangle: fire, infectious disease, a woman's biological clock, accidents, birth interval between children, the cost of a big brain, violence. To make sense of why we are the longest-living apes, it helps to come down from the trees, so scientists are studying the humans whose lifestyle most closely resembles chimpanzees, hunter-gatherers, and the chimpanzees that live in captivity and benefit from human caretakers who intervene when there are fights, provide medicine when there's disease, and protect newborns from harm until they're old enough to fend for themselves.

THE Hadza of Tanzania. The !Kung of Kalahari. The Aché of Paraguay, the Agta of the Philippines, the Hiwi of Venezuela. The members of these keenly observed hunter-gatherer societies led brief lives "precontact," which is anthropological jargon for the time before colonists arrived wielding the double-edged sword of modernity that cleaves cultures but staves off death with medicines and farm-grown food in markets. Recently, Michael Gurven and Hillard Kaplan, anthropologists at the University of California at Santa Barbara and the University of New Mexico, respectively, investigated just how brutish and short those lives really were.[9]

In their sweeping review of mortality, Gurven and Kaplan compared traditional hunter-gatherer groups to their own acculturated relatives and to wild and captive chimpanzees, revealing fascinating distinctions at a variety of ages. They concluded that human bodies "are designed to function well for about seven decades in the environment in which our species evolved." To work out the life trajectory of wild chimps, they relied on the longevity data collected by Wrang-

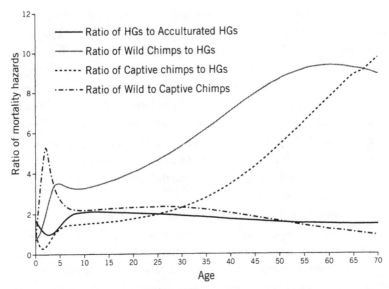

When Chimps and Hunter-Gatherers Die

The first spike on the left shows that many more wild chimps die than captive chimps before the age of five (a ratio of 1 on the y-axis would mean they were the same). This reflects the extent to which captivity protects animals from disease (through veterinary care) and violence (including infanticide). The first big dip, strikingly, reveals that captive chimps suffer less infant mortality than traditional hunter-gatherer populations (HGs). But as time passes (moving right on the chart), the degree of benefit to being a captive chimp steadily wanes, and humans win the mortality sweepstakes.

ADAPTED FROM MICHAEL GURVEN AND HILLARD KAPLAN, "LONGEVITY AMONG HUNTER-GATHERERS: A CROSS-CULTURAL EXAMINATION," *POPULATION AND DEVELOPMENT REVIEW* 33:321–65 (JUNE 2007).

ham, Goodall, Boesch, and others, while for captive chimps they turned to an analysis of some one thousand animals that had lived in research colonies such as Yerkes in the United States and in Europe.[10] Whether living in the wild or in captivity, by age thirty, chimps begin to die far more frequently than hunter-gatherers. And by their fourth decade, they stop functioning well—three decades earlier than humans. "The lifespan of chimps, even under the best of circumstances in captivity, doesn't look anything like humans," Gurven said.

But how long can the body truly last? Analyzing life expectancy from birth obscures the answer because there are great differences in early-life mortality rates. A much better measure is to look at what happens to individuals that make it past the critical first few years. Wild chimps have an average life expectancy of thirteen, whereas chimpanzees raised in captivity live about forty-two years. Yet a chimp in the wild that survives to fifteen likely will live until thirty, and one that makes it to thirty will probably see thirty-eight. Gurven and Kaplan believe traditional hunter-gatherers only have a life expectancy from birth that ranges from twenty-one to thirty-seven, but those who make it to their twentieth birthday party on average will die at age seventy-two.

The human body, then, can take twice as much wear and tear as the corporeal components of our closest relative. That point now is clear. But figuring out which evolutionary forces allowed us to keep on keeping on decades after our closest relatives had literally bit the dust continues to keep theorists in business.

—◦◦◦—

EVOLUTIONARY theory holds that as individuals age and become less capable of reproducing, they become less valuable, and the body will not work as hard to maintain itself and repair the problems that develop. Women, by this logic, should die around the time of menopause, which on average occurs at fifty-one years of age, plus or minus roughly ten years.[11] On the surface, the life expectancy of hunter-gatherers from birth seems to confirm this, as would the same analysis for chimpanzees. But more than two-thirds of hunter-gatherer girls who live long enough to bear children survive beyond menopause.[12] "In chimpanzees, if they make it to adulthood, their chances of living for several years beyond the age of their last birth is 1 percent," the University of Utah anthropologist Kristen Hawkes told me. "Nobody does."

Hawkes has spent two decades refining a theory to explain why so many women live beyond menopause, and why humans—men and women alike—enjoy such spectacular longevity compared to other great apes. To test her hypothesis, she has examined the lives of female chimpanzees, analyzing their behavior, menstrual cycles, age at maturity, intervals between births, relationship to males, and eggs inside their ovaries. "I don't think chimpanzees are australopithecines," she joked, referring to our hominin ancestors. "But chimpanzees are a useful model for life history studies." She points to the Miocene apes, which include the common ancestor to chimps and humans and stretch back nearly 18 million years to the oldest ape fossil, Proconsul. "The burden of evidence is strongly on the side that says the average life spans for Miocene apes in general and australopithecines probably weren't that different. And then something happens in genus *Homo*, and it happens both to the boys and the girls."

When Hawkes was starting her career, she believed the something that happened was hunting. The hunting hypothesis, which dates back to Darwin, held that bipedalism made it easier to hold a weapon and kill big game ("Hit 'Em Where It Hurts"), and meat provided extra energy for the ravenous large brain, which began to grow larger 1.8 million years ago because the smartest *Homo erectus* or *ergaster* hunted with the most skill and efficiency. But, the thinking goes, the narrow female pelvis could not fit the bigger head, so brain growth was delayed until after birth, which in turn led women to spend more time caring for children and converted hunting into a male-dominated activity. With the gender division of labor, nuclear families formed in which men provided meat to a family in exchange for women raising the children, providing sex, and caring for the shelter. "Originally, I really had the hunting hypothesis firmly in mind," said Hawkes. "It didn't even occur to me to question it."

After doing research on the Binumarien language of the highlands of Papua New Guinea, in 1980 Hawkes began to study foraging among the Aché, hunter-gatherers who had only begun interacting peacefully with outsiders in the early 1970s. Her entry to the Aché came through a student studying under her who had worked in Paraguay during a Peace Corps stint and had built strong bonds with the group.[13] Although she did not realize it at the time, their early observations laid the foundation for what would become her own theory of human origins and the centerpiece of her research today.[14] "The stuff that men acquired, their wives and kids were no more likely to eat that than anybody else," said Hawkes. "And the fraction that went to their own kids was very small. This was so different from the hunting hypothesis."

Hawkes next shifted to field studies of the Hadza in Tanzania. "I was looking at sex differences and resource choices and what the hell men are up to," she said. The Hadza men, unlike the Aché, routinely hunt big game, and again, the hunting hypothesis seemed like bunkum. "One of the things you see is a big animal comes down, everybody comes to make claims," she said. "They say, 'This is our meat.' It isn't the private property of the hunter." Further chipping away at the idea that meat drove the evolution of modern humans was the reality that there just wasn't much of it: the Hadza hunters, who had bows and poisoned arrows with big game aplenty, only succeeded in killing an animal about once out of every thirty tries.

But what really twisted Hawkes's head were the grandmothers. "These old ladies, man, they were just amazing," she said. The grandmothers worked hard, particularly in helping their daughters raise and feed their children. This led Hawkes to formulate the Grandmother Hypothesis, an XX challenge to the XY-centric "Man-the-Hunter." In an earlier permutation of the idea, first floated in 1957 by George Williams, menopause evolved to allow a woman to "stop dividing her declining faculties between the care of extant offspring and the production of new ones."[15] Menopause would spare the woman the dangers of pregnancy and childbirth, increasing the likelihood that she would live long enough to raise her children—and her grandchildren after them—and be able to invest her energy, undiluted, into their success. Hawkes said hooey. Menopause, to her mind, is not an adaptation; it is the ancestral state.

In Hawkes's view, grandmothers eased the burden on their daughters, allowing them to wean earlier and have more babies at shorter intervals. The grandchildren that were cared for and well fed would, in an evolutionary sense, become more fit. Hadza grandmothers, she noted, could pull tubers from the ground, a staple food source that young children were not strong enough to forage for; children similarly could not crack seeds and nuts. Natural selection would have thus favored a longer life span for postmenopausal women, and, in something of a trickle-down theory, their children and grandchildren also would be more robust, and, over the generations, their life spans would increase.

The Grandmother Hypothesis provoked a lot of chin pulling from colleagues, who questioned whether the theory adequately explained long-lived men, noted that it failed to discuss the effects of a grandmother caring for her son's children, and suggested that longer life spans might simply reflect the prolonged childhood in humans. Or it could be that people began to live longer because the young started to care for and feed the old. These reservations have merit.[16] But the Grandmother Hypothesis continues to receive a serious scientific hearing because Hawkes and others steadily have accrued evidence for it—some of which involves chimpanzees.[17] "I realized from the outset that the differences between us and chimpanzees are really central to it," Hawkes told me. "And that meant really trying to improve what we knew about aging in chimpanzees."

First, Hawkes looked at differences in chimpanzee and human life histories. In early chimpanzee life, Hawkes noted a marked difference in humans that reduces the need for a grandmother's help: after weaning, chimps typically feed themselves, save for difficult to obtain foods such as the meat of monkeys, or nuts. Hawkes also contended that her theory explained why chimpanzee females and other apes reached maturity earlier than humans.[18]

Hawkes calculated the age of maturity based on reproduction. Females become mature, by her definition, when they first give birth, less the gestation period (chimp gestation is about thirty-one weeks versus forty weeks in humans).[19] Based on the data sets she consulted for one of her first papers on the Grandmother Hypothesis, in 1998, chimps become mature at age thirteen, and subsequent studies in both captive and wild chimps put the age even younger.[20] Aché and !Kung females, in contrast, become mature at 17.3 years of age. There is a trade-off between how long a juvenile will grow, and the risk of dying before reproducing, as Eric Charnov, one of Hawkes's mentors, has put forward. It follows, then, that as the risk of dying drops—such as when grandmothers help provision and care for juveniles—it pays to delay becoming an adult because an individual can grow larger. No great apes have grandmothers that provide anything like the support seen in humans. "The life history contrast between us and the other living apes is the standout thing," she said. Indeed, as she showed, maturity also comes relatively early to orangutans (14.3 years) and gorillas (9.3 years).

The biology of chimpanzees further supports Hawkes's theory. If menopause is not an adaptation but a consequence of living longer, the reason menopause is not visible in chimpanzees is simply because they die too young for anyone to witness it. So she decided to examine the ovaries of nineteen chimpanzees that had died between three months and forty-seven years of age, the first such study of its kind.[21]

Menopause is not, as people commonly think, defined by age. It occurs when a female's pool of eggs, or oocytes, inside her ovaries diminish to about one thousand. Mammalian females develop their entire pool of oocytes while they are still in utero, and, for mysterious reasons, they begin self-destructing en masse even before birth. Human fetuses have about 7 million eggs at their fifth month in utero, and that figure plummets to 2 million or less at birth, with fewer than three hundred thousand surviving to puberty. A woman typically only menstruates one egg per cycle, so she'll only need five hundred or so during her reproductive years.[22] As she approaches menopause, the quality of the eggs declines, and it becomes increasingly difficult to conceive a healthy baby and carry it to term.

If menopause is an ancestral state and not a derived trait, as Hawkes suspected, then chimpanzees' oocyte depletion should mirror what occurs in humans. That is precisely what she and her colleagues discovered: although chimps began with fewer oocytes than humans, they declined at the same rate, with the oldest chimp studied, at forty-seven, having hardly any eggs left. The findings meshed well with a study of captive chimps that found statistically significant increases in miscarriages and stillbirth as the mother's age increased.[23]

A few months after Hawkes reported these findings, Melissa Emery Thompson, a postdoctoral anthropologist at Harvard working with Richard Wrangham, published a comprehensive study of aging and fertility in wild chimpanzees.[24] Thompson and Wrangham's collaborators included a who's who of chimpology, including Jane Goodall and Anne Pusey at Gombe, Yukimaru Sugiyama and Tetsuro Matsuzawa at Bossou, Vernan Reynolds at Budongo, and Toshisada Nishida at Mahale. All told, they followed three hundred females over the course of several decades, and 165 of the chimps had babies—including Tongo, the mother of three that I watched in Kibale.

Of course these mother chimps began bearing offspring earlier

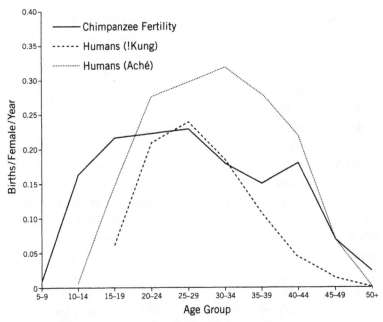

When Chimps and Hunter-Gatherers Give Birth

Although chimpanzees give birth at a younger age, their fertility peaks at a similar age to that of hunter-gatherers. But, in contrast to humans, the decline in chimpanzee fertility is linked to their lifespan: few individuals live to forty, and the ones that do likely are the most fit and capable of reproducing at an older age.

ADAPTED FROM MELISSA EMERY THOMPSON, JAMES H. JONES, ANNE E. PUSEY, ET AL., "AGING AND FERTILITY PATTERNS IN WILD CHIMPANZEES PROVIDE INSIGHTS INTO THE EVOLUTION OF MENOPAUSE," *CURRENT BIOLOGY* 17:2150–56 (DEC. 18, 2007).

than humans, but their "reproductive performance" started to decline at nearly the same time as the Aché and !Kung, with healthy chimps experiencing a 20 percent drop in fertility after thirty-five, the age which humans use as the cutoff for "advanced maternal age." Compared to humans, though, wild chimpanzees' fertility seems to senesce at the same pace as the rest of the body. "We're seeing a pattern in decline in fertility that would lead to infertility if the animals lived long enough," Emery Thompson said. "It's a completely normal mammalian pattern, just like cardiac function will decline with age." To Emery Thompson and Hawkes, it was compelling confirmation that menopause is the ancestral state. "Do most adult female chimpanzees go through menopause?" asked Hawkes. "No, they die first. It's all a definitional thing."

In chimpanzees, it appears, menopause is a signal that not just the oocyte inventory, but the whole body, has worn out. Chimpanzees that continue to have babies late in life—in this study, thirty-four females gave birth after age forty, and one mother was fifty-five—are the ones that have the hardiest somatic, or nonreproductive, cells. Human females, in contrast, have a disconnect between the life span of their somatic and reproductive cells.

One of the practical consequences of this chimp/human difference is that it may help explain why many men prefer younger women, while many chimpanzee males have a thing for older females. Emery Thompson, Wrangham, and Harvard's Martin Muller published a hugely entertaining study on this point that explored what makes for an attractive female chimp.[25] No matter what they measured—male-male battles for an estrous female, the preferences of the population's alpha male, and a variety of other parameters—the males went for age before beauty. The researchers asserted that the human male hankering for young-looking females with low waist-to-hip ratios, doe eyes, and slender jaws—all paedomorphic, or juvenile, characteristics—was not a quirk introduced by advertising agencies, but a derived trait, likely driven by the desire to have a long-term mate who did not bring along the baggage of previous kids and could reproduce longer; chimp males, in contrast, are more promiscuous than humans, and fathers don't even know which children are theirs. "Chimpanzee males may not find the wrinkled skin, ragged ears, irregular bald patches, and elongated nipples of their aged females as alluring as human men find the full lips and smooth complexions of young women, but they are clearly not reacting negatively to such cues," they concluded.

Hawkes's work does not clarify why it is that human somatic cells live longer. She posits that menopause allowed surviving females to spend more energy on repair and maintenance of their somatic cells, a benefit that evolutionary forces would have extended to their descendants. Humans may also have adapted better mechanisms to deal with oxidative stress, a key metabolic factor that damages cells, as a recent study that compares men to male chimps found.[26] We may have evolved strategies to delay neurologic damage, too, staving off dementias such as Alzheimer's disease until our sixth or seventh decade.

Persuasive as the Grandmother Hypothesis may be, it is but one insight into the forces behind the extraordinary longevity of humans, which eclipses the life span not only of apes but of every extant terrestrial mammal. To wit: it may not be who provides the food—the macho hunter or the little ole grandma—but what we eat and how we eat it that is key to our longevity.

—∞—

WHILE I was sitting under the sprawling *Ficus capensis* following old man Makoku, Tongo, and her three children as they scampered around the tree and gorged themselves on figs, I asked Paco Bertolani whether I could eat one. He advised against it: there are often wasp larvae inside them.

A gray-cheeked mangabey swung into the tree and began to share the bounty of fruit. At that point, I had been watching the researchers watch the chimps eat figs for a few hours, and, frankly, was excited by the prospect that they would move on to the entrée. But the chimps ignored the mangabey. Bertolani joked that the monkey had an evil-looking face and was not only a demonic male but would make a demonic meal. A few minutes later, a blue monkey arrived for a fig breakfast, and again the chimps paid no heed to the potential steak and ribs on offer.

The Gombe chimpanzees are more carnivorous than those at any other site studied, yet meat makes up only half a kilogram, or 2 percent, of their weekly diet, which mainly consists of five to seven kilograms of plant material each day. Based on observations of the Kanyawara chimps in Kibale, Richard Wrangham calculated that meat consists of just 0.1 percent to 0.3 percent of their more vegetarian diets. And an analysis of the chimps' monkey meals conducted between 1988 and 1997 pinpointed that eighty-four of their ninety-two kills were red colobus, with the balance all being black and white colobus.[27] Although other studies have shown that chimps in the region occasionally do eat gray-cheeked mangabey and blue monkey, those species either are too difficult to catch or are too unpalatable to attract much interest.[28] Any way about it, meat for chimpanzees is a luxury item, and not a major source of protein, which also exists in the leaves, nuts, and oil palm fruit they eat.

If the waspy figs seemed like an odd breakfast by human standards, when Makoku came down from the fig tree, the first thing he did was start eating flowers. As we followed him around for the next seven hours, he would stop at a plant and eat for fifteen or twenty minutes and then take a short siesta or amble to his next feeding, which never was more than twenty minutes away. He had a broad palate, snacking on leaves from *Bridelia micanthra*, *Celtis africana*, and *Celtis durandii* trees, and the pith of an herb called *Aframomum mala*. He made few sounds all day, other than his flatulence and the rumble of moving through an undulating rainforest, and his existence reminded me of the movie *La Grand Bouffe* (*Blow-Out*), in which the characters attempt to eat themselves to death.

But in fact Makoku was eating himself to life.

When Wrangham first entered the chimpanzee field, he focused on feeding behavior. After his year working with Goodall at Gombe, he finished his undergraduate degree at Oxford and then completed his Ph.D. on the behavioral ecology of the Gombe chimps at Cambridge University. In a paper he published in 1977, only the second of more than a hundred he would go on to write about chimps, he calculated how much time the Gombe chimps spent chewing. He recorded their every activity, minute by minute from sunup to sundown, and by his tally, the chimps passed more than six hours a day using their teeth to rip and grind the tough, fibrous raw leaves and fruit that are the staples of their diet.[29] "When you add up the amount of time spent digesting their food, and the rest of the time traveling, you realize they're spending all of their time concerned about eating," Wrangham told me. "It's an astonishing difference when we think about humans."

Wrangham noted several studies that have calculated how long humans spend eating. Nine- to twelve-year-old children in the United States dedicate little more than one hour out of their twelve-hour day, so about 10 percent, to food. Children in subsistence societies in Venezuela, Samoa, and Kenya average about the same. As adults, we even take in more calories per day—between 2,000 and 2,500 for a human versus 1,800 in a chimp. How can we feed ourselves at least five times more quickly than chimpanzees? Wrangham's answer: cooking.[30]

Only humans have mastered fire, which means only humans cook. Not only does cooking soften food and make it taste better, it also aids digestion, making it easier to extract nutrients. As Wrangham argues in his book *Catching Fire*, "the transformative moment that gave rise to the genus *Homo*, one of the great transitions in the history of life, stemmed from the control of fire and the advent of cooked meals."

Even Darwin recognized the singular importance of controlling fire, a discovery he ranked as "probably the greatest ever made by man, excepting language."[31] Still, the Cooking Hypothesis, first published by Wrangham and four colleagues in 1999, was initially viewed by many as an audacious, radical claim. Indeed, it purports to explain everything from our unique longevity to the shrinking of our guts and teeth, the expansion of our brains, the loss of our hair, the reduced weaning time for our babies, the division of labor between the sexes, and the way we cooperate and flirt with each other.[32] Some colleagues rejected the Cooking Hypothesis out of hand, punningly complaining that the idea is "half-baked wishful thinking," "too many cooks spoiling the broth," and "a good case of indigestion." Others slammed it as "anthropological folklore," "absolutely charming," and "little more than a just-so story." But the most brutal critiques used the restrained scientific lingo for caca, calling it "unconvincing." The main hole in the theory: the earliest evidence of humans controlling fire dates back only 790,000 years, and even that is scant, leading many anthropologists to conclude that we started hitting flint to stones in the last 250,000 years. *Homo erectus* emerged about 1.8 million years ago.

Wrangham conceded up front that the critics are right about that timing, but said only a "pretty extreme archeologist" would dismiss an idea out of hand because of the absence of evidence. For now, the cooking theory must rely on a biological argument, he said, not the physical remains of controlled fire from a few million years ago. "Humans are biologically adapted to cook food, to judge from all the evidence," he told me.

The supporting facts include people who restrict their diets to raw food, be it to live what they perceive as a more natural lifestyle or because they are shipwrecked. Eating a raw food diet leads to significant weight loss, and people constantly feel hungry. They survive, but their reduced energy intake means that half the women who are raw foodists temporarily stop menstruating, and some men report reduced sex drives. Wrangham is careful, in his droll way, not to dismiss the potential virtues of eating raw, natural food. "It's healthful—if you want to lose weight," he said. Notably, vegetarians and vegans who eat cooked food have no problems with menstruation or excessive weight loss.[33]

Wrangham illustrates the limitations of a raw food diet by examining how much a person would need to eat each day to remain healthy. His case study is a theoretical 120-pound, five-foot-four-inch hunter-gatherer woman who menstruates regularly, which is a reasonable measure of "healthy" from an evolutionary vantage since fitness, by definition, is the capability to reproduce.[34] By his calculations, this woman would need 2,000 calories a day if she had a relatively sedentary lifestyle. A balanced diet of raw fruits, vegetables, and greens that would give her both the calories and protein she needs would require more than nine pounds of food each day, that is, 7.6 percent of her body weight. This is about the weight of food that an average American eats only on Thanksgiving Day. If she had cooked the same foods, 4.2 pounds would yield the same calories. He then factors in the likelihood that a hunter-gatherer would have had meat, so he adds half a pound of raw, ground venison and removes the equivalent in fruits and leaves. This would reduce her total daily food intake to a more reasonable 5.4 percent of her body weight. But another problem surfaces: it takes an extraordinarily long time to chew raw meat. Wrangham turns to chimpanzees for evidence and describes a feeding on baboon and bushbuck antelope timed by Goodall and a red colobus feast timed by him. Using this gauge, it would take forty-five minutes to chew half a pound of meat. The nonmeat on the plate would still total six pounds of food, and though he does not calculate how long it would take to chew that, the simple act of eating—let alone hunting and gathering—would clearly have occupied much more of the day than it now does. Wrangham asserts that a hunter who went out for the day would not have had enough time to eat a nutritionally sufficient raw food dinner. He goes a step further and suggests that "after cooking was adopted, humans lost the ability to survive on raw food except under unusual circumstances."

Cooking makes it easier to extract calories from food in several ways. It denatures the proteins in meats and plants, altering their three-dimensional structure to open the way for enzymes that can cut the bites into more digestible bits. Raw starches like potatoes and plantains are gelatinized by heat, which similarly allows enzymes and acids to aid digestion by breaking complex carbohydrates into simpler sugars. Cooking also destroys the cell walls of plants and gelatinizes the proteins in meat, softening foods and making them easier to tear and chew.

Many people assume that food has the same number of calories regardless of how it is prepared, but Wrangham convincingly shows this not to be the case. The vast majority of studies on the amount of energy people extract from different types of food analyze human feces. "There are endless studies of fecal digestibility, which are meaningless," Wrangham told me. When food leaves the stomach, it enters the small intestine. This, in his language, is where digestion by the body takes place. The rest of digestion occurs in the large intestine, or colon, where hordes of bacteria and protozoa ferment the food—and keep most of the nutrients for themselves. Rather than looking at feces, Wrangham believes it makes more sense to analyze the intestinal contents higher up the pipe, which is possible because some people have their large colons removed, a process called an ileostomy, and wear a bag.

Contents of ileostomy bags show that cooked wheats, cereals, and bread are 95 percent digested by the time they leave the small intestine. Digestibility drops to about 50 percent with raw potatoes or plantains. The body similarly does not digest about half the protein in raw eggs. "This is amazingly underappreciated," said Wrangham.

Experiments in animals further make the case that as food is made more tender, calories become more available. In one study, rats were fed food pellets that carried the exact same content but one version was hard and the other was puffed. Within twenty-two weeks, the rats that ate the soft pellets were 6 percent heavier—they had become obese. Wrangham's lab teamed up with a group that studies digestion in Burmese pythons, which can be fed food directly into the esophagus—no chewing involved—and then lie around afterward during digestion. They gave some snakes whole rats and others round steak that was either whole or ground, and then either raw or cooked. The amount of oxygen the snakes used after a meal indirectly indicates the amount of energy used, a proxy for the cost of digestion. Cooking and grinding the meat reduced the cost of digestion by a little over 12 percent each, and the effects were additive, so that a cooked, ground piece of round steak had a 23.4 percent drop in energy consumption.[35]

All of this makes a devastating case against the claim that raw foodism is "natural," but, more important, cooking stands out as a critical difference for humans—and offers yet another perspective on how we alone among the apes evolved such long life spans. Wrangham thinks that cooking primarily impacts the human life span by providing more energy to the immune system. He has no proof for this, but he does have a testable hypothesis: humans invest more energy in maintaining their immune systems than do chimpanzees. The increased energy derived from cooking also means that during times of food shortages, humans can do more with less, properly fueling the immune system and thereby reducing vulnerability to disease. Cooking kills many toxins in plants, as well as ones released by pathogens that make their home in raw meat, including *E. coli* and *Salmonella*. Wrangham and his graduate student Rachel Carmody calculated that the average amount of meat an American eats would make a person sick forty-two times a year and would raise body temperature 145 days annually, which in a seventy-five-year-old translates to 6.9 years of an increased metabolic rate necessary to fuel an elevated immune response.[36]

Now add to this the likelihood that fire itself altered our lifestyles and longevity. Based on the anatomy of *Homo erectus*, Wrangham surmises that this archaic human did not climb trees well and likely slept on the ground. "Can you imagine curling up and going to sleep without a fire on the African plains?" he asked me. "It would be terrifying. As long as you've got a fire, then lions will respect you." Not only can fire scare off predators; it also can help people see what menaces lurk in the dark. And fire, especially when it is used to cook, attracts groups of humans to gather around, as anyone who has ever camped can attest. This may have contributed to important behavioral differences between humans and chimps, including our unusual levels of cooperation and reduced impulsive aggression. Wrangham goes so far as to suggest that fire may in effect have domesticated humans, rewarding the better behaved individuals with more cooked victuals—much in the way the kindest wolves received more scraps and evolved into dogs—increasing their chance of survival and passing on those calmer genes.

When it comes to the family way, Wrangham asserts that cooking was part of a "primitive protection racket" in which men chased off robbers who wanted to filch food stored or cooked by their wives in exchange for two hots and a cot. In this view it is Man-the-Security-Guard, not Man-the-Hunter, that creates pair bonding, households, and the division of labor between the sexes. Cooked food also made it possible to wean babies at 2.8 years of age versus 4.8 years for chimps, accounting for the fact that human birth intervals are much shorter than chimps'—and offering support for the distinct but complementary Grandmother Hypothesis.

Fire finally may have altered our physiology. Compared to chimps, humans have much smaller mouths, molars, jaws, stomachs, and large intestines, each of which reflects that cooked food is easier to chew and digest. Over evolutionary time, the warmth of the fire, Wrangham imagines, may have allowed us to shed our hair, which opens the door to endurance running. And of course the more efficient provision of calories could have allowed our brains, fuel hogs that they are, to expand.

The critics chortle at many of these bold propositions, noting that antelopes sleep on the ground without the safety of fire, no tools from 2 million years ago appear to have been used to harvest tubers, rats on restricted diets live longer, chimpanzees exchange raw meat for sex, raw meat alone could have led to the expansion of the brain and the shrinking of the guts, and, most damningly, that there *should* be evidence of fire because humans would have made many millions of them between 1.8 million and 1 million years ago if only one *Homo erectus* and his or her descendants cooked just once a week.[37] Wrangham has compelling rebuttals for most of these, which he claims are based on flimsy data (regarding sex for meat), misguided logic (calorie-restricted rats may simply benefit from fortified immune systems), or overly exacting standards (traces of fire and wood tools vanish quickly).

But from the first article that Wrangham and his colleagues published on the role of cooking in human evolution, they stressed they were not claiming proof for this idea; "one person's just-so story is another's hypothesis," they wrote. Wrangham, to his credit, subsequently has tested the theory in creative ways. He behaves, in short, like a scientist—curious, skeptical, intellectually honest, welcoming of criticism, and bound by data. He even has thanked critics of the Cooking Hypothesis for their "horrendous puns." Wrangham is an unusually humble cook of hearty food for thought, a remarkable quality in the field of human origins, where credulous theories all too often are overdressed as the Be-All-End-All Hypothesis.

—◦◦◦—

THE Grandmother Hypothesis, the Cooking Hypothesis, and even Man-the-Hunter each point out that our uniqueness, be it longevity or brain power, has strong ties to our diets. It follows that our genes further should reveal how we are what we eat.

Genes typically adapt by mutation, with natural selection favoring the version that makes an individual most fit. But to adapt to a diet that had an increasing amount of roots and tubers—which possibly was linked to the advent of cooking, more grandmothers digging up plants, or both—humans took a different route: the human genome simply made more copies of a gene that helps us digest starch.

Humans usually have two copies of each gene, one from each parent. Yet a landmark study published in 2004 reported that hundreds of genes in the human genome appear more than two times, or fewer.[38] A shortage or an abundance of copies can lead to less or more of the gene's product (typically a protein). The gene that stood out as having the most variation in humans coded for a version of the amylase enzyme in saliva, which helps the body chew up starches.

Three years later, a group led by Nathaniel Dominy, an evolutionary anthropologist at the University of California, Santa Cruz, and George Perry at Arizona State University in Tempe, analyzed the copy number of that amylase gene, *AMY1*, in different populations of humans, chimpanzees, and bonobos, and found persuasive evidence that the amount of starch in our diets differentiates us from our closest primate relatives as well as from one another.[39] The team analyzed

AMY1 in high-starch eaters such as Americans of European descent, Japanese, and the Hadza. They then compared these groups to people who don't eat many roots and tubers, such as the Biaka of the Central African Republic and the Mbuti from Congo, both rainforest hunter-gatherers, and Tanzania's Datog and Siberia's Yakut pastoralists. In all, the researchers studied samples from more than two hundred people. Rather than having mutations that boosted *AMY1*'s activity, the high-starch eaters had seven copies of the gene, whereas the low-starch populations had only five. "If you have a gene that's working well, why not just copy it over and over again?" Dominy asked when I spoke with him. "Why wait for evolution to just roll the dice?"

Their analysis of the gene copy number in fifteen chimpanzees, which eat relatively little starch, furthered the case that *AMY1* had evolved in a special way to adapt to the new humans' diets. In each instance, the chimps only had two copies of the gene. Three bonobos studied had four copies each, but they had mutations in the gene that suggested to the researchers that *AMY1* no longer even functioned in them. (As occurs all too often in the study of endangered species, the researchers needed bonobo saliva to confirm this, but they could not find a zoo with a captive population that would provide a sample.)

If the human genome adapted to handle increased levels of starch, our DNA similarly should somehow capture our fondness for flesh. Caleb "Tuck" Finch, who studies longevity at the University of Southern California, argues that precisely such an adaptation occurred. Finch focuses on the gene for apolipoprotein E, or *apoE*, which carries cholesterol in the blood and regulates how it and other components of fat move into cells. Humans have three main versions, or alleles, of *apoE*. The most common allele, *apoE3*, appears in anywhere from 65 percent to 85 percent of the population, whereas *apoE2* is relatively rare. The third allele, *apoE4*, plays the villain in the bloodthirsty story.[40]

ApoE4 has strong links to Alzheimer's and cardiovascular disease. It typically appears in no more than 20 percent of a population, though its frequency climbs to as high as 40 percent in some subgroups. In Caucasian people who have a copy of *apoE4* from both parents, the risk of developing Alzheimer's disease jumps more than ten times, and half of them will develop the characteristic dementia symptoms if they live to be ninety. Finch also notes that studies have shown that eating two eggs will shoot cholesterol levels in the blood four times higher in *E4/E4* people than in those with *E3/E3*. Cardiovascular disease is increased, too, in the *E4/E4* population, though not as dramatically as Alzheimer's. *E4*, in this scenario, is retained in humans because it does help fight some infectious diseases, like hepatitis C.

Finch, who specializes in reproduction and gerontology, became intrigued by a paradox in our diets, and he thought the *E3* versus *E4* distinction might solve it. Basically, meat cuts both ways: it can extend life by providing loads of energy but can shorten it by exposing us to increased levels of fats, toxins, and pathogens. "In thinking about evolution of human life spans, you have two enormously opposed processes," he told me in October 2006, sitting in his venerable USC office. "Life span got longer by twenty, thirty years, at the same time our diet shifted in a direction that should have inhibited the lengthening of life, accelerating vascular pathology and Alzheimer's disease. So how did we pull this off?" His answer: *apoE3* is what he calls a "meat-adapted gene."

Finch, who collaborated with the primate researcher Craig Stanford to formulate this hypothesis, believes that *apoE4* is the ancestral gene, and that the *E3* allele became prominent as human meat-eating increased.[41] They suggest that chimpanzees do not get Alzheimer's or humanlike cardiovascular disease because they have only one *apoE* allele and it functions more like the *E3* than the *E4*. And *E3* played a critical role in allowing humans to more safely eat meat, extending our life spans.[42]

ApoE is just one of several putative meat-adaptive genes that Finch and Stanford have floated, and the availability of entire genome sequences from humans, chimpanzees, and other species now allows researchers to deeply probe such a hypothesis. These types of studies promise to clarify how our unusual diets, complex family structures, unique genes, peculiar disease patterns, and extraordinarily long life spans all gather at the intersection of nature and nurture.

Then again, conducting this type of research is becoming an endangered species itself.

12

BORN TO BE WILD

WHEN THE VETERINARIAN WHEELED IN THE GURNEY HOLD-ing a sedated Melinda, a twenty-six-year-old chimpanzee, I thought, now that looks almost human.

It was November 2008, and I was at the Yerkes Imaging Center in Atlanta. Melinda was being prepared to enter an MRI machine to have her brain scanned for aging studies being conducted by Todd Preuss and Jim Rilling. She had been sedated with an injection of Telazol, a short-acting anesthetic that also relaxes the muscles. Save for the extra hair and the hulking body, it could have been a lightly anesthetized human on the white-sheeted gurney. Melinda looked beatific, her arms folded across her torso, her eyes gazing toward the ceiling the way a child's do upon waking to the first light of day. The thick muscles in her massive 166-pound body were at ease. An intravenous drip was slipped into a vein in case she needed emergency fluid, and monitors taped to her chest tracked her heartbeat.

Although Preuss, Rilling, and I were not touching Melinda, we had all gowned up from head to toe, including face masks and latex gloves, making it appear as though we were about to perform surgery on her. In reality, we mainly would sit in front of a bank of computer monitors in a glass-walled room adjacent to the massive Siemens Trio 3 Tesla MRI scanner that would make images of her brain for Rilling and Preuss to study later. And no one would be cutting into Melinda's skin or bones. "This is completely noninvasive," said Preuss, noting that MRI creates magnetic fields to temporarily alter the hydrogen protons in water molecules inside the body, which then can be scanned—a safe technique that does not involve any radiation. "It's the kind of procedure we'd do with humans."

They did have to intubate Melinda before putting her in the scanner, sliding a tube down her trachea so that they could continue to sedate her during the procedure with an anesthetic gas. I have reported on many stories that take me inside hospital emergency rooms, intensive care units, and operating rooms, so I have watched several humans being intubated, and for the minute or so it takes to snake the tube into their windpipes, they can appear to be choking to death, gasping and writhing. Melinda was no different, and while people who do not believe that chimpanzees should be used in research might be outraged, I was so familiar with human intubation that it furthered my sense of Melinda being treated like one of us. Then, in another genuinely human touch, they placed a green heating pad over her torso to make sure she stayed warm in the MRI during the two hours of scans.

Before the scans started, they took a few extra steps that are unnecessary with humans. Because the two hemispheres of the human brain typically are asymmetrical, it usually is easy to tell left from right. To discern Melinda's left hemisphere from the right, the MRI team taped a vitamin E capsule to one side of Melinda's head. They also placed several pads around her massive cranium in order to position it properly in a helmet called a head coil, which is designed for the geometry of a flat-faced human. "You've got this multimillion-dollar machine and very high-tech software and computer systems, and the quality of the images all comes down to the foam and Styrofoam to fit the head in there just so," Preuss explained.

As the scans started, the machine filled the room with pulses, buzzes, hums, and honks that changed depending on whether they were doing diffusion tensor imaging or a variety of other sophisticated scanning sequences, which reveal such subtle details as the direction in which signals move through axons in the brain's white matter, the density of the white matter, the amount of myelin that wraps around the axons to insulate them, the ratio of gray to white matter, and the blood vessels feeding different regions. "No one has ever compared human brain aging with brain aging in our closest living relative to identify what's really distinctive about humans," said Rilling over the machine, which alternately sounded like an especially annoying alarm clock buzzer, boop-boops from a fork lift backing up, a foghorn, and the unh-unh-unh that signals a security breach.

Rilling and Preuss feel strongly that their work not only treats the animals humanely but also meets the ethical standards of any clinical research, as it aims to help answer critical medical questions that ultimately could benefit both species. In this study, Melinda was the twenty-ninth of thirty-six chimpanzees they planned to scan, and they were conducting similar scans of humans and macaque monkeys for comparison. The humans would include Alzheimer's patients and people with a pre-Alzheimer's condition called mild cognitive impairment. The researchers particularly wanted to see whether they could link aging to differences in the integrity of the brain's white matter, including declines in its volume, the frequency of lesions, and demyelinization of axons—each of which may be linked to the onset of age-related dementias.

On a more fundamental level, the scientists hope to map the chimpanzee's brain more precisely than ever before. "We have aspirations to make an atlas of chimp brain connectivity—all the white matter pathways in chimps—and compare that with humans and try to identify unique features of human brain wiring," said Rilling. "We know next to nothing." Preuss noted that the field relied primarily on a study of chimpanzee brain architecture that was published in 1950.[1] "Very, very little modern work has been done on the chimpanzee brain," he said.

But the window to conduct research on the brains of living chimpanzees is rapidly shrinking, both because their numbers are dwindling and because the precepts of what constitutes "ethical" experiments continue to become ever more restrictive. Until 1973, there were virtually no limits on the use of chimpanzees in research. But that year, the United States and seventy-nine other countries signed the Convention on International Trade in Endangered Species of Wild Fauna and Flora, CITES, which explicitly restricted the importation of wild chimpanzees. The convention's rules went into full effect two years later. Wildlife advocacy groups around the world soon began to lobby governments to increase protections for captive chimpanzees and to discourage—if not outright ban—their use in research. In 1986, NIH, facing increased demands for chimpanzees because of the urgency of slowing HIV's global march, had begun its breeding program, but it wasn't long before the model fell out of favor. At that time, most every other country in the world was starting to ban invasive chimpanzee experiments. The philosophers Paola Cavalieri and Peter Singer gathered primatology celebrities including Jane Goodall, Toshisada Nishida, the evolutionary biologist Richard Dawkins, and the polymath Jared Diamond under an umbrella organization, called the Great Ape Project, which demanded more equality and rights—"personhood" itself—be extended to our closest relatives, and freeing all "imprisoned" chimpanzees, bonobos, orangutans, and gorillas.[2] By 1995, NIH, the world's main financial backer of biomedical research with chimpanzees—and indeed the funder of Preuss and Rilling's aging experiment—decided that, for an unspecified period, it would not pay centers such as Yerkes to breed.

Preuss and many of the other investigators who conducted studies on chimpanzees worried that the NIH moratorium would effectively doom Yerkes and the five other U.S. research colonies, which had only a limited number of female chimps that could reproduce. "If we don't start breeding chimpanzees soon, they're going to go away," Preuss complained to me when I first visited his lab in 2006. "And they're going to be gone for good. They're going to be extinct in the wild in fifty years, and our research colonies are within about five years of reaching the point where they won't be sustainable anymore. And if we lose the chimpanzees, future generations will never forgive us. These are essential for us to make determinations about the nature of the human/chimpanzee comparisons. It's one of the most fundamental comparisons in biology. I tell that to everybody I meet. It's a really bad situation."

There were, of course, many people who thought the breeding moratorium was a good idea. Foremost among them was Geza Teleki, an alumnus of Goodall's Gombe, who for many years ran the Committee for Conservation and Care of Chimpanzees, a pioneering attempt to prevent humans from wiping chimpanzees off the planet.

—◆◆◆—

"YOU'VE never heard of Geza Teleki?"

The woman who asked me this question in November 1990 happened to be sitting behind me at a Washington, D.C., press conference to announce that human trials would soon begin in the United States for an AIDS vaccine made by an Austrian company, Immuno-AG.[3] The Food and Drug Administration had approved a clinical trial of the vaccine in part because of promising data from HIV studies in a few chimpanzees, which were then still considered by many the best animal for AIDS vaccine tests. The woman, it turned out, was Geza Teleki's wife, Heather, and she had come to the press conference to gather intelligence and goad Immuno, which her husband, a former professor at George Washington University, insisted was illegally using wild-born chimpanzees for research.

Several weeks later, I phoned Geza, who invited me to their D.C. home to review reams of documents he had collected about Immuno's chimps. He described his days at Gombe working under Goodall, and then his stint in Sierra Leone, where he battled the animal trade and was in charge of the national parks. He was especially livid about what he called NIH's malfeasance, and about a $4-million libel suit brought by Immuno against a medical journal for publishing a letter to the editor that criticized the company for conducting tests on wild-caught chimps. (The case made it to the U.S. Supreme Court.)[4] He was exotically handsome but had a wandering left eye, which he could control independently, giving him a somewhat feral look. Our conversation unfailingly circled around the tragic plight of chimpanzees.

In 1986 Teleki, Goodall, and a few dozen other people who studied chimps banded together to form the Committee for Conservation and Care of Chimpanzees (CCCC).[5] The scientists had gathered in Chicago for a symposium, "Understanding Chimpanzees," that was the precursor to the Mind of the Chimpanzee meeting. Teleki would chair the thirty-member CCCC—initially part of the Jane Goodall Institute but soon spun off—and Goodall would lend her name and presence as necessary. "She's the public persona," Teleki explained to me. "People are afraid of Jane's access to the media. I'm the person who scurries around in the background and collects information."

A few months later, Teleki would invite me to his house to meet Goodall, who was staying with him while she worked on a new book about chimpanzee conservation. The Understanding Chimpanzees symposium had been the turning point in her life, the moment she decided to leave research for care and conservation, she told me, emphasizing that Teleki played a central role in her new incarnation. "Geza's specialty is collecting facts, organizing them in a crystal clear way, and delivering them seemingly at random to anyone who asks," said Goodall, sitting on his sofa under an expansive painting of the Ngorongoro Crater in Tanzania. "He's very lucid. People have said we make a great team because he comes pounding like the breakers on a beach, and then I come in like the tide. And then they've had it." She smiled.

While the charming, poised Goodall would be elevated to something of a saint for championing the cause of chimpanzee conservation, behind the scenes Teleki was ethicist-in-chief, the bite that followed the bark, the unashamedly obsessive and ferocious campaigner who would call bullshit *bullshit* and cared so passionately about his mission that he could alienate adversaries and admirers alike. He was a strong cup of coffee, a triple espresso. He had a unique way of pushing the conservation agenda forward, relying less on heart-tugging anecdote than head-turning fact. And in many ways, he was my odd entrée to the world of chimpanzees beyond AIDS vaccine research, introducing me to the animal trafficking trade, issues of housing and care, the very real possibility that humans would soon drive chimps to extinction, and Goodall herself.

As one of CCCC's first actions, Teleki documented more thoroughly than ever before the status of wild chimpanzee populations. He conducted the study as part of a petition in 1987 to convince the U.S. Fish and Wildlife Service to upgrade the status of chimpanzees and bonobos from "threatened" to "endangered." The petition was filed by the Jane Goodall Institute, the Humane Society of the United States, and the World Wildlife Fund, and the agency took it seriously, conducting a status review of the chimpanzee and the bonobo.

Teleki's analysis of the world's wild chimpanzee population became the standard reference for the field.[6] By his accounting, chimpanzees once lived in twenty-five sub-Saharan African countries that spanned from Senegal in the northwest to Tanzania in the southeast. He noted that four of these countries no longer had chimps, and five had populations so small that he predicted they would disappear soon. In all, he estimated there were at most 235,440 chimpanzees in the wild, and possibly as few as 150,900. (A 2003 update of his estimates had a high of 299,700 and a low of 172,200.)[7]

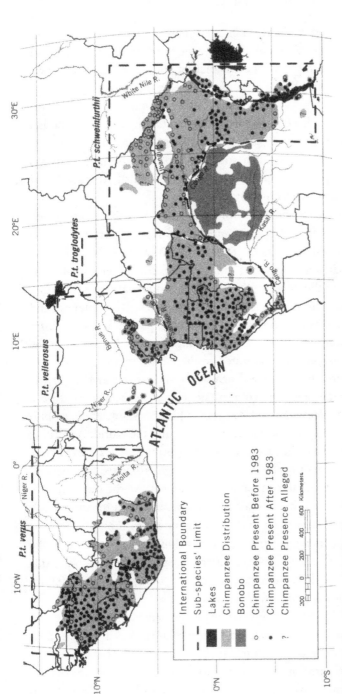

The Shrinking Range of Wild Chimpanzees

Wild chimpanzees live in a few dozen countries across sub-Saharan Africa, but their range has decreased because humans have destroyed their habitats, poached them for zoos and researchers, kept them as pets, and hunted them for food.

Teleki's report explicitly did not estimate the number of chimpanzees in the "original" population, say, one hundred years ago. He had done a back-of-the-envelope calculation, however, based on their original range and the density of observed chimpanzees per square kilometer, which pegged the number at 1 to 2 million. That guesstimate persists to this day, bandied about as fact by the Jane Goodall Institute and the Humane Society of the United States, among others. "I did not want to use such soft numbers, but journalists always pressed us to give a lump historical figure for all of Africa," Teleki explained. "They have no great meaning."

Whatever the size of the historical population, it had plummeted, and Teleki exhaustively documented the forces behind this "extermination." Most important, humans had destroyed chimp habitat through logging, mining, and farming, fragmenting their populations and restricting the flow of genes between individuals. As inbreeding increased, we reduced their fitness as a species. Deforestation and building of roads for mining or logging also put chimps into closer contact with many more disease-carrying humans. Locals—and the miners and loggers who migrated to these once remote areas—also hunted chimps, both to feed themselves and to sell as part of the bushmeat trade. Poachers sold infants to people who wanted pets, and for many years, animal dealers could legally ship them to biomedical research institutions and zoos around the world. According to Teleki's calculations, for each infant captured, at least five adults were killed, including the mother and other protective members of the community—and many infants did not survive the shipping process. He estimated that at least ten chimpanzees died for every wild-caught infant that made it to the United States or Europe.

As part of its review, the Fish and Wildlife agency sought public comment, and received 54,212 letters, mostly preprinted postcards supplied by the Humane Society, supporting the upgrade of the chimpanzee to endangered status. Six were against.[8]

NIH and biomedical research institutions including Yerkes urged the U.S. government not to lump captive chimpanzees into the endangered category, and instead to list them as threatened. They argued that research with captive chimps had nothing to do with the decline in wild populations. "The allegations are absolutely untrue," said

George Galasso, a top NIH official who oversaw the Public Health Service AIDS Animal Model Committee. Galasso further claimed that "making them endangered would endanger our research."[9]

In essence, threatened status cut down on researchers' paperwork and bureaucracy—and it also left open the door to continue to import more chimps from captive colonies in Africa for biomedical research. Frederick King, then head of Yerkes, predicted that "species loyalty" sooner or later would provide a wake-up call to humans who objected to this research. "When the pandemic of AIDS becomes a truly frightening thing, humans will not stand by and watch their own species reduced while they protect animals that could help test vaccines and drugs," King said. "It's not a very popular thing to say, but I think it's true."[10]

Fish and Wildlife made a final ruling on March 12, 1990, to reclassify chimpanzees and bonobos as endangered. Captive chimpanzees, to Teleki's dismay, would still be deemed threatened. According to the agency, there were then between 1,100 and 1,450 chimpanzees in U.S. biomedical research institutions, another 240 in facilities that were members of the American Association of Zoological Parks and Aquariums, and maybe two hundred more that were pets or in the entertainment industry. Europe then had about one thousand captive chimps, three hundred or so that lived in biomedical research facilities. Japan had another three hundred, one-third of which were set aside for research, and Australia and New Zealand had about sixty. Fewer than one hundred bonobos were thought to live in captivity. No mention was made of chimps living in captivity in Africa.

When I first met Teleki, the agency's decision was still fresh, and it rankled him. "We view NIH as our enemy," he told me. It struck me as a remote possibility that NIH would actually try to import captive chimps from Africa for biomedical research, but Teleki persuaded me with his voluminous documentation that some had been brought in after the CITES rules went into effect.[11] There also were compelling technical reasons for giving captive chimps an endangered status, he said, noting that the government would document the fate of each individual animal more closely, which would allow him and other advocates to monitor their well-being. But the bottom line was that Teleki simply did not trust humans to do the right thing—with chimpanzees or with each other. "Human beings are a difficult species, and I suppose part of the reason I went into wildlife work is I didn't have much faith in human beings since I was four years old," he said. "The irony is I have tremendous faith in individuals. But as a class, I don't perceive humans as being a gift to the world."

Alison Brooks, an anthropology professor at George Washington University who recruited Teleki to teach there, offered this assessment of her longtime friend: "He's seen so many disgusting people and so few disgusting chimps."

—◁◦▷—

WHILE Communist secret police repeatedly thunked the four-year-old Geza Teleki with their gun butts, he gazed up at the hill behind the Hungarian courtyard in which he was lying and caught sight of some flowers. "Those flowers I can still see today," Teleki told me, sitting in the manicured garden behind his house. It was a few years after the end of World War II, and the police were hunting for his father, who during wartime had served as Hungary's minister of education and minister of foreign affairs. Though his father had reluctantly dealt with the Communists, he was a staunch nationalist and had gone into hiding. "The Hungarian and Russian Communists wanted to find out where my father was," says Teleki. "That's how they found out—by taking women and children and beating the crap out of them."

Politics and science were in Teleki's blood. His Transylvanian clan was elevated to nobility in 1409 and eventually acquired vast tracts of land. "I'm carrying on the tradition but in a different context: I have no power base and no money," he joked. His great-grandfather, Count Geza Teleki, served as a cabinet minister and headed the Hungarian Historical Society. His great-uncle, Samuel Teleki, trekked through a vast region of unexplored Africa in the late 1800s—shooting hordes of large game along the way—discovering Ethiopia's Lake Stefanie and Kenya's Lake Rudolph, near which Teleki's Volcano stands today. His grandfather Paul, who had a Ph.D. in geography and was a noted cartographer, twice served as Hungary's prime minister (he killed himself when Hitler pressured Hungary to support the German invasion of Yugoslavia). In addition to holding government posts, Geza's father was a geologist.

The Telekis remained in Hungary after the war so his father could try to help prevent the Communists from running riot, but the Hungarian underground warned him in 1949 that his next posting would be in Siberia. "We just picked up our toothbrushes and left," remembered Geza. "All we had left was a knapsack of Maria Theresa coins that we used to pay our Russian drivers. We crashed through the border crossing in Vienna, with us in the back of the truck. They were shooting at us from the Russian side." The U.S. government found Geza's father a teaching job at the University of Virginia, and, later in 1949, they moved there.

Teleki found Jane Goodall in 1968. Enrolled in a master's program in anthropology at Pennsylvania State University, Teleki wrote Louis Leakey hoping to obtain interesting fieldwork. Leakey recognized the name because of Geza's great-uncle Samuel and, having no slot for the young man, arranged for a job with Goodall. "Jane was not very sociable," remembered Teleki of their first meeting. "It was hard to talk to her if you didn't know her, partly because she's British and partly because of who she is. And for another thing, she didn't pick me to work with her." Teleki, truth be told, also had little interest in chimpanzees. "I had never observed anything living," he said. "The closest I had come was looking at bones in the Smithsonian." But during the next year, he fell in love with Gombe.

One day while tracking chimps, Teleki had forgotten his lunch. Stick in hand, he vainly swung at fruit hanging from a tree. A chimp noticed his dilemma, picked a bunch of fruit, and dropped it at Teleki's feet. "These are not pets," he told me. "They're not under your direction. These are voluntary actions." He became especially close to a chimp named, of all things, Leakey. "When I couldn't keep up with him, he'd stop and wait for me on the trail," he said. One day, Teleki nodded off while Leakey and the other chimps he was following took a nap, and when he woke, Leakey was sitting next to him, the chimp's hair against Teleki's leg. "He put his head on my tennis shoe and went to sleep," said Teleki. "On the shores of Lake Tanganyika, sitting in the woods with a wild chimp with his head on my foot. What an incredible relationship. I don't feed him, mess him around. Yet he feels comfortable enough to use my tennis shoe as his pillow. An hour later, he got up and walked off. It's not as though it was a human relationship.

But a trust built up, and, in a way, I was a chimp for that short period, and when I sometimes want to give up these days of fighting for a cause, I remember days like that. That's what drives all of us."

Teleki stayed at Gombe for the next two and a half years, publishing studies about chimps' predatory behavior and range, crossing paths with a young Richard Wrangham, and becoming engaged to a fellow researcher who tragically fell off a cliff while chimping and died.[12] In 1981, he moved to Sierra Leone and became head of what became known as the Outamba-Kilimi National Park, which he ran until 1984. He had a brief tenure in academia at George Washington University before helping to launch the Committee for Conservation and Care of Chimpanzees.

Teleki and CCCC's efforts went beyond protecting wild chimps and the captive ones used for biomedical research. He worked to return four chimps to Uganda that in September 1990 had been illegally shipped to the Soviet Union to join an ice-skating circus.[13] He went after the owner of the circus, a Soviet-born U.S. citizen named Victor Shulman, using both CITES and the new Fish and Wildlife endangered status, which made it illegal for Americans to trade in wild-bred chimps anywhere.

Named Katya, Kolya, Vanya, and Emisha, the chimps learned to skate with the circus's bears, goats, geese, dogs, and a cow. Their owner, who lived in Long Island, New York, insisted he had no idea the animals were wild born. "A child should have as much care," Shulman told me. "These animals are tremendously loved. The Soviet people don't live as well as those animals." What about the ethics of putting chimps on ice skates? "We're not the first or the last people to do that," he said. "It's not something we invented. And they liked it. They enjoyed it. When they didn't have rehearsals, they were sad. Is it better that they live in some zoo and God knows how they live? Or is it better that they have someone hugging and kissing them?"

Shulman shuttled the chimps and the rest of his circus act from Moscow to Yugoslavia to Rome, and then to Austria, with CITES taking interest in the case and confiscating the chimps at the Hungarian-Soviet border in September 1991. When Teleki learned of the seizure, he phoned the CITES Hungary office, where the person answering the phone asked, "Are you related to the Telekis?" The Hungarians immediately cooperated with him. "If an official had requested their assistance, they probably would have responded less, if at all," he said. "It was a fortuitous accident." A few weeks later, the Uganda Four were repatriated to the Entebbe Zoo.

In 1995, Teleki quit CCCC, which then evaporated. He was exhausted, frustrated, angry, and depressed, he told me, pointing a finger at the duplicity and stupidity of many conservationists, the difficulty of dealing with Goodall and her institute's seemingly constant turnover of directors, and the continuing destruction of the world. "Add it all up and it became more than I could face for another year," he wrote in an e-mail. "I never was one of those (like Jane) who preferred a feel-good approach to saving species. Roses were ok with me, but only if backed up by guns. I know of no human society that was/ is kind to nature by choice rather than force. (And remember that I was trained as an anthropologist, so I'm not just gassing off here.)"

—◆◇◆—

In February 2008, three of the Uganda Four—renamed Megan, Sunday, and Maisko—were alive and well at the Ngamba Island sanctuary, which the Jane Goodall Institute helped establish a decade earlier near their former home, Entebbe.[14]

Chimpanzee sanctuaries blossomed in the years after I first spoke with Geza Teleki, so much so that in 2000 the Pan African Sanctuary Alliance was founded. According to PASA records, there were 851 chimpanzees in PASA sanctuaries in 2009, a jump from 407 since the group started.[15] And that was just in Africa.

In the United States, sanctuaries received a boost in the 1990s, when the rise in opposition to conducting biomedical research on chimpanzees intertwined with the declining interest in using the chimp model, for both scientific and financial reasons. The AIDS vaccine field had all but stopped testing on chimpanzees. The U.S. Air Force, which in 1993 owned about 120 chimponauts from the days of the space program, wanted out of the business, as did New Mexico State University, which cared for those animals and 235 others, making it host to the largest colony in the country.[16] The Coulston Foundation—run by the toxicologist Frederick Coulston and long the bane of CCCC, In Defense of Animals, and other advocacy groups because it supplied chimps to research labs and conducted its own invasive studies—obtained those chimpanzees. To the anthropologist Carole Noon, that was an outrage. In 1997, she formed the Center for Captive Chimpanzee Care (later renamed Save the Chimps) and sued the U.S. Air Force for possession of its chimpanzees. That same year, the National Research Council issued a game-changing report, *Chimpanzees in Research*, that said biomedical institutions in the United States that had chimps, which then numbered about fifteen hundred, should transfer "surplus" animals no longer needed for research or breeding to government-owned or -supported sanctuaries.[17] In an interesting sign of the shifting moral tides, the committee that wrote the report debated the ethics of using euthanasia to kill excess chimps as part of a cost-saving population control strategy; they weighed in against the idea, although one member did write a minority statement that said "in the face of limited financial resources, euthanasia is an appropriate mechanism." The NRC report paved the way for a bill, the Chimpanzee Health Improvement, Maintenance, and Protection Act, which Congress easily passed and President Bill Clinton signed on December 20, 2000.[18] The law forbids the euthanasia of chimps and promises that the government will provide long-term care for any surplus chimps that it used in research. Research on the chimpanzees living in a sanctuary could only be done if it involved information taken from routine care or was noninvasive, provided benefit to chimpanzees, and only involved "minimal" mental or physical harm.

The CHIMP Act did include some exceptions to these strict research limitations, which riled Teleki and other advocates. Individuals could be called up for research duty if they were needed for a study because of their unique medical or research history, "technological or medical advancements," or if there was "an important public health need." Any exceptions, however, would have to be approved by a sanctuary board—which had to include at least one person with experience in animal protection—and public input, making it unlikely this would ever happen.

Carole Noon won her lawsuit against the air force, and in 2001 twenty-one former chimponauts moved to a three-acre island in Fort Pierce, Florida. The next year, the Coulston Foundation was on the verge of bankruptcy, and it offered to "donate" its remaining 266 chimps to Noon if she purchased the labs and buildings where they lived in Alamogordo, New Mexico. Save the Chimps made the deal, becoming the world's largest sanctuary of its kind, constructing eleven additional three-acre islands (with indoor housing) in Fort Pierce for the chimps.[19] As envisioned, soon after the passage of the CHIMP Act, a nonprofit, Chimp Haven, won a contract to set up a system of sanctuaries, with the first one opening on two hundred forested acres near Keithville, Louisiana, in 2005. Chimp Haven features indoor housing and allows residents to form their own groups, roaming and climbing through five acres of forest. (Congress in 2007 amended the CHIMP Act to prevent any of the animals from being returned to research.)[20] At the end of 2009, Chimp Haven housed more than one hundred retired biomedical chimpanzees and hoped to expand to accommodate up to three hundred animals.

In concert with these changes in the United States, biomedical research with chimpanzees fell out of favor most everywhere else in the world. The United Kingdom in 1986 passed the Animals (Scientific Procedures) Act, which its Home Office boasted was "the most rigorous piece of legislation of its type in the world," but a loophole technically allowed most any type of research with great apes if the government granted a license.[21] Although this never happened, the loophole was tied shut in 1997 by a new law that expressly banned any biomedical research with great apes.[22] That same year, Immuno was acquired by Baxter, and in December 2002 the company moved its forty-one chimpanzees to Home of Primates–Europe, a safari park in Gänserndorf, Austria.[23] In 2004, the Biomedical Primate Research Center in Rijswijk, the Netherlands—the only European site that conducted this type of research with chimpanzees—stopped its program after a government-initiated review by the Dutch Royal Academy of Science found that the colony "was not big enough to sustain a long-term scientific research programme."[24] Some of the hundred or so chimps living at the facility in Rijswijk were moved to the AAP Sanctuary for Exotic Animals in Almere, while others went to the Safaripark Beekse Bergen in Hilvarenbeek, both in Holland.[25] The New York Blood Center, which conducted hepatitis research with chimpanzees in Liberia, in the fall of 2006 released its remaining seventy-four animals onto island sanctuaries there. In Japan, a group started by Tetsuro

Matsuzawa, Support for African/Asian Great Apes, in October 2006 brought an end to the last invasive research with chimpanzees in that country when Sanwa Kagaku Kenkyusho, a pharmaceutical company, agreed to donate its sixty-seven chimpanzees to the newly formed Uto Sanctuary in Kumamoto, to be run by Kyoto University.[26]

By the end of 2009, according to the New England Antivivisectionist Society's Project R&R—which advocates for the release and restitution of chimpanzees from U.S. laboratories—Sweden, Belgium, and Austria had completely outlawed research with chimpanzees. New Zealand and Australia also either banned or severely restricted the use of chimpanzees in biomedical research.[27] Outside of the United States, the only country that still had captive chimpanzees and allowed them to be used for research was Gabon.

—◦◦◦—

In November 2006, several of the leading scientists who studied chimpanzees in an effort to better understand humans gathered at the National Academy of Sciences' outpost in Irvine, California, for an unusual cross-disciplinary meeting on the "New Comparative Biology of Human Nature." After a presentation by Evan Eichler, the University of Washington geneticist who took a lead role in the chimpanzee genome project, Danny Povinelli raised a provocative question about the future of chimpanzee research. It was then still unclear whether the NIH would once again decide to extend the breeding moratorium beyond its sunset date of December 2007, and anxiety about that prospect was high. Povinelli, a psychologist at the University of Louisiana at Lafayette—which had 350 chimpanzees at New Iberia, the world's largest colony available for research—asked his colleagues what questions would never be addressed if chimpanzees disappeared in captivity and the wild.

"It's incalculable," said Kristen Hawkes, the anthropologist who coined the Grandmother Hypothesis. Hawkes noted that future generations would only have chimpanzee bones to study—and they all knew how difficult it was to make sense of human fossils. "We actually have a living critter and to not take advantage of that would be astonishing," she said. "Imagine if we had a Neandertal or we had an *Australopithecus*. We have chimpanzees now, but for how long?"

"Veterinarians view us geneticists as vultures," Eichler said. "All this genome sequencing isn't worth spit unless you can do something with it at the end of the day," he continued, noting that many facilities that do have chimps did not have trained people on staff to preserve the animals when they die.

Bernard Wood, a surgeon who became a prominent paleoanthropologist and was based at George Washington University, said the people assembled in part had themselves to blame for the dwindling interest in captive chimpanzee research. "In the history of science, the confirmation that modern humans are part of the tree of life and we know where they fit in will be seen like the splitting of the atom and the discovery of DNA," said Wood. "The evidence has been accruing incrementally—there wasn't a huge breakthrough—and we're responsible for not going out there and explaining that this is a really profound piece of information."

It was true that the public had come to the conclusion that past generations of scientists had not been respectful enough of chimpanzees. As I have recounted, researchers had conducted some horrific experiments, harvesting chimp organs for human transplantation, slamming chimponauts to death in sleds that tested G-forces, boring into their brains and implanting electrodes, intentionally infecting them with lethal pathogens, and trying to inseminate females with human sperm. "There has to be some fairly bright line that's drawn about what type of work can be done," Povinelli said.

UCSD's Pascal Gagneux had described that bright line in an editorial—essentially, Do Unto Others—he had published with his collaborators Ajit Varki and Jim Moore when the chimp genome was unveiled. They wrote, "The study of great apes should follow ethical principles generally similar to those currently used in studies on human subjects who cannot give informed consent."[28] In other words, treat humans and chimpanzees the same. They also emphasized the need for improved housing and care that respected chimps' social nature and the need for activities that engaged them intellectually and physically.

Despite these attempts to redefine chimp research ethics, a renewal of the moratorium was threatening. The leaders of the various primate centers had gathered to discuss the likely fate of their populations. John VandeBerg, the director of Southwest National Primate Research Center in San Antonio, Texas, tallied the chimps at all six centers: they had 1,133 animals, only 200 of which were females that potentially could breed. If the breeding moratorium were not lifted, VandeBerg calculated that there would be no research chimps left by 2037.[29]

In May 2007, the director of the NIH branch in charge of chimpanzee research announced that the institute was not just extending the breeding moratorium but making it permanent. The researchers in the field were stunned.[30] Although the Humane Society of the United States claimed a victory for its public letter-writing campaign, the official who headed primate resources at the institution, John Harding, told me that it was purely a fiscal decision.[31]

Chimpanzee care and feeding is expensive: it cost NIH $10.9 million in 2006, and it only owned or supported 650 of the 1,100 or so "biomedical research chimps." NIH estimates that over a chimp's lifetime, it spends five hundred thousand dollars per animal. But many researchers in the field do not believe it simply came down to money: the NIH annual budget in 2009 was more than $30 *billion*.

NIH for many years wanted out of the business. It had tired of being assailed by Saint Jane in the media, of responding to endless Freedom of Information requests from the Humane Society, of making itself vulnerable to attacks from extreme animal activists, who threatened scientists like Frank Novembre with razor-blade-laced envelopes and worse. "Most of what they do is fund mass spectrometers and clinical research centers," said Ajit Varki about NIH's National Center for Research Resources, NCRR. "I think the chimpanzees are a thorn in their side." Even researchers who did biomedical studies with primates had misgivings about working with chimpanzees. "If you talk to a lot of primate researchers, they're not comfortable with it," the virologist Jonathan Allan, who conducted AIDS research with monkeys at the Southwest National Primate Research Center, told me. "You shouldn't be comfortable with it. You should have to search your soul as to the balance between the research and the good that comes from it and the bad part, which is what happens to the animals." And not that many researchers actually requested to do studies with the chimpanzees. According to the veterinarian William Morton, who ran a primate consulting company in Edmonds, Washington, and sat on a working group that advised NCRR about chimpanzees, there was "a huge number of chimps that weren't being used" for research. "They were just sitting there."

Yet the real death knell was that NIH "developed" the resource to test drugs and vaccines for major diseases including AIDS, hepatitis, and respiratory syncytial virus. It had scant interest in supporting expensive studies to better understand the fundamental differences between us and them. Researchers like Preuss, Hawkes, Eichler, and Varki were just piggybackers, an epiphenomenon.

——◦∞◦——

"EXCLUSIVE: Ex-Employees Claim 'Horrific' Treatment of Primates at Lab," read the headline of a story posted on ABC News's Web site the morning of March 4, 2009. "Hidden-Camera Investigation Goes Behind Closed Doors at New Iberia Research Center." As the story breathlessly explained, *Nightline* had "obtained the results of a nine-month undercover investigation by the Humane Society of the United States."[32]

The allegations that the University of Louisiana's research center mistreated its chimpanzees and monkeys appeared serious. The undercover video showed a chimpanzee falling from a perch and smacking the floor after being darted by a tranquilizer gun, improper transportation of a sedated chimp, an anesthetized monkey rolling off a table, a baby monkey writhing while receiving a feeding tube, and other strong images of caged primates. A 108-page complaint filed by the Humane Society of the United States with the Department of Agriculture alleged 338 "possible violations" of the Animal Welfare Act. Jane Goodall, who was given the videos to review, issued a statement, lamenting that "in no lab I have visited have I seen so many chimpanzees exhibit such intense fear." The chimpanzee screaming she heard "was, for me, absolutely horrifying." A "tragedy" was occurring at New Iberia Research Center, said an ex-employee. "It's about the money," the employee claimed. "There's big bucks in this research, especially chimp research. We're talking millions. Millions of dollars."

I sensed there was something overblown about the story from the get-go, and as I looked into the charges, I became less than convinced that a horrific tragedy was under way. The Humane Society refused to share with me its complaint. The preceding January, USDA had received a separate complaint regarding New Iberia from Stop Animal Exploitation NOW, but an investigation had revealed no violations, nor had a routine USDA inspection in September 2008. The society's charge that NIH had broken its own moratorium against breeding chimps ignored the fact that the rule did not apply to New Iberia, which wasn't an NCRR facility.[33] The university's president, E.Joseph Savoie, called a press conference to denounce the images as a product of "interpretation and impression."[34] New Iberia employees contended that many of the most disturbing images were distorted. For example, they alleged that the employee who shot the video of the monkey falling from a table appeared to be a technician responsible for the care of anesthetized animals—the person who could have tended to the monkey's safety.

On closer inspection, the story was mainly about the Humane Society's push to pass legislation banning all invasive research on great apes.[35] The group wanted to give a boost to a bill introduced the next day to the House of Representatives by Edolphus Towns, a Democrat from New York. Kathleen Conlee, the society's director of program management, conceded as much. "We figured this was a good opportunity," Conlee told me. "He agreed to introduce the legislation on a timetable we asked [for]." Towns's bill, which became the Great Ape Protection Act of 2009, banned any tests that could cause "death, bodily injury, pain, distress, fear, injury, or trauma to a great ape." No research could restrain, tranquilize, or anesthetize animals. The legislation, a version of which had died in committee a year earlier, stipulated that each violation could cost up to ten thousand dollars per day.

I share the Humane Society's outrage about the mistreatment of chimpanzees—or any other animals. But campaigns like the one against New Iberia, slick as they may be, do little if anything to stop the already limited—and progressively decreasing—biomedical research with chimpanzees.

Similarly, the movement to grant chimpanzees "personhood," which aims not just to stop research but to give them the legal rights of humans, has an absurd flavor: I understand the aspiration of treating apes better by affording them our legal status, but I cannot see how it will measurably improve the way that humans treat chimpanzees. Take the case of Matthias "Hiasl" Pan. Born in 1981 in Sierra Leone, Hiasl was separated from his mother by an ex-Nazi animal trader whom Geza Teleki battled for years, and was shipped off to Austria, headed for Immuno.[36] But Austria signed the CITES agreement the day before Hiasl and eleven other chimps arrived at the Schwechat Airport outside of Vienna. They were confiscated by customs. Hiasl went to a local animal shelter, before a caretaker moved him into his own home.

Immuno fought for Hiasl in court, and after some initial setbacks, prevailed. But when Immuno went to retrieve him in November 1984, animal rights activists blocked the company's representatives. Immuno returned to court and demanded that Austria intervene and remove Hiasl by force. The company again prevailed, and in 1987 the government began a protracted struggle with the shelter to take possession of Immuno's "property." The shelter was steadfast and eventually wore the company down. When Baxter acquired Immuno, it donated the chimp to the shelter.

By 2006, the shelter had run into financial straits. A donor offered an animal rights group, the Association Against Animal Factories, or VGT, money to care for Hiasl if the chimp had a legal guardian, which by law required that he be mentally handicapped and be facing an imminent threat. A court in February 2007 ruled that he did not meet these conditions. VGT argued that Hiasl was mentally fit but had suffered through the trauma of being abducted as a baby. And he clearly was facing an imminent threat because his keepers were broke and he might be shipped anywhere. The court turned down the appeal, contending that Hiasl had no legal standing. That's when the activists turned to the personhood argument. When Austria's Supreme Court ruled in January 2008 that Hiasl was not a person, the activists vowed to appeal the case to the European Court of Human Rights.[37] That summer, Spain granted legal rights to apes, banning their use in research (which was not actively being done in Spain) and the entertainment industry, but, highlighting the absurdity of the law, not springing the 350 apes in the country's zoos.[38]

Philosophical arguments against granting personhood range from a strict speciesist perspective to the animal activist abolitionist one. The speciesist, who believes that human life is more important than chimp, chimp life more than monkey, all the way down the chain—says if chimps are granted personhood, they will be subject to all of our laws: Frodo would have stood trial in Tanzania for killing a human baby. The "abolitionist" perspective says captivity should be illegal for all sentient creatures. The legal scholar and animal rights theorist Gary Francione argues that singling out the most cognitively developed nonhumans for special treatment "is like having a human rights campaign that focuses on giving rights to the 'smarter' humans first in the hope that we will extend rights to less intelligent ones later on." (Francione contributed an essay on personhood for *The Great Ape Project* book, and in a stunning reversal, wrote that he regretted taking part in the effort because it reinforced the speciesist hierarchy.)[39]

Debates about legal status or the propriety of conducting biomedical research with chimpanzees raise the most essential questions about where to draw the dividing line between us and them. But they do not, unfortunately, directly address the most pressing dilemma of all. Unless something dramatically changes, teenagers today who live a normal human lifespan may see the last chimpanzee disappear from the wild, and maybe even captivity, before they are laid to rest themselves. Why, then, are so few people using that most human of skills—comprehending a complicated message—and taking appropriate action?

SEVERAL hours after following Makoku through the Kibale Forest, I noticed that his left hand only had four digits, one of which looked as fat as two swollen together. Makoku, as with 25 percent or more of the chimpanzees in Kibale, had suffered an injury from a poacher's trap. When I visited Budongo, I learned that the problem there might be worse still.

The poachers typically make traps from sticks and the brake wire from motorbikes or rope, and they are hunting for duikers and bush pigs, not chimpanzees. But the young male chimps in particular step into them, typically when they are first going on patrol with the elder males and do not know enough to stay away from strange-looking objects. The wire or rope nooses around a limb or a digit, and instead of simply unloosening it, they pull on it, cinching off blood supply. The limb or digit rots off and heals—or the chimp dies.

Emily Otali, the researcher who runs Kibale, showed me a box full of traps that her team had collected. On average, they find thirty each month. We were sitting on the covered porch of Richard Wrangham's house in Kibale, where Otali lives when he is not there. A rain fell as she told me the story of Max, a male who at eight years old lost one leg to a snare and six months later a second. "When Max lost his first leg, he was a sport about it," said Otali, the first African woman to ever earn a Ph.D. for studying chimpanzees. Max limped on his good leg, and the stump quickly healed. "Then he lost his other leg, and he had to use his former stump, and he bruised it and it was bleeding, and the new one was bleeding," Otali said. "He sat on the tree whimpering."

Max was still alive—and so was Otali's frustration and rage. "When Max lost his second leg, I cried," she said. "And I just thought, if we caught the poacher, should we also cut off his limbs?"

Actually, Otali has hired a few ex-poachers to remove traps and to educate the local villagers about the harm they cause to chimpanzees. The "snare removal team" also receives a bonus if they catch a poacher who is later convicted. When chimpanzees have snares on their limbs, Otali and her team send out veterinarians to dart and treat them (which has only limited success). The group is also trying to provide local children with a better education—a project spearheaded by Wrangham's wife, Elizabeth Ross—in the hopes this will raise their standard of living and provide them with more money to purchase meat, rather than to hunt for it. One of the difficulties that project faces is that many of the villagers have little concern about the well-being of chimpanzees. "People look at chimps as belonging to the white people," Otali explained. "Chimps are *mzungu*" (Swahili for white person). A few chimps, like Saddam, have attacked and killed local children. "How do you go out into the community and tell them we need to save these chimps that they know go out and kill their children?" asked Otali. "That makes life really hard."

Indirect poaching has taken an enormous—if rarely documented—toll on the chimpanzee population.[40] "Most of our young males are dying or are getting injured, which gives a really gloomy picture for our community's future," Otali said. "The old males are growing older and will all die away. So then in ten, twenty years' time we'll be left with a community of half-limbed males that cannot fight another community." When I visited, the Kanyawara community only had forty chimps left, a drop of 25 percent since the time when Otali began studying them, only seven years earlier.

Otali's efforts at Kibale are not singular. The Jane Goodall Institute in Uganda has a snare removal program that works in the Kibale, Budongo, and Kalinzu forests. In Gombe, the institute is linking "village-managed" forest reserves to increase habitat, and is promoting ecotourism with local villagers. An ambitious Roots & Shoots project teaches kids in some one hundred countries how to address local environmental, animal, and humanitarian issues. In Guinea, Tetsuro Matsuzawa's team is working with local villagers and the government to plant trees and create a "green passage" that allows the disappearing community in Bossou—where there were reportedly only thirteen chimps left in 2009—to connect with the much larger population in the nearby Nimba Mountains.[41] Toshisada Nishida and Jun'ichiro Itani helped persuade the Tanzanian government to set aside Mahale as a national park in 1985, and later, a conservation society was formed.[42] Christophe Boesch formed the Wild Chimpanzee Foundation to protect the wild populations he follows in the Taï National Park and branched out to several other parks in Côte d'Ivoire, as well as sites in Sierra Leone and Guinea; activities range from educating children about chimpanzees to biomonitoring populations. Vernon Reynolds created the Budongo Conservation Field Station to, in part, document the impact of logging and hunting. Jill Pruetz more recently started Neighbor Ape, an effort to protect the savanna chimps she monitors in Senegal, and Crickette Sanz and David Morgan launched Goualougo Triangle Ape Project to help protect the chimpanzees and gorillas they study. Smaller efforts are under way to protect chimpanzees in Nigeria, Cameroon, and Rwanda.

On a more global scale, the United Nations Environment Programme in 2001 launched the Great Ape Survival Partnership. It set itself the challenging goal of uniting "all the principal institutional actors in great ape conservation—UN agencies, biodiversity-related multilateral environmental agreements, great ape range state and donor governments, nongovernmental organizations, scientists, local communities and the private sector—in an internationally concerted approach to a major extinction crisis." Practically speaking, it provides technical support and attempts to shape government policies. The World Wildlife Fund also has an African Great Apes Program.

As much as conservation-minded scientists, professional conservationists, NGOs, and governments work to prevent wild chimpanzee populations from declining, many at the forefront have a bleak outlook. The International Union for Conservation of Nature's much respected Red List of Threatened Species, which ranks *Pan troglodytes* as endangered, in 2009 said "this species is estimated to have experienced a significant population reduction in the past 20 to 30 years . . . and it is suspected that this reduction will continue for the next 30 to 40 years. The maximum population reduction over a three-generation (i.e., 60-year) period from the 1970s to 2030 is suspected to exceed 50 percent."[43] (Bonobos similarly are endangered, with estimates suggesting there are somewhere between 29,500 to 50,000 in the wild.)[44]

These dire predictions have not won over everyone in the field. The anthropologist John Oates from Hunter College in New York City in 2006 published a thoughtful analysis that questioned whether chimpanzees even deserve to be deemed "endangered," especially since they live in protected parks in many countries. "If relatively modest improvements are made (and sustained) in the management of many of the national parks in which chimpanzees occur, it seems very unlikely that the species will be extinct by the year 2100," wrote Oates.[45]

Oates, however, is a minority voice, and the anticipated steep decline in wild chimpanzees has led many researchers who study the species to conclude that maintaining a robust captive population provides the best hedge against extinction. But the number of captive chimps approaches the strictest definition of "endangered" (fewer than 250 mature individuals). The American Zoo and Aquarium

Association, which sponsors chimpanzee and bonobo species survival plans, reported that as of July 2009, there were just 270 chimpanzees and eighty-four bonobos in North American zoos. Deciding how to breed these animals to keep the population genetically diverse is a formidable challenge. Worse, the European Association of Zoos and Aquaria has wrestled with a "surplus" of chimpanzees, based on a lack of available housing in the wake of increasing public interest in captive bonobos and other great apes. It now encourages its members not to breed chimps.[46] Japan, Australia, and New Zealand have even smaller populations. No hard numbers exist, but UCSD's Pascal Gagneux estimated that captive chimps in Europe number seven hundred, and in Asia, five hundred.[47]

Some researchers simply do not trust zoos to protect chimps from extinction. "The zoo community doesn't have a good track record of keeping things alive," said Gagneux, who once worked for the renowned San Diego Zoo. "Even some of the best zoos in the world are notoriously bad at long-term guarantees of the persistence of a species."

One last-ditch strategy for the survival of the species is the return of captive animals to the wild, known as reintroduction, which Ben Beck, of Iowa's Great Ape Trust, pushed at the Mind of the Chimpanzee Conference. Beck had headed an effort in Brazil that successfully reintroduced to the forest the golden lion tamarin, a monkey species that subsequently moved from critically endangered to merely endangered. Beck stressed that sanctuaries already held as many great apes as all the world's zoos, and they could not afford to maintain large chimpanzee populations indefinitely. "Something has to give, or we need to get the financial support to increase sanctuary space," Beck told me.

Beck pointed to the one well-documented attempt to reintroduce chimpanzees, Project HELP (Habitat Ecologique et Liberté des Primates), in the Republic of Congo. As of 2009, Project HELP had released forty orphaned chimpanzees that had been raised in a sanctuary run by a French ex-pat, Aliette Jamart, who lived in Pointe Noire.[48] Not all of the reintroductions went well, but Jamart and a scientific team radio-collared the released chimps in order to provide veterinary care when individuals were attacked by other chimps—their biggest threat. Three of the chimps released in the project's early years had been killed by wild chimpanzees, and one was killed by another released chimp.[49] (One other had drowned, and others had disappeared.) But on a more heartening note, three infants birthed by reintroduced females had survived, and released females regularly interacted with wild chimps without incident. Ongoing genetic analyses will reveal whether the chimps became pregnant by wild or released males. "Although we are not at the stage yet where the wild chimpanzee population is so low that it needs to be supplemented with captive chimpanzees, knowledge of how to release chimpanzees successfully in the future should be obtained now, before it is urgently needed," the project researchers wrote. It was not a resounding triumph, but Beck "would not characterize it as a disaster."

Humans will determine the fate of chimpanzees. Chimpanzees of course will have no say in the fate of humans. And that may be the single most conspicuous difference between the two species.

―◦◦◦―

IN *The Great Ape Project* book, Geza Teleki contributed the chapter "They Are Us." He described interactions with Gombe chimpanzees that had led him to devote most of his career to chimpanzee conservation. "I had seen my species inside the skin of another," he wrote. He did not idealize chimpanzees, but he also had great disdain for the skepticism that some people brought to claims that another species could have emotions or thoughts on the level of a human. "My view of chimpanzees is much like my view of humans: some are scoundrels and some are saints, and most are somewhere between those extremes." And he was enraged that "hoes, saws and guns" would likely drive chimpanzees to extinction.

When I first started my research for this book, I checked with Teleki to hear his thoughts about my planned journey into his world. I explained to him that I wanted to explore our differences. He warned me that I would anger many conservationists who had dedicated their lives to emphasizing the similarities. Of course, I replied, emphasizing similarities made great sense in the days of Yerkes and Goodall—even in the days of Teleki—but times had changed. No real debates persist about whether chimpanzees have feelings, use tools, can learn words and numbers, or cooperate. "Be careful," he said.

When I reached the end of my expedition, I contacted Teleki again. He was, as usual, both good-natured and gloomy. "I think the situation for chimps has drastically worsened since the time I started my work," he said. "I don't think it's controllable. I think only stop-gap measures are working. On the whole, I think the species is doomed. I don't have much room for positive thinking here."

What did he think of the idea that breeding captive chimpanzees might save the species from extinction? I asked. "I've heard this bullshit all my life: They're at risk, lock 'em up," he said. "The survival of that species isn't going to help anyone. That's breeding everyone in a penitentiary. That's going to save humanity? Those chimps are ruined. Behaviorally, biologically, medically. And transpose it to your world. Someone comes to your door with a gun and aims at you and says, 'I've got just the women you should be living with, come with me. And you should produce this many people in this amount of time.' "

By the time we turned to similarities and differences, Teleki was good and riled. "I couldn't give a damn whether one addresses similarities or differences," he said. "For me, the whole reason to focus on similarities was to leave these beings alone." Then he launched into one of the most beautiful descriptions I have ever heard not about our differences, but about why chimpanzees are different: "There's no possible way anyone with an ounce of logic could argue that chimps and humans are the same or that individual chimps are the same as each other," he said. "What's crucial at an ethical level and a moral level—and that's a whole different issue—is that there is no animal anywhere in the world where I've experienced what I've experienced with a chimp. To walk into a situation absolutely cold and within minutes understand what's going on around me because all of my life I've seen what's going on with the people around me. That understanding which is absolutely intuitive. And that feeling of knowing what's happening does not exist with any other species."

I agree, and it became obvious to me on the second day I saw wild chimpanzees. It was early afternoon, and I was sitting against a tree and resting from a long morning of chimping while more than a dozen chimpanzees scattered about me in a midday siesta, reclining with one hand behind the head, picking through one another's hair, playing with their babies, quietly digesting food and thoughts from a busy morning foraging. It was as though I had stumbled into a group of ancient humans. It was as though I was almost a chimpanzee myself.

NOTES

Quotations are from interviews I conducted unless otherwise noted. I have covered primate research since 1989, but began this project in 2006. I live in San Diego and made many research visits for the book between 2006 and 2009 to the University of California at San Diego, La Jolla; the Salk Institute for Biological Studies; and the San Diego Zoo. The main out-of-town trips I made to do research for this book were:

University of Southern California, Los Angeles, California, October 19, 2006

Yerkes National Primate Research Center and Emory University, Atlanta, Georgia, November 14, 2006, and November 11, 2008

Stanford University, Palo Alto, California, November 16, 2006

National Academy of Science, Sackler Colloquium, "The New Comparative Biology of Human Nature," Irvine, California, November 17–18, 2006

Oxford University, Oxford, United Kingdom, November 27–28, 2006

Kirill Rossiianov, Moscow, Russia, November 30–December 1, 2006

Institute for Child Health, London, England, December 4, 2006

Max Planck Institute for Evolutionary Anthropology, Leipzig, Germany, December 5–8, 2006

University of Washington, Seattle, January 2, 2007

The Institute of Greatly Endangered and Rare Species, Myrtle Beach, South Carolina, and Miami, Florida, January 22–25, 2007

Primate Foundation of Arizona, Mesa, Arizona, February 15, 2007

Broad Institute and Harvard University, Cambridge, Massachusetts, March 7, 2007, and February 7, 2008

Lincoln Park Zoo, "Mind of the Chimpanzee," Chicago, Illinois, March 23–25, 2007

University of California, Los Angeles, September 11 and October 2, 2007

National Anthropological Archives, Smithsonian Institution, Suitland, Maryland, September 27, 2007

Great Ape Trust, Des Moines, Iowa, March 12–13, 2008

Primate Research Institute, Inuyama, Japan, March 20–22, 2008

Kibale National Park, Kibale, Uganda, April 30–May 1, 2008

Budongo Forest Reserve, Budongo, Uganda, May 2–4, 2008

16th Conference on Retroviruses and Opportunistic Infections, Montreal, Canada, February 8–11, 2009

Introduction

1. The Dutch ethologist Adriaan Kortlandt observed wild chimpanzees in the Belgian Congo at a banana and papaya plantation a few months before Goodall began her observations in Gombe. But Kortlandt hid in observation stations that he built. See Adriaan Kortlandt, "Chimpanzees in the Wild," *Scientific American* 5:128–38 (Sept. 1962). Robert Yerkes also had sent Henry Nissen to French Guinea to collect chimpanzees in 1930, and Nissen studied them in the wild, moving about with six porters, guides, and scouts. On the forty-nine days he spotted chimps, the observations were fleeting and from a distance. Not surprisingly, Nissen's observations were riddled with inaccurate assumptions about their behavior. Henry W. Nissen, "A Field Study of the Chimpanzee: Observations of Chimpanzee Behavior and Environment in Western French Guinea," *Comparative Psychology Monograph* 8:1–122 (1931).

2. Paco Bertolani and Christophe Boesch, "Habituation of Wild Chimpanzees (*Pan troglodytes*) of the South Group at Taï Forest, Côte d'Ivoire: Empirical Measure of Progress," *Folia Primatologica* 79:162–71 (Apr. 2008).

3. Robert M. Yerkes, *Almost Human* (New York and London: Century, 1925), p. xii. Displaying his own prejudices, Yerkes also opined about the way that the "primitive varieties of the human species" who lived in Africa viewed chimpanzees and vice versa (p. 26). "Negro and chimpanzee seem to recognize in each other similarities which attract and differences which repel," claimed Yerkes. "The feelings of the negro are pretty generally shared by mankind, for the appearance and behavior of monkeys and apes offend while they fascinate most of us."

1. The Family Tree

1. *Untamed and Uncut*, a show on the Animal Planet TV network, aired the video, which I viewed on its Web site, http://animal.discovery.com/videos/untamed-uncut -animal-rescues/, accessed Feb. 22, 2010.

2. Jane Goodall, *Beyond Innocence: An Autobiography in Letters*, Dale Peterson, ed. (Boston: Houghton Mifflin Harcourt, 2001), p. 356.

3. Jeannie Williams, "Liz Not at Loss for Words on Weight," *USA Today*, May 8, 1991.

4. Jane Goodall, "Learning from the Chimpanzees: A Message Humans Can Understand," *Science* 282:2184–85 (Dec. 18, 1998). Also see Jane Goodall, "Chimpanzees: Bridging the Gap," in Paola Cavalieri and Peter Singer, eds., *The Great Ape Project: Equality Beyond Humanity* (New York: St. Martin's Griffin, 1993), pp. 10–18.

5. Robert M. Yerkes, *Almost Human* (New York and London: Century, 1925), p. 263.

6. Robert Yerkes, "Provision for the Study of Monkeys and Apes," *Science* 43:231–34 (Feb. 18, 1916).

7. "The First 100 Chimpanzees," http://first100chimps.wesleyan.edu/, accessed Sept. 16, 2008.

8. W. C. Osman-Hill, "The Discovery of the Chimpanzee" in G. H. Bourne, ed., *The Chimpanzee*, vol. 1 (New York: S. Karger, 1969), pp. 1–21.

9. Thomas Henry Huxley, *Evidence as to Man's Place in Nature* (New York: D. Appleton, 1863), pp. 85–86.

10. George Gaylord Simpson, "The Principles of Classification and a Classification of Mammals," *Bulletin of the American Museum of Natural History* 85:187–88 (1945).

11. A translation of the letter, written to Johann Georg Gmelin and dated Feb. 25, 1747, appears on the Linnaeus Wikipedia entry, http://en.wikipedia.org/wiki/Carolus_ Linnaeus, accessed Sept. 18, 2008.

12. Vincent M. Sarich and Allan C. Wilson, "Immunological Time Scale for Hominid Evolution," *Science* 158:1200–1203 (Dec. 1, 1967).

13. Ann Gibbons, *The First Human: The Race to Discover Our Earliest Ancestors* (New York: Doubleday, 2006), p. 75.

14. Peter Andrews and J. E. Cronin, "The Relationships of *Sivapithecus* and *Ramapithecus* and the Evolution of the Orang-utan," *Nature* 197:541–46 (June 17, 1982).

15. Charles G. Sibley and Jon E. Ahlquist, "The Phylogeny of the Hominoid Primates, as Indicated by DNA-DNA Hybridization," *Journal of Molecular Evolution* 20:2–15 (1984).

16. Maryellen Ruvolo, Todd R. Disotell, Marc Allard, et al., "Resolution of the African Hominoid Trichotomy by Use of a Mitochondrial Gene Sequence," *Proceedings of the National Academy of Sciences* 88:1570–74 (Feb. 1991). As late as 2000, at least one group of researchers wrote, "The question of the closest living relative of the human species has not been settled." See Yoko Satta, Jan Klein, and Naoyuki Takahata, "DNA Archives and Our Nearest Relative: The Trichotomy Problem Revisited," *Molecular Phylogenetics and Evolution* 14:259–75 (Feb. 2000).

17. Edwin McConkey and Morris Goodman, "A Human Genome Evolution Project Is Needed," *Trends in Genetics* 13:350–51 (Sept. 1997).

18. Evan E. Eichler, Catherine B. Kunst, Kellie A. Lugenbeel, et al., "Evolution of the Cryptic *FMR1* CGG Repeat," *Nature Genetics* 11:301–8 (Nov. 1995).

19. Elaine A. Muchmore, Sandra Diaz, and Ajit Varki, "A Structural Difference between the Cell Surfaces of Humans and the Great Apes," *American Journal of Physical Anthropology* 107:187–98 (Oct. 1998).

20. Geoff Spencer, who works in the public affairs department at NIH's National Human Genome Research Institute, told me the $42 million went to the NHGRI Sequencing Research Network through cooperative agreements, not through specific grants.

21. International Human Genome Sequencing Consortium, "Finishing the Euchromatic Sequence of the Human Genome," *Nature* 431:931–45 (Oct. 21, 2004).

22. Katherine S. Pollard, Sofie R. Salama, Bryan King, et al., "Forces Shaping the Fastest Evolving Regions in the Human Genome," *PLoS Genetics* 2:1599–1611 (Oct. 13, 2006). See also Katherine S. Pollard, Sofie R. Salama, Nelle Lambert, et al., "An RNA Gene Expressed during Cortical Development Evolved Rapidly in Humans," *Nature* 443:167–72 (Sept. 14, 2006).

23. Shyam Prabhakar, Axel Visel, Jennifer A. Akiyama, et al., "Human-Specific Gain of Function in a Developmental Enhancer," *Science* 321:1346–50 (Sept. 5, 2008). The table here is adapted from figure S6 of the supporting online material.

24. Matthew W. Hahn, Jeffery P. Demuth, and Sang-Gook Han, "Accelerated Rate of Gene Gain and Loss in Primates," *Genetics* 177:1941–49 (Nov. 2007).

25. Nick Patterson, Daniel J. Richter, Sante Gnerre, et al., "Genetic Evidence for Complex Speciation of Humans and Chimpanzees," *Nature* 441:1103–8 (June 29, 2006).

26. When Patterson and Reich factored in a "more realistic" estimate of the divergence between the orangutan and human genome, they concluded that human speciation occurred less than 5.4 million years ago.

27. Alain Beauvilain and Yves Le Guellec, "Further Details Concerning Fossils Attributed to *Sahelanthropus tchadensis* (Toumaï)," *South African Journal of Science* 100:142–44 (Spring 2004).

28. Elizabeth Pennisi, "Genomes Throw Kinks in Timing of Chimp-Human Split," *Science* 312:985–86 (May 19, 2006).

2. Two Become One

1. Charles Darwin, *On the Origin of Species* (London: John Murray, 1859), chap. 8.
2. "Hominin," short for "Hominini," refers to the tribe of humans from *homo sapiens* back to the common ancestor with chimpanzees. It replaces "hominid," or "hominidae," a term that fell out of favor for humans and our extinct relatives when researchers started to include chimpanzees and other great apes in that family because of genetic similarities. Much confusion still exists about which terminology to use, and many scientists still prefer "hominids" to refer to what others insist should be called "hominins."
3. John Wakeley, "Complex Speciation of Humans and Chimpanzees," *Nature* 452:E3–4 (Mar. 13, 2008). Also see the reply from Patterson, Reich, et al. that follows.
4. Nick H. Barton, "Evolutionary Biology: How Did the Human Species Form?" *Current Biology* 16: R647–50 (Aug. 22, 2006).
5. Charles Darwin, *The Variation of Animals & Plants Under Domestication*, vol. 2 (London: John Murray, 1868), chap. 13.
6. Ernst Mayr, *Animal Species and Evolution* (Cambridge, Mass.: Belknap Press, 1963).
7. Andrei Soficaru, Adrian Dobos, and Erik Trinkaus, "Early Modern Humans from the Pestera Muierii, Baia de Fier, Romania," *Proceedings of the National Academy of Sciences* 103:17196–201 (Nov. 14, 2006). Richard Green, Anna-Sapfo Malaspinas, Johannes Kraus, et al., "A Complete Neandertal Mitochondrial Genome Sequence Determined by High-throughput Sequencing," *Cell* 134:416–26 (Aug. 8, 2008).
8. Björn Kurtén, *Dance of the Tiger: A Novel of the Ice Age* (New York: Pantheon, 1980).
9. Ibid., p. 65.
10. Catherine Turleau, Jean de Grouchy, and M. Klein, "Phylogénie chromosomique de l'homme et des primates hominiens (*Pan troglodytes, Gorilla gorilla* et *Pongo pygmaeus*)," *Essai de reconstitution du caryotype de l'ancêtre commun, Annales de Génétique* 15: 225–40 (1972).
11. Debates swirl around the role that inversions have played in human/chimp speciation, and few clear answers exist. A prominent publication in 2003 showed that the real estate where these "rearrangements" have occurred in the human chromosomes has also been a hotbed of evolution itself. A difference in the DNA between humans and chimps often has no meaning, merely reflecting that a random mutation occurred that neither helped nor hurt the species and, by chance, stuck around. But sometimes—and this is gold to evolutionary biologists—a mutation occurs and persists because it makes the species more fit. Arcadi Navarro and Nick Barton, evolutionary biologists working in Spain and Scotland, respectively, looked for "positive selection" in the chromosomes of humans and found that it occurred twice as frequently in rearranged regions. They argued that these DNA changes created a barrier to gene flow, allowing humans to gradually become more fit. In this line of thought, differences in the structure of chromosomes then at the very least greased the wheels that separated humans and chimps, and may well have started the wheels spinning in the first place. See Arcadi Navarro and Nick H. Barton, "Chromosomal Speciation and Molecular Divergence—Accelerated Evolution in Rearranged Chromosomes," *Science* 300:321–24 (Apr. 11, 2003). But, in a stunning turnabout, Navarro's group revisited the question with more data and came to precisely the opposite conclusion. "More experimental and theoretical knowledge needs to be gathered before the debate can be satisfactorily settled," they wrote. See Tomàs Marques-Bonet, Jesús

Sànchez-Ruiz, Lluís Armengol, et al., "On the Association between Chromosomal Rearrangements and Genic Evolution in Humans and Chimpanzees," *Genome Biology* 8: R230 (Oct. 30, 2007).

12. Kirill Rossiianov, "Beyond Species: Il'ya Ivanov and His Experiments on Cross-Breeding Humans with Anthropoid Apes," *Science in Context* 15:277–316 (June 2002).

13. David Rhees, "A New Voice for Science: Science Service under Edwin E. Slosson, 1921–29," master's thesis, University of North Carolina at Chapel Hill, 1979.

14. "Soviet Backs Plan to Test Evolution," *New York Times*, June 17, 1926.

15. J. Michael Bedford, "Sperm/Egg Interaction: The Specificity of Human Spermatozoa," *Anatomical Record* 188:477–87 (Aug. 1977).

16. Sally McBrearty and Nina Jablonsky, "First Fossil Chimpanzee," *Nature*, 437:105–8 (Sept. 1, 2005).

3. In Sickness and Health

1. Hsun-hua Chou, Hiromu Takematsu, Sandra Diaz, et al., "A Mutation in Human CMP-Sialic Acid Hydroxylase Occurred after the Homo-Pan Divergence," *Proceedings of the National Academy of Sciences* 95:11751–56 (Sept. 1998). A separate lab in Japan actually published this discovery a few months earlier but did not explore its uniqueness to humans. Atsushi Irie, Susumu Koyama, Yasunori Kozutsumi, et al., "The Molecular Basis for the Absence of N-Glycolylneuraminic Acid in Humans," *Journal of Biological Chemistry* 273:15866–71 (June 19, 1998).

2. *CMAH* stands for cytidine monophosphate-N-acetylneuraminic acid hydroxylase, which is the enzyme that synthesizes Neu5Gc.

3. Pam Tangvoranuntakul, Pascal Gagneux, Sandra Diaz, et al., "Human Uptake and Incorporation of an Immunogenic Nonhuman Dietary Sialic Acid," *Proceedings of the National Academy of Sciences* 100:12045–50 (Oct. 14, 2003).

4. Michael L. Lammey, Gary B. Baskin, Andrew P. Gigliotti, et al., "Interstitial Myocardial Fibrosis in a Captive Chimpanzee (*Pan troglodytes*) Population," *Comparative Medicine* 58:389–94 (Aug. 2008). Tom P. Meehan and Linda J. Lowenstine, "Causes of Mortality in Captive Lowland Gorillas: A Survey of the SSP Population," *Proceedings of the Annual Conference of the American Association of Zoo Veterinarians* 216–18 (Nov. 1994).

5. Frans B. M. de Waal, *Our Inner Ape: A Leading Primatologist Explains Why We Are Who We Are* (New York: Riverhead Books, 2005), pp. 236–37. The hybrids were described in more detail by Hilde Vervaecke and Linda van Elsacker, "Hybrids between Common Chimpanzees (*Pan troglodytes*) and Pygmy Chimpanzees (*Pan paniscus*) in Captivity," *Mammalia* 56:667–69 (Dec. 1992).

6. Kamala and Amala, the wolf girls of Midnapore, were discovered in 1920 living with wolves in northern India, according to a reverend of a local orphanage who took them in. Many controversies surround whether they truly were raised by wolves. For a thorough debunking of the myth, see the French surgeon Serge Aroles's book, *L'enigme des enfants-loups: une certitude biologique mais un déni des archives, 1304–1954* (Paris: Publibook, 2007). I revisit the wolf girls' story in chap. 5, "Talking Apes."

7. Paul de Kruif, *Microbe Hunters* (New York: Harcourt, Brace, 1926). See chap. 7, "Metchnikoff."

8. Richard M. Krause, "Metchnikoff and Syphilis Research during a Decade of Discovery, 1900–1910," *ASM News* 62:307–10 (2002).

9. Félix Bosch and Laia Rosich, "The Contributions of Paul Ehrlich to Pharmacology: A Tribute on the Occasion of the Centenary of His Nobel Prize," *Pharmacology* 82:171–79 (Aug. 2008).

10. Philip D. Curtin, "The End of the 'White Man's Grave'?: Nineteenth-Century Mortality in West Africa," *Journal of Interdisciplinary History* 21:63–88 (Summer 1990).

11. Donald Blacklock and Saul Adler, "A Parasite Resembling *Plasmodium Falciparum* in a Chimpanzee," *Annals of Tropical Medicine and Parasitology* 160:99–106 (1922). See also B. Feldman-Muhsan, "Centenary Biographical Note: Saul Adler, 1895–1966," *International Journal of Parasitology* 25:1141–43 (Oct. 1995).

12. Jérôme Rodhain, "Les plasmodiums des anthropoïdes de l'Afrique centrale et leurs relations avec les plasmodiums humains," *Annales de la Société Belge de médecine tropicale* 190:563–72 (1939).

13. World Health Organization, "World Malaria Report, 2005," (Geneva: WHO and UNICEF, 2005).

14. A. P. Waters, D. G. Higgins, and T. F. McCutchan, "*Plasmodium falciparum* Appears to Have Arisen as a Result of Lateral Transfer between Avian and Human Hosts," *Proceedings of the National Academy of Sciences* 88:3140–44 (Apr. 1991).

15. Stephen M. Rich, Monica C. Licht, Richard R. Hudson, and Francisco J. Ayala, "Malaria's Eve: Evidence of a Recent Population Bottleneck throughout the World Populations of *Plasmodium falciparum*," *Proceedings of the National Academy of Sciences* 95:4425–30 (Apr. 1998).

16. Gagneux's aside might have come off as trifling to me before I had immersed myself deeply in the chimpanzee research world. Among other things, it starkly highlighted the fact that remarkably few scientists, maybe a few hundred total, study chimpanzees. Fewer still examine what they tell us about human origins, and many of the leaders in the field were with me in the room for Ayala's talk. In contrast, scientific conferences focused on HIV/AIDS, the neurosciences, and clinical oncology each regularly fill convention centers with more than twenty thousand researchers.

17. Maria J. Martin, Julian C. Rayner, Pascal Gagneux, et al., "Evolution of Human–Chimpanzee Differences in Malaria Susceptibility: Relationship to Human Genetic Loss of N-glycolylneuraminic Acid," *Proceedings of the National Academy of Sciences* 102:12819–24 (Sept. 6, 2005).

18. Weimin Liu, Yingying Li, Gerald H. Learn, et al., "Origin of the human malaria parasite *Plasmodium falciparum* in gorillas," *Nature* 467:420–25 (Sept. 23, 2010).

19. William D. Hillis, "An Outbreak of Infectious Hepatitis among Chimpanzee Handlers at a United States Air Force Base," *American Journal of Hygiene*, 73:316–28 (May 1961).

20. Richard Witkin, "Chimpanzees Pass Space Speed Test: Withstand 100 G's and More in Abrupt-Stop Rides in Ground Rocket Sled," *New York Times*, Jan. 31, 1958.

21. Hillis, "An Outbreak of Infectious Hepatitis."

22. National Research Council, *Chimpanzees in Research: Strategies for Their Ethical Care, Management, and Use* (Washington, D.C.: National Academy Press, 1997).

23. World Health Organization, "Hepatitis B Fact Sheet," http://www.who.int/mediacentre/factsheets/fs204/en/, accessed Jan. 5, 2010.

24. Pascal Gagneux and Elaine Muchmore, "The Chimpanzee Model: Contributions and Considerations for Studies of Hepatitis B Virus," in R. K. Hamatake and J. Y. N. Lau, eds., *Methods in Molecular Medicine*, vol. 95: *Hepatitis B and D Protocols*, vol. 2 96:289–318 (2004).

25. Scientists only recently uncovered persuasive evidence that hepatitis B virus infects chimps in the wild. See Maria Makuwa, Sandrine Souquière, Olivier Bourry, et al.,

"Complete-Genome Analysis of Hepatitis B Virus from Wild-born Chimpanzees in Central Africa Demonstrates a Strain-specific Geographical Cluster," *Journal of General Virology* 88:2679–85 (Oct. 2007). Nevertheless, the "origin and evolution of HBV remains elusive," according to this thoughtful review: Betty H. Robertson and Harold S. Margolis, "Primate Hepatitis B Viruses—Genetic Diversity, Geography and Evolution," *Reviews in Medical Virology* 12:133–41 (May–June 2002).

26. Qui-Lim Choo, George Kuo, Amy J. Weiner, et al., "Isolation of a cDNA Clone Derived from a Blood-Borne Non-A, Non-B Viral Hepatitis Genome," *Science* 244:362–64 (Apr. 21, 1989).

27. Daniel Lavanchy, "The Global Burden of Hepatitis C," *Liver International* 29(s1): 74–81 (Jan. 2009).

28. Raija H. Bettauer, "Chimpanzees in Hepatitis C Virus Research: 1998–2007," *Journal of Medical Primatology*, electronic publication ahead of print (Nov. 9, 2009).

29. For arguments that chimpanzees are not needed in hepatitis research, see *Alternatives to Animal Testing and Experimentation*, Special Issue, Proceedings of the 6th World Congress on Alternatives and Animal Use in the Life Sciences, Aug. 21–25, 2007, Tokyo, Japan. In particular, see the essays by Kathleen Conlee of the Humane Society of the United States, "Chimpanzees in Research and Testing Worldwide: Overview, Oversight and Applicable Laws" (pp. 111–18), and Andrew Knight of the UK's Animal Consultants International, "Chimpanzee Experiments: Questionable Contributions to Biomedical Progress" (pp. 119–24).

30. Janette Wallis and D. Rick Lee, "Primate Conservation: The Prevention of Disease Transmission," *International Journal of Primatology* 20:803–26 (Dec. 1999).

31. Sophie Köndgen, Hjalmar Kühl, Paul K. N'Goran, et al., "Pandemic Human Viruses Cause Decline of Endangered Great Apes," *Current Biology* 18:1–5 (Feb. 26, 2008).

4. Of Epidemic Proportions

1. D. Carleton Gajdusek, Clarence J. Gibbs Jr., and Michael Alpers, "Experimental Transmission of a Kuru-like Syndrome to Chimpanzees," *Nature* 209:794–96 (Feb. 19, 1966).

2. D. Carleton Gajdusek, Herbert L. Amyx, Clarence J. Gibbs Jr., et al., "Transmission Experiments with Human T-lymphotropic Retroviruses and Human AIDS Tissue," *Lancet* 1(8391):1415–16 (June 23, 1984). Also see D. Carleton Gajdusek, Clarence J. Gibbs Jr., Pamela Rodgers-Johnson, et al., "Infection of Chimpanzees by Human T-lymphotropic Retroviruses in Brain and Other Tissues from AIDS Patients," *Lancet* 1(8419):55–56 (Jan. 5, 1985).

3. Committee on Long-Term Care of Chimpanzees, National Research Council, *Chimpanzees in Research: Strategies for Their Ethical Care, Management, and Use* (Washington, D.C.: National Academy Press, 1997), p. 15.

4. I delve into this subject in detail in my book *Shots in the Dark: The Wayward Search for an AIDS Vaccine* (New York: Norton, 2001), chap. 5, "Animal Illogic."

5. Raymond V. Gilden, Larry O. Arthur, W. Gerard Robey, et al., "HTLV-III Antibody in a Breeding Chimpanzee Not Experimentally Exposed to the Virus," *Lancet* 1(8482):678–79 (Mar. 22, 1986).

6. Martine Peeters, Cécile Honoré, Thierry Huet, et al., "Isolation and Partial Characterization of an HIV-related Virus Occurring Naturally in Chimpanzees in Gabon," *AIDS* 3:625–30 (Oct. 1989).

7. Martine Peeters, Katrien Fransen, Eric Delaporte, et al., "Isolation and Characterization of a New Chimpanzee Lentivirus (Simian Immunodeficiency Virus Isolate cpz-ant) from a Wild-captured Chimpanzee," *AIDS* 6:447–51 (May 1992).

8. Beatrice Hahn, "The Origin of HIV-1: A Puzzle Solved?" 6th Conference on Retroviruses and Opportunistic Infections, Chicago, Ill. (Jan. 31, 1999).

9. Researchers debate how many subspecies of chimpanzees actually exist, with some recent DNA studies showing that there are only two, the west African *Pan troglodytes vellerosus* and *P. t. troglodytes* in central and eastern Africa. Mary Katherine Gonder, Todd R. Disotell, and John F. Oates, "New Genetic Evidence on the Evolution of Chimpanzee Populations and Implications for Taxonomy," *International Journal of Primatology* 27:1103–27 (Aug. 2006).

10. Lawrence K. Altman, "H.I.V. Is Linked to a Subspecies of Chimpanzee," *New York Times*, Feb. 1, 1999.

11. Bette Korber, "Timing the Origin of the AIDS Epidemic," 7th Conference on Retroviruses and Opportunistic Infections, San Francisco, Calif. (Jan. 30–Feb. 2, 2000).

12. Sandrine Souquire, Pierre Roques, Ahidjo Ayouba, et al., "Newly Derived HIV-1 Group N and SIVcpz (*P.t.t.*) Strains Cluster Together in the HIV-1/SIVcpz Lineage," abstract 213, 7th Conference on Retroviruses and Opportunistic Infections, San Francisco, Calif. (Jan. 30–Feb. 2, 2000).

13. Edward Hooper, *The River: A Journey to the Source of HIV and AIDS* (Boston: Little, Brown, 1999).

14. I wrote a detailed critique of the oral polio vaccine hypothesis, "The Hunt for the Origin of AIDS," *Atlantic Monthly*, Oct. 2000.

15. Elizabeth Bailes, Feng Gao, Frederic Bibollet-Ruche, et al. "Hybrid Origin of SIV in Chimpanzees," *Science* 300:1713 (June 13, 2003).

16. Vernon Reynolds, *The Chimpanzees of the Budongo Forest: Ecology, Behaviour, and Conservation* (Oxford and New York: Oxford University Press, 2005), p. 22.

17. Mario L. Santiago, Cynthia M. Rodenburg, Shadrack Kamenya, et al., "SIVcpz in Wild Chimpanzees," *Science* 295:465 (Jan. 18, 2002).

18. Mario L. Santiago, Frederic Bibollet-Ruche, Elizabeth Bailes, et al., "Amplification of a Complete Simian Immunodeficiency Virus Genome from Fecal RNA of a Wild Chimpanzee," *Journal of Virology* 77:2233–42 (Feb. 2003). Mario Santiago, Magdalena Lukasik, Shadrack Kamenya, et al., "SIVcpz Prevalence and Genetic Diversity in Wild Communities of Eastern Chimpanzees," 10th Conference on Retroviruses and Opportunistic Infections, abstract 501, Boston, Mass. (Feb. 10–14, 2003).

19. Michael Worobey, Mario L. Santiago, Brandon F. Keele, et al., "Contaminated Polio Vaccine Theory Refuted," *Nature* 428:820 (Apr. 22, 2004).

20. Donald Francis, Paul Feorino, J. Roger Broderson, et al., "Infection of Chimpanzees with Lymphadenopathy-Associated Virus," *Lancet* 2(8414):1276–77 (Dec. 1, 1984).

21. Francis J. Novembre, Michelle Saucier, Daniel C. Anderson, et al., "Development of AIDS in a Chimpanzee Infected with Human Immunodeficiency Virus Type 1," *Journal of Virology* 71:4086–91 (May 1997).

22. Francis J. Novembre, Juliette de Rosayro, Soumya Nidtha, et al., "Rapid CD4+ T-Cell Loss Induced by Human Immunodeficiency Virus Type 1_{NC} in Uninfected and Previously Infected Chimpanzees," *Journal of Virology* 75:1533–39 (Feb. 2001). See also Shawn P. O'Neil, Francis J. Novembre, Anne Brodie Hill, et al., "Progressive Infection in a Subset of HIV-1-positive Chimpanzees," *Journal of Infectious Diseases* 182:1051–62 (Oct. 2000).

23. Norman L. Letvin, "Progress in the Development of an HIV-1 Vaccine," *Science* 280:1875–80 (June 19, 1998).

24. Alfred M. Prince and Linda Andrus, "AIDS Vaccine Trials in Chimpanzees," *Science* 282:2194 (Dec. 18, 1998).

25. Alfred M. Prince, Jonathan Allan, Linda Andrus, et al., "Virulent HIV Strains, Chimpanzees, and Trial Vaccines," *Science* 283:1115 (Feb. 19, 1999).

26. Patricia Fultz infected two females and a male chimpanzee with a viral strain known as HIV-1$_{JC499}$. The animals, housed at the Laboratory for Experimental Medicine and Surgery in Primates in New York, were infected by different routes (IV, vagina, penis). The virus soon disappeared from the male, and one of the females died of a non-AIDS disease twenty-one months after becoming infected, making it impossible to determine whether the virus caused her harm. In 2001 the third chimp was moved to the Southwest Foundation of Biomedical Research in San Antonio, Texas, and I could not determine her fate. Fultz published several papers about HIV-1$_{JC499}$, the last in 2002. See Qing Wei and Patricia N. Fultz, "Differential Selection of Specific Human Immunodeficiency Virus Type 1/JC499 Variants after Mucosal and Parenteral Inoculation of Chimpanzees," *Journal of Virology* 76: 851–64 (Jan. 2002).

27. Jon Cohen, "Researchers Urged Not to Inject Virulent HIV Strain Into Chimps," *Science* 283:1090–91 (Feb. 19, 1999).

28. Weiss, who later earned a law degree, cofounded the Laboratory Primate Advocacy Group, and she posted her writings about Jerom on its Web site (http://www.lpag.org) under the "Memorials" section.

29. Jon Cohen, "Chimps and Lethal Strain a Bad Mix," *Science* 286:1454–55 (Nov. 19, 1999).

30. Laure Y. Juompan, Karen Hutchinson, David C. Montefiori, et al., "Analysis of the Immune Responses in Chimpanzees Infected with HIV Type 1 Isolates," *AIDS Research and Human Retroviruses* 24:573–86 (Apr. 2008).

31. Dzung H. Nguyen, Nancy Hurtado-Ziola, Pascal Gagneux, and Ajit Varki, "Loss of Siglec Expression on T Lymphocytes during Human Evolution," *Proceedings of the National Academy of Sciences* 103:7765–70 (May 16, 2006).

32. Frederic Bibollet-Ruche, Brett A. McKinney, Alexandra Duverger, et al., "The Quality of Chimpanzee T-Cell Activation and Simian Immunodeficiency Virus/Human Immunodeficiency Virus Susceptibility Achieved via Antibody-Mediated T-Cell Receptor/CD3 Stimulation Is a Function of the Anti-CD3 Antibody Isotype," *Journal of Virology* 82:10271–78 (Oct. 2008). In February 2010, Varki and coworkers had a cogent rebuttal in press at the *Journal of Immunology*, "Relative Over-reactivity of Human versus Chimpanzee Lymphocytes: Implications for the Human Diseases Associated with Immune Activation."

33. João Dinis de Sousa, Philippe Lemey, Viktor Müller, and Anne-Mieke Vandamme, "Date to a Unique Window of Opportunity for the Initial Spread of SIV Infections Transmitted to Human Populations," 16th Conference on Retroviruses and Opportunistic Infections, paper 293, Montreal, Canada (Feb. 8–11, 2009).

34. As but one example of the argument that the chimp immune system has evolved to handle SIVcpz, see Natasja G. de Groot, Nel Otting, Gaby G. M. Doxiadis, et al., "Evidence for an Ancient Selective Sweep in the MHC Class I Gene Repertoire of Chimpanzees," *Proceedings of the National Academy of Sciences* 99:11748–53 (Sept. 3, 2002). The paper shows that a molecule on immune cells that typically comes in many forms is not as diverse in chimpanzees, which they suggest may have occurred because of a "selective sweep" for protective genes that followed a widespread viral infection. Specifically, they write: "The relative resistance of HIV-1-infected chimpanzees to the development of AIDS may be the consequence of an effective immune response controlled, at least in part, by the present set of MHC class I molecules, which are the result of positive selection."

35. Dale Peterson, *Eating Apes* (Berkeley: University of California Press, 2003), p. 92.
36. Pascale Ondoa, Luc Kestens, David Davis, et al., "Longitudinal Comparison of Virus Load Parameters and CD8 T-cell Suppressive Capacity in Two SIVcpz-infected Chimpanzees," *Journal of Medical Primatology* 30:243–52 (Oct. 2001). Also see Jonathan L. Heeney, Erik Rutjens, Ernst J. Verschoor, et al., "Transmission of Simian Immunodeficiency Virus SIVcpz and the Evolution of Infection in the Presence and Absence of Concurrent Human Immunodeficiency Virus Type 1 Infection in Chimpanzees," *Journal of Virology* 80:7208–18 (July 2006).
37. Luc Kestens, Pascale Ondoa, Kim Vereecken, et al., "SIVcpz-ant Causes a Non-escalating Immune Activation and Uses Multiple Co-receptors in Chimpanzees," abstract 11210, 12th World AIDS Conference, Geneva, Switzerland (June 28–July 3, 1998). This singular study received virtually no comment from the AIDS intelligentsia, which all too often intently searches for clues to finding a vaccine using the already established compass points and as a result misses potentially exciting leads and truly novel insights.
38. Thushan I. de Silva, Matthew Cotton, and Sarah L. Rowland-Jones, "HIV-2: The Forgotten AIDS Virus," *Trends in Microbiology* 16:588–95 (Dec. 2008).
39. Marc van den Eerenbeemt, "Proefchimpansee Niko overleden in onderzoekscentrum," *de Volkskrant*, Jan. 25, 2006.

5. Talking Apes

1. Sue Savage-Rumbaugh, Stuart G. Shanker, and Talbot J. Taylor, *Apes, Language, and the Human Mind* (New York: Oxford University Press, 1998), p. v.
2. Charles Darwin, *Descent of Man, and Selection in Relation to Sex* (London: John Murray, 1871), chap. 3, "Comparison of the Mental Powers of Man and the Lower Animals."
3. "Mr. Garner Has Mastered the Ape Language," *Baltimore Sun*, Feb. 28, 1909.
4. Richard L. Garner, *The Speech of Monkeys* (London: Heinemann, 1892); Garner, *Gorillas & Chimpanzees* (London: Osgood, McIlvaine, 1896); and Garner, *Apes and Monkeys: Their Life and Language* (London: Ginn, 1900).
5. These quotes are excerpted from the following correspondences and his diary, all in box 1 of the Richard Lynch Garner Collection at the Smithsonian Institution's National Anthropological Archives in Suitland, Maryland: "I have heard it said" from diary, Feb. 19, 1905; "mere mechanics" from letter to Professor W. H. Holmes, Dec. 6, 1900; "a certain contempt for the greater portion of the genus homo" and "unscrupulous little miscarriage" from letter to Dear Son, May 18, 1909. The "unscrupulous little miscarriage," Ida Vera Simonton, came to Gabon to work with Garner, but they had a falling-out. She later wrote *Hell's Playground* (New York: Moffat, Yard, 1912), based on her Africa travels; in 1942 it was made into a movie, *White Cargo*, that starred Hedy Lamarr and Frank Morgan (most famous for his role as the Wizard of Oz).
6. Gregory Radick, *The Simian Tongue: The Long Debate about Animal Language* (Chicago: University of Chicago Press, 2007), p. 89.
7. Richard Lynch Garner, "The Simian Tongue," *New Review* 4:555–62 (June 1891).
8. Radick, *Simian Tongue*, p. 96.
9. Richard L. Garner, "What I Expect to Do in Africa," *North American Review*, 154:713–18 (June 1892).
10. Garner, *Gorillas & Chimpanzees*, p. 16.
11. The Edison papers are available online at http://edison.rutgers.edu. There are twelve

related letters dated between July 27, 1891, and Jan. 11, 1893. The last one is from Garner's agent, Samuel Sydney McClure, and includes Edison's tart handwritten note at the bottom. A Mar. 18, 1892, note from Grover Cleveland calls Garner's Africa plans "manly and straightforward."

12. Garner, *Apes and Monkeys*, pp. 137, 143.

13. "Says Monkeys Answer Him: Prof. Garner Declares They Reply to His Calls in the Jungle," *New York Times*, Aug. 26, 1906.

14. Yerkes, *Almost Human*, pp. 168–69. As Radick notes in *Simian Tongue*, four years after *Almost Human* came out, Yerkes had a "distinctly less generous" take on Garner, writing that "his publications indicate serious lack of scientific competence." See Robert Yerkes, *The Great Apes: A Study of Anthropoid Life* (New Haven, Conn.: Yale University Press, 1929).

15. Yerkes, *Almost Human*, pp. 173–74.

16. William H. Furness III, "Observations on the Mentality of Chimpanzees and Orang-utans," *Proceedings of the American Philosophical Society*, 55:281–90 (1916). All quotations by Furness are taken from this talk.

17. Yerkes, *Almost Human*, p. 180.

18. Paul C. Squires, "The 'Wolf Children' of India," *American Journal of Psychology*, 38:313–15 (Apr. 1927).

19. W. N. Kellogg and L. A. Kellogg, *The Ape and the Child: A Study of Environmental Influence upon Early Behavior* (New York: McGraw-Hill, 1933).

20. Ludy T. Benjamin Jr., and Darryl Bruce, "From Bottle-fed Chimp to Bottlenose Dolphin: A Contemporary Appraisal of Winthrop Kellogg," *Psychological Record* 32:461–82 (1982).

21. "Babe and Ape," *Time*, Jun. 19, 1933.

22. For an overview, see Paul Bloom, *How Children Learn the Meaning of Words* (Cambridge, Mass.: MIT Press, 2000). The first- through fifth-grade word acquisition figures are from Jeremy Anglin, "Vocabulary Development: A Morphological Analysis," *Monographs of the Society for Research in Child Development* 58:i–186 (1993). The commonly cited estimate of sixty thousand words by high school graduation probably comes from Seashore and Eckerson's 1940 studies, as noted in Anglin's monograph, and also George A. Miller, *The Science of Words* (New York: Scientific American Library/Freeman, 1996). See also Paul Bloom and Lori Markson, "Capacities Underlying Word Learning," *Trends in Cognitive Sciences* 2:67–73 (Feb. 1998).

23. Cathy Hayes, *The Ape in Our House* (New York: Harper and Brothers, 1951).

24. Philip H. Lieberman, Dennis H. Klatt, and William H. Wilson, "Vocal Tract Limitations on the Vowel Repertoires of Rhesus Monkey and Other Nonhuman Primates," *Science* 164:1185–87 (June 6, 1969). Lieberman has written extensively about this elsewhere, but this paper quickly lays out the basic vocal limitation in chimpanzees and other nonhuman primates.

25. David Premack and Ann James Premack, *The Mind of an Ape* (New York: Norton, 1983).

26. R. Allen Gardner and Beatrice T. Gardner, "Teaching Sign Language to a Chimpanzee," *Science* 165:664–72 (Aug. 15, 1969).

27. R. Allen Gardner and Beatrice T. Gardner, "Early Signs of Language in Child and Chimpanzee," *Science* 187:752–53 (Feb. 28, 1975).

28. Herbert S. Terrace, Laura-Ann Petitto, Richard J. Sanders, and Thomas G. Bever, "Can an Ape Create a Sentence?" *Science* 206:891–902 (Nov. 23, 1979).

29. R. Allen Gardner, Beatrix T. Gardner, and Thomas E. Van Cantfort, eds., *Teaching Sign Language to Chimpanzees* (Albany: State University of New York Press, 1989), pp. 21 and 22. The book also covers a study conducted with Nim Chimpsky two years after Project Nim ended, showing improved performance when the researchers tried a more "conversational" approach than the training protocol Herbert Terrace used (see chap. 7).

30. Nicholas Wade, "Does Man Alone Have Language?: Apes Reply in Riddles, and a Horse Says Neigh," *Science* 208:1349–51 (June 20, 1980).

31. D. R. Shaffer, *Developmental Psychology* (Pacific Grove, Calif.: Brooks/Cole, 1989), p. 295.

32. Steven Pinker, *The Language Instinct* (New York: Morrow, 1994), pp. 340–42.

33. E. Sue Savage-Rumbaugh, Jeannine Murphy, Rose A. Sevcik, et al., *Language Comprehension in Ape and Child*, Monographs of the Society for Research in Child Development 58:v–255 (1993).

34. Sue Savage-Rumbaugh, Duane M. Rumbaugh, and Kelly McDonald, "Language Learning in Two Species of Apes," *Neuroscience and Biobehavioral Reviews* 9:653–65 (Winter 1985); Sue Savage-Rumbaugh, Kelly McDonald, Rose A. Sevcik, et al., "Spontaneous Symbol Acquisition and Communicative Use by Pygmy Chimpanzees (*Pan paniscus*)," *Journal of Experimental Psychology: General* 115:211–35 (Sept. 1986); Sue Savage-Rumbaugh, Rose A. Sevcik, and William D. Hopkins, "Symbolic Cross-modal Transfer in Two Species of Chimpanzees," *Child Development* 59:617–25 (June 1988).

35. Mark S. Seidenberg and Laura-Ann Petitto, "Communication, Symbolic Communication, and Language: Comment on Savage-Rumbaugh, McDonald, Sevcik, Hopkins, and Rupert (1986)," *Journal of Experimental Psychology: General* 116: 279–87 (Sept. 1987). Petitto coauthored the famous 1979 *Science* article with Terrace, and Seidenberg, who did his Ph.D. under another of the coauthors, Tom Bever, lived with Nim for a time.

36. Sue Savage-Rumbaugh, "Communication, Symbolic Communication, and Language: Reply to Seidenberg and Petitto," *Journal of Experimental Psychology: General* 116:288–92 (Sept. 1987).

37. The one review I found of the monograph was published three years later and argued that the researchers, contrary to their claims, had reinforced and trained Kanzi's behavior. See Mark L. Sundberg, "Analysis of Behavior Toward Granting Linguistic Competence to Apes: A Review of Savage-Rumbaugh et al.'s *Language Comprehension in Ape and Child*," *Journal of the Experimental Analysis of Behavior* 65:477–92 (Mar. 1996).

38. Juliane Kaminski, Josep Call, and Julia Fischer, "Word Learning in a Domestic Dog: Evidence for 'Fast Mapping,'" *Science* 304:1682–83 (June 11, 2004).

6. The Fox in the Chimp House

1. Pierre Paul Broca, "Loss of Speech, Chronic Softening and Partial Destruction of the Anterior Left Lobe of the Brain," *Bulletin de la Société Anthropologique* 2:235–38 (1861).

2. Carl Wernicke, "The Symptom-Complex of Aphasia," translated into English in R. S. Cohen and M. W. Wartofsky, eds., *Boston Studies in the Philosophy of Science*, 4:34–97 (Dordrecht, the Netherlands: Reidel, 1969). •

3. Jane A. Hurst, Michael Baraitster, Elizabeth Auger, et al., "An Extended Family with a Dominantly Inherited Speech Disorder," *Developmental Medicine and Child Neurology* 32:352–55 (Apr. 1990).

4. Myrna Gopnik, "Feature-blind Grammar and Dysphasia," *Nature* 344:715 (Apr. 19, 1990).

5. Paul Fletcher, and separately, Faraneh Vargha-Khadem and Richard Passingham, "Speech and Language Defects," *Nature* 346:226 (July 19, 1990).

6. Myrna Gopnik, "Genetic Basis of Grammar Defect," *Nature* 347:26 (Sept. 6, 1990).

7. Steven Pinker, "Rules of Language," *Science* 253:530–35 (Aug. 2, 1991).

8. James K. Kilpatrick, "Better Grammar through Genetics," *St. Petersburg Times*, Feb. 25, 1992; Thomas H. Maugh II, "Language Defects Linked to the Grammar of Genes Science," *Los Angeles Times*, Feb. 11, 1992; Erma Bombeck, "Poor Grammar? It Are in the Genes," cited in Steven Pinker, *The Language Instinct* (New York: Morrow, 1994), p. 303.

9. Faraneh Vargha-Khadem, Kate Watkins, Katie Alcock, et al., "Praxic and Nonverbal Cognitive Deficits in a Large Family with a Genetically Transmitted Speech and Language Disorder," *Proceedings of the National Academy of Sciences* 92:930–33 (Jan. 1995).

10. Simon E. Fisher, Faraneh Vargha-Khadem, Kate E. Watkins, et al., "Localisation of a Gene Implicated in a Severe Speech and Language Disorder," *Nature Genetics* 18:168–70 (Feb. 1998).

11. Cecilia S. L. Lai, Simon E. Fisher, Jane A. Hurst, et al., "A Forkhead-domain Gene Is Mutated in a Severe Speech and Language Disorder," *Nature* 413:519–23 (Oct. 4, 2001).

12. Nicholas Wade, "Researchers Say Gene Is Linked to Language," *New York Times*, Oct. 4, 2001.

13. Steven Pinker, "Talk of Genetics and Vice Versa," *Nature* 413:465–66 (Oct. 4, 2001).

14. Simon Baron-Cohen, *Mindblindness: An Essay on Autism and Theory of Mind* (Cambridge, Mass.: MIT Press, 1995).

15. Gang Li, Jinhong Wang, Stephen J. Rossiter, et al., "Accelerated *FoxP2* Evolution in Echolocating Bats," *PLoS ONE* 2:e100 (Sept. 2007).

16. Johannes Krause, Carles Lalueza-Fox, Ludovic Orlando, et al., "The Derived *FOXP2* Variant of Modern Humans Was Shared with Neandertals," *Current Biology* 17:1908–12 (Nov. 6, 2007).

17. Nicholas Wade, "Neanderthals Had Important Speech Gene, DNA Evidence Shows," *New York Times*, Oct. 19, 2007.

18. Amy S. Pollick and Frans B. M. de Waal, "Ape Gestures and Language Evolution," *Proceedings of the National Academy of Sciences* 104:8184–89 (May 8, 2007). See also Lisa Parr and Bridget Waller, "Understanding Chimpanzee Facial Expression: Insights into the Evolution of Communication," *Social Cognitive and Affective Neuroscience* 1:221–28 (Dec. 2006).

19. Catherine Crockford, Ilka Herbinger, Linda Vigilant, and Christophe Boesch, "Wild Chimpanzees Produce Group-Specific Calls: A Case for Vocal Learning?" *Ethology* 110:221–43 (Mar. 2004).

20. A subsequent study found that captive chimps produced a raspberry to grab the attention of humans. William D. Hopkins, Jared P. Taglialatela, and David A. Leavens, "Chimpanzees Differentially Produce Novel Vocalizations to Capture the Attention of a Human," *Animal Behaviour* 73:281–86 (Feb. 2007).

21. Katie E. Slocombe and Klaus Zuberhüler, "Food-associated Calls in Chimpanzees: Responses to Food Types or Food Preferences?" *Animal Behaviour* 72:989–99 (Nov. 2006).

22. Robert M. Seyfarth, Dorothy L. Cheney, and Peter Marler, "Monkey Responses to Three Different Alarm Calls: Evidence of Predator Classification and Semantic Communication," *Science* 210:801–3 (Nov. 14, 1980).

23. Roman M. Wittig, Catherine Crockford, Eva Wikberg, Robert M. Seyfarth, and Dorothy L. Cheney, "Kin-mediated Reconciliation Substitutes for Direct Reconciliation in Female Baboons," *Proceedings of the Royal Society: Biological Sciences* 274:1109–15 (Apr. 22, 2007).

24. Other researchers, including some of Crockford and Wittig's close colleagues, were similarly conducting careful studies of chimp communications. Ilka Herbinger's doctoral dissertation used playback experiments with Taï chimpanzees to simulate intrusions of one community's territory by another. See Ilka Herbinger, "Inter-group Aggression in Wild West African Chimpanzees (*Pan troglodytes verus*): Mechanisms and Functions," Ph.D. dissertation, University of Leipzig, Germany, Apr. 23, 2004. One group of scientists used playback experiments to show that Budongo chimps could distinguish different types of screams. See Katie E. Slocombe, Simon Townsend, and Klaus Zuberbühler, "Wild Chimpanzees (*Pan troglodytes schweinfurthii*) Distinguish between Different Scream Types: Evidence from a Playback Study," *Animal Cognition* 12:441–49 (May 2009).

25. For an extensive discussion of the social intelligence hypothesis and its history, see Dorothy L. Cheney and Robert M. Seyfarth, *Baboon Metaphysics: The Evolution of a Social Mind* (Chicago: University of Chicago Press, 2007), chap. 7.

26. Norman Geschwind and Walter Levitsky, "Human Brain: Left-Right Asymmetries in Temporal Speech Region," *Science* 161:186–87 (July 12, 1968).

27. Albert M. Galaburda, Marjorie LeMay, Thomas L. Kemper, and Norman Geschwind, "Right-Left Asymmetries in the Brain: Structural Differences between the Hemispheres May Underlie Cerebral Dominance," *Science* 199:852–56 (Feb. 24, 1978).

28. Daniel H. Geschwind, Bruce L. Miller, Charles DeCarli, and Dorit Carmelli, "Heritability of Lobar Brain Volumes in Twins Supports Genetic Models of Cerebral Laterality and Handedness," *Proceedings of the National Academy of Sciences* 99:3176–81 (Mar. 5, 2002).

29. Philip Lieberman, *Toward an Evolutionary Biology of Language* (Cambridge, Mass.: Belknap Press, 2006), p. 87.

30. Clyde Francks, Shinji Maegawa, Juha Laurén, et al., "*LRRTM1* on Chromosome 2p12 Is a Maternally Suppressed Gene That Is Associated Paternally with Handedness and Schizophrenia," *Molecular Psychiatry* 12:1129–39 (Dec. 2007).

31. Brett S. Abrahams, Dmitri Tentler, Julia V. Perederiy, et al., "Genome-wide Analyses of Human Perisylvian Cerebral Cortical Patterning," *Proceedings of the National Academy of Sciences* 104:17849–54 (Nov. 6, 2007).

32. Kevin A. Strauss, Erik G. Puffenberger, Matthew J. Huentelman, et al., "Recessive Symptomatic Focal Epilepsy and Mutant Contactin-Associated Protein-like 2," *New England Journal of Medicine* 354:1370–77 (Mar. 30, 2006).

33. Maricela Alarcón, Brett S. Abrahams, Jennifer L. Stone, et al., "Linkage, Association, and Gene-Expression Analyses Identify *CNTNAP2* as an Autism-Susceptibility Gene," *American Journal of Human Genetics* 82:150–59 (Jan. 2008).

34. Sonja C. Vernes, Dianne F. Newbury, Brett S. Abrahams, et al., "A Functional Genetic Link between Distinct Developmental Language Disorders," *New England Journal of Medicine* 359:2337–45 (Nov. 27, 2008).

35. Fisher and Geschwind did not find *CNTNAP2* in their initial hunt for *FOXP2* targets because of their methodology. FOXP2 typically regulates a gene by binding to a

stretch of DNA called a promoter region (which, in turn, recruits RNA polymerase and allows transcription from DNA into mRNA to occur). The researchers first looked for FOXP2 targets by determining which promoter regions bound the protein. They later used a more random approach, chopping up chunks of DNA and asking which of them attached to FOXP2. This uncovered that FOXP2 bound an intron—a piece of a gene that is removed before translation—of *CNTNAP2*.

36. Matthias Groszer, David A. Keays, Robert M. J. Deacon, et al., "Impaired Synaptic Plasticity and Motor Learning in Mice with a Point Mutation Implicated in Human Speech Deficits," *Current Biology* 18:354–62 (Mar. 11, 2008).

37. Weiguo Shu, Julie Y. Cho, Yuhui Jiang, et al., "Altered Ultrasonic Vocalization in Mice with a Disruption in the *Foxp2* Gene," *Proceedings of the National Academy of Sciences* 102:9643–48 (July 5, 2005).

38. The species comparison mainly comes from Constance Scharff. See Sebastian Haesler, Kazuhiro Wada, A. Nshdejan, et al., "*FoxP2* Expression in Avian Vocal Learners and Non-Learners," *Journal of Neuroscience* 24:3164–75 (Mar. 31, 2004). For elephants, see Joyce H. Poole, Peter L. Tyack, Angela S. Stoeger-Horwath, and Stephanie Watwood, "Elephants Are Capable of Vocal Learning," *Nature* 434:455–56 (Mar. 24, 2005).

39. Sebastian Haesler, Christelle Rochefort, Benjamin Georgi, et al., "Incomplete and Inaccurate Vocal Imitation after Knockdown of *FoxP2* in Songbird Basal Ganglia Nucleus Area X," *PLoS Biology* 5:e321 (Dec. 2007).

40. Julie E. Miller, Elizabeth Spiteri, Michael C. Condro, et al., "Birdsong Decreases Protein Levels of FoxP2, a Molecule Required for Human Speech," *Journal of Neurophysiology* 100:2015–25 (Oct. 2008). Constance Scharff and Sebastian Haesler, "An Evolutionary Perspective on *FoxP2*: Strictly for the Birds?" *Current Opinion in Neurobiology* 15:694–703 (Dec. 2005).

41. Michael T. Ullman, "A Neurocognitive Perspective on Language: The Declarative/ Procedural Model," *Nature Reviews Neuroscience* 2:717–26 (Oct. 2001).

7. Mind the Gap

1. Nobuyuki Kawai and Tetsuro Matsuzawa, "Numerical Memory Span in a Chimpanzee," *Nature* 403:39–40 (Jan. 6, 2000).

2. For a detailed description of these experiments, see Sana Inoue and Tetsuro Matsuzawa, "Working Memory of Numerals in Chimpanzees," *Current Biology* 17:R1004–5 (Dec. 4, 2007).

3. Matsuzawa writes about the trade-off in more depth in "Comparative Cognitive Development," *Developmental Science* 10:97–103 (Jan. 2007).

4. Tetsuro Matsuzawa, "The Ai Project: Historical and Ecological Contexts," *Animal Cognition* 6:199–211 (Oct. 18, 2003).

5. Matsuzawa also discussed his climbing days in a Q&A in *Current Biology* 19:R310–13 (Apr. 28, 2009).

6. Tetsuro Matsuzawa and William McGrew, "Kinji Imanishi and 60 Years of Japanese Primatology," *Current Biology* 18:R587–91 (July 22, 2008).

7. Frans B. M. de Waal, "Silent Invasion: Imanishi's Primatology and Cultural Bias in Science," *Animal Cognition* 6:293–99 (Dec. 2003).

8. Hiroyuki Takasaki, "Traditions of the Kyoto School of Field Primatology in Japan," chap. 7 in Shirley Carol Strum and Linda Marie Fedigan, eds., *Primate Encounters* (Chicago: University of Chicago Press, 2000). See also Tetsuro Matsuzawa, "Koshima Monkeys and Bossou Chimpanzees: Long-Term Research on Culture in Nonhuman

Primates," chap. 14 in Frans B. M. de Waal and Peter Tyack, eds., *Animal Social Complexity* (Cambridge, Mass.: Harvard University Press, 2003).

9. John C. Mitani, William McGrew, and Richard Wrangham, "Toshisada Nishida's Contributions to Primatology," *Primates* 47:2–5 (Jan. 2006).

10. Yukimaru Sugiyama and Jeremy Koman, "Tool-Using and -Making Behavior in Wild Chimpanzees at Bossou, Guinea," *Primates* 20:513–24 (Oct. 1979).

11. Frans B. M. de Waal and Frans Lanting, *Bonobo: The Forgotten Ape* (Berkeley: University of California Press, 1997).

12. Tetsuro Matsuzawa, "Use of Numbers by a Chimpanzee," *Nature* 315:56–59 (May 2, 1985). Because Matsuzawa has studied both wild and captive chimpanzees for more than two decades (as well as several monkey species), he has one of the most diverse publication records of anyone in the chimp research field.

13. Frans B. M. de Waal, *The Ape and the Sushi Master: Cultural Reflections of a Primatologist* (New York: Basic Books, 2001).

14. Dora Biro, Cláudia Sousa, and Tetsuro Matsuzawa, "Ontogeny and Cultural Propagation of Tool Use by Wild Chimpanzees at Bossou, Guinea: Case Studies in Nut Cracking and Leaf Folding," chap. 28 in Tetsuro Matsuzawa, Masaki Tomonaga, and Masayuki Tanaka, eds. *Cognitive Development in Chimpanzees* (Tokyo: Springer-Verlag, 2006). See also Dora Biro, Noriko Inoue-Nakamura, Rikako Tonooka, et al., "Cultural Innovation and Transmission of Tool Use in Wild Chimpanzees: Evidence from Field Experiments," *Animal Cognition* 6:213–23 (Dec. 2003).

15. Christophe Boesch, "Teaching among Wild Chimpanzees," *Animal Behaviour* 41:530–32 (Mar. 1991).

16. Gaku Ohashi, "Behavioral Repertoire of Tool Use in the Wild Chimpanzees at Bossou," chap. 26 in Matsuzawa, Tomonaga, and Tanaka, eds. *Cognitive Development in Chimpanzees.*

17. Matsuzawa conducted the study of mutual gaze with a research group that had followed mother-infant pairs at Yerkes. The pairs at the Primate Research Institute indulged in significantly more mutual gazing, which may reflect differences in their conditions or the mothers' childbearing histories. See Kim A. Bard, Masako Myowa-Yamakoshi, Masaki Tomonaga, et al., "Group Differences in the Mutual Gaze of Chimpanzees (*Pan troglodytes*)," *Developmental Psychology* 41:616–24 (July 2005).

18. Masako Myowa-Yamakoshi, Masaki Tomonaga, Masayuki Tanaka, and Tetsuro Matsuzawa, "Imitation in Neonatal Chimpanzees (*Pan troglodytes*)," *Developmental Science* 7:437–42 (Sept. 2004).

19. The only place I have seen Matsuzawa spell out this theory is in a chapter of a book he coedited. See Tetsuro Matsuzawa, "Evolutionary Origins of the Human Mother-Infant Relationship," chap. 8 in Matsuzawa, Tomonaga, and Tanaka, eds., *Cognitive Development in Chimpanzees.*

20. William McGrew, *The Cultured Chimpanzee: Reflections on Cultural Primatology* (New York: Cambridge University Press, 2004), p. 1.

21. William McGrew and Caroline E. G. Tutin, "Evidence for a Social Custom in Wild Chimpanzees?" *Man* 13:234–51 (June 1978).

22. Andrew Whiten, Jane Goodall, William McGrew, et al., "Culture in Chimpanzees," *Nature* 399:682–85 (June 17, 1999).

23. Tatyana Humle and Tetsuro Matsuzawa, "Ant-dipping Among the Chimpanzees of Bossou, Guinea, and Some Comparisons with Other Sites," *American Journal of Primatology* 58:133–48 (Nov. 2002).

24. David A. Leavens, Autumn B. Hostetter, Michael J. Wesley, and William D. Hopkins, "Tactical Use of Unimodal and Bimodal Communication by Chimpanzees, *Pan troglodytes*," *Animal Behaviour* 67:467–76 (Mar. 2004). See also William D. Hopkins, Jared P. Taglialatela, and David A. Leavens, "Chimpanzees Differentially Produce Novel Vocalizations to Capture the Attention of a Human," *Animal Behaviour* 73:281–86 (Feb. 2007).

25. Andrew Whiten, Victoria Horner, and Frans B. M. de Waal, "Conformity to Cultural Norms of Tool Use in Chimpanzees," *Nature* 437:737–40 (Sept. 29, 2005).

26. Ken Sayers and C. Owen Lovejoy, "The Chimpanzee Has No Clothes: A Critical Examination of *Pan troglodytes* in Models of Human Evolution," *Current Anthropology* 49:87–114 (Feb. 2008).

27. Esther Herrmann, Josep Call, María Victoria Hernández-Lloreda, et al., "Humans Have Evolved Specialized Skills of Social Cognition: The Cultural Intelligence Hypothesis," *Science* 317:1360–66 (Sept. 7, 2007).

28. Michael Tomasello, Josep Call, and Brian Hare, "Chimpanzees Understand Psychological States—The Question Is Which Ones and to What Extent," *Trends in Cognitive Sciences* 7:153–56 (Apr. 2003).

29. Alicia P. Melis, Brian Hare, and Michael Tomasello, "Chimpanzees Recruit the Best Collaborators," *Science* 311:1297–1300 (Mar. 3, 2006).

30. Felix Warneken, Brian Hare, Alicia P. Melis, et al., "Spontaneous Altruism by Chimpanzees and Young Children," *PLoS Biology* 5:e184 (July 2007).

31. Brian Hare, Alicia P. Melis, Vanessa Woods, et al., "Tolerance Allows Bonobos to Outperform Chimpanzees on a Cooperative Task," *Current Biology* 17:1–5 (Apr. 3, 2007).

32. David Premack, "Human and Animal Cognition: Continuity and Discontinuity," *Proceedings of the National Academy of Sciences* 104:13861–67 (Aug. 28, 2007).

33. Derek C. Penn, Keith J. Holyoak, and Daniel J. Povinelli, "Darwin's Mistake: Explaining the Discontinuity between Human and Nonhuman Minds," *Behavioral and Brain Sciences* 31:109–78 (Apr. 2008).

34. Daniel J. Povinelli, "Behind the Ape's Appearance: Escaping Anthropocentrism in the Study of Other Minds," *Dædalus* 133:29–41 (Winter 2004).

8. Head to Head

1. Sandra F. Witelson, Debra L. Kigara, Thomas Harvey, et al., "The Exceptional Brain of Albert Einstein," *Lancet* 353:2149–53 (June 19, 1999).

2. Harry J. Jersion, *Evolution of the Brain and Intelligence* (New York Academic Press, 1973).

3. Min S. Park, Andrew D. Nguyen, Henry E. Aryan, et al., "Evolution of the Human Brain: Changing Brain Size and the Fossil Record," *Neurosurgery* 60:555–62 (Mar. 2007).

4. Karl Zilles, Este Armstrong, Axel Schleicher, and Hans-Joachim Kretschmann, "The Human Pattern of Gyrification in the Cerebral Cortex," *Anatomy and Embryology* 179:173–79 (Nov. 1988).

5. Chet C. Sherwood, Cheryl D. Stimpson, Mary Ann Raghanti, et al., "Evolution of Increased Glia–neuron Ratios in the Human Frontal Cortex," *Proceedings of the National Academy of Sciences* 103:13606–11 (Sept. 12, 2006).

6. Francis Crick and Edward Jones, "Backwardness of Human Neuroanatomy," *Nature* 361:109–10 (Jan. 14, 1993).

7. M. P. Crawford and John F. Fulton, "Frontal Lobe Ablation in Chimpanzee: A

Resumé of Becky and Lucy," *Association for Research in Nervous and Mental Disease* 27:3–58 (1948). See also W. Ross Adey, Raymond T. Kado, and John M. Rhodes, "Sleep: Subcortical and Cortical Recordings," *Science* 141:932–33 (Sept. 6, 1963).

8. Todd M. Preuss, "Who's Afraid of *Homo sapiens?*" *Journal of Biomedical Discovery and Collaboration* 1:17 (Nov. 29, 2006).

9. Katerina Semendeferi, Hanna Damasio, Randall Frank, and Gary W. Van Hoesen, "The Evolution of the Frontal Lobes: A Volumetric Analysis Based on Three-dimensional Constructions of Magnetic Resonance Scans of Human and Ape Brains," *Journal of Human Evolution* 32:375–88 (Apr. 1997).

10. Katerina Semendeferi and Hanna Damasio, "The Brain and Its Main Anatomical Subdivisions in Living Hominoids Using Magnetic Resonance Imaging," *Journal of Human Evolution* 38:317–32 (Feb. 2000).

11. Karen Emmorey, John S. Allen, Joel Bruss, et al., "A Morphometric Analysis of Auditory Brain Regions in Congenitally Deaf Adults," *Proceedings of the National Academy of Sciences* 100:10049–54 (Aug. 19, 2003).

12. Nicole Barger, Lisa Stefanacci, and Katerina Semendeferi, "A Comparative Volumetric Analysis of the Amygdaloid Complex and Basolateral Division in the Human and Ape Brain," *American Journal of Physical Anthropology* 134:392–403 (Nov. 2007).

13. Natalie M. Schenker, Daniel P. Buxhoeveden, William L. Blackmon, et al., "A Comparative Quantitative Analysis of Cytoarchitecture and Minicolumnar Organization in Broca's Area in Humans and Great Apes," *Journal of Comparative Neurology* 10:117–28 (Sept. 2008).

14. Michael Balter, "Brain Evolution Studies Go Micro," *Science* 315:1208–11 (Mar. 2, 2007).

15. Todd M. Preuss, Huixin Qi, and Jon H. Kass, "Distinctive Compartmental Organization of Human Primary Visual Cortex," *Proceedings of the National Academy of Sciences* 96:11601–6 (Sept. 28, 1999).

16. "The New Comparative Biology of Human Nature," Arthur M. Sackler Colloquia of the National Academy of Sciences, Irvine, Calif., Nov. 16–18, 2006.

17. Esther A. Nimchinsky, Brent A. Vogt, John H. Morrison, and Patrick Hof, "Spindle Neurons of the Human Anterior Cingulate Cortex," *Journal of Comparative Neurology* 355:27–37 (Apr. 24, 1995).

18. Esther A. Nimchinsky, Emmanuel Glissen, John M. Allman, et al., "A Neuronal Morphologic Type Unique to Humans and Great Apes," *Proceedings of the National Academy of Sciences* 96:5268–73 (Apr. 27, 1999).

19. Camilla Butti, Chet C. Sherwood, Atiya Y. Hakeem, et al., "Total Number and Volume of Von Economo Neurons in the Cerebral Cortex of Cetaceans," 515:243–59 (July 10, 2009). Also see Atiya Y. Hakeem, Chet C. Sherwood, Christopher J. Bonar, et al., "Von Economo Neurons in the Elephant Brain," *Anatomical Record* 292:242–48 (Feb. 2009).

20. Carrolee Barlow's group at the Salk Institute for Biological Studies contributed substantially to this work. Mario Cáceres, Joel Lachuer, Matthew A. Zapala, et al., "Elevated Gene Expression Levels Distinguish Human from Non-human Primate Brains," *Proceedings of the National Academy of Sciences* 100:13030–35 (Oct. 28, 2003). See also this earlier study, Wolfgang Enard, Philipp Khaitovich, Joachim Klose, et al., "Intra- and Interspecific Variation in Primate Gene Expression Patterns," *Science* 296:340–43 (Apr. 12, 2002).

21. Karen S. Christopherson, Erik M. Ullian, Caleb C. A. Stokes, et al., "Thrombospon-

dins Are Astrocyte-Secreted Proteins that Promote CNS Synaptogenesis," *Cell* 120: 421–33 (Feb. 11, 2005).

22. Mario Cáceres, Carolyn Suwyn, Marcelia Maddox, et al., "Increased Cortical Expression of Two Synaptogenic Thrombospondins in Human Brain Evolution," *Cerebral Cortex* 17:2312–21 (Oct. 2007).

23. Eric J. Vallender, Nitzan Mekel-Bobrov, and Bruce T. Lahn, "Genetic Basis of Human Brain Evolution," *Trends in Neurosciences* 31:637–44 (Dec. 2008).

24. Magdalena C. Popesco, Erik J. MacLaren, Janet Hopkins, et al., "Human Lineage–Specific Amplification, Selection, and Neuronal Expression of DUF1220 Domains," *Science* 313:1304–7 (Sept. 1, 2006).

25. Researchers do not pretend to know what all of this means, but journalists all too often do, wrote Elizabeth Pennisi, another of my *Science* colleagues who regularly covers these advances, in a trenchant article. Elizabeth Pennisi, "Mining the Molecules That Made Our Mind," *Science* 313:1908–11 (Sept. 29, 2006).

26. James K. Rilling, Matthew F. Glasser, Todd M. Preuss, et al., "The Evolution of the Arcuate Fasciculus Revealed with Comparative DTI," *Nature Neuroscience* 11:426–28 (Apr. 2008).

27. James K. Rilling, Sarah K. Barks, Lisa A. Parr, et al., "A Comparison of Resting-state Brain Activity in Humans and Chimpanzees," *Proceedings of the National Academy of Sciences* 104:17146–51 (Oct. 23, 2007).

28. Lisa A. Parr, Erin Hecht, Sarah K. Barks, et al., "Face Processing in the Chimpanzee Brain," *Current Biology* 19:50–53 (Jan. 13, 2009).

29. Stephen Jay Gould, *Ontogeny and Phylogeny* (Cambridge, Mass.: Belknap Press, 1985), p. 399.

30. Helene Coqueugniot, Jean-Jacques Hublin, Francis Veillon, et al., "Early Brain Growth in *Homo erectus* and Implications for Cognitive Ability," *Nature* 431:299–302 (Sept. 16, 2004).

31. Jean-Jacques Hublin and Helene Coqueugniot, "Absolute or Proportional Brain Size: That Is the Question. A Reply to Leigh's (2006) Comments," *Journal of Human Evolution* 50:109e113 (2006).

32. Eric Courchesne used magnetic resonance imaging and postmortem examinations to establish these comparative measurements. Eric Courchesne, Karen Pierce, Cynthia M. Schumann, et al., "Mapping Early Brain Development in Autism," *Neuron* 56:399–413 (Oct. 25, 2007).

9. Walk This Way

1. Jonathan Kingdon, *Lowly Origin: Where, When, and Why Our Ancestors First Stood Up* (Princeton, N.J.: Princeton University Press, 2003).

2. Russell H. Tuttle, David M. Webb, and Nicole I. Tuttle, "Laetoli Footprint Trails and the Evolution of Hominid Bipedalism," *Origine(s) de la Bipédie chez les Hominidés* (Paris: CNRS Editions, 1991), pp. 187–98. The paper formed the basis of Russell Tuttle's presentation at the Colloque International de la Fondation Singer-Polignac, June 5–8, 1990.

3. Ann Gibbons, *The First Human* (New York: Doubleday, 2006), p. 148.

4. Susannah K. S. Thorpe, Roger L. Holder, and Robin H. Crompton, "Origin of Human Bipedalism as an Adaptation for Locomotion on Flexible Branches," *Science* 316:1328–31 (June 1, 2007).

5. Paul O'Higgins and Sarah Elton, "Walking on Trees," *Science* 316:1292–94 (June 1, 2007).

6. David R. Begun, Brian G. Richmond, and David S. Strait, "Comment on 'Origin of Human Bipedalism as an Adaptation for Locomotion on Flexible Branches,'" *Science* 318:1066d (Nov. 16, 2007). Crompton and Thorpe rebut these arguments in "Response to 'Comment on "Origin of Human Bipedalism as an Adaptation for Locomotion on Flexible Branches,"'" *Science* 318:1066e (Nov. 16, 2007). See also the back-and-forth between Jeffrey Schwartz and Thorpe, Holder, and Crompton on the letters-to-the-editor page, "The Origins of Human Bipedalism," *Science* 318:1065 (Nov. 16, 2007).

7. C. Richard Taylor and V. J. Rowntree, "Running on Two or on Four Legs: Which Consumes More Energy?" *Science* 179:186–87 (Jan. 12, 1973).

8. Michael D. Sockol, David A. Raichlen, and Herman Pontzer, "Chimpanzee Locomotor Energetics and the Origins of Human Bipedalism," *Proceedings of the National Academy of Sciences* 104:12265–69 (July 24, 2007).

9. Herman Thieme, "Lower Palaeolithic Hunting Spears from Germany," *Nature* 385:807–10 (Feb. 27, 1997).

10. Dennis M. Bramble and Daniel E. Lieberman, "Endurance Running and the Evolution of *Homo*," *Nature* 432:345–52 (Nov. 18, 2004).

11. Nina G. Jablonski, *Skin: A Natural History* (Berkeley: University of California Press, 2006).

12. Desmond Morris, *The Naked Ape: A Zoologist's Study of the Human Animal* (New York: McGraw-Hill, 1967).

13. Nina G. Jablonski and George Chaplin, "The Evolution of Human Skin Coloration," *Journal of Human Evolution* 39:57–106 (July 2000).

14. McBrearty and Jablonski described their findings in the same issue of *Nature* that included the first draft of the chimpanzee genome. Sally McBrearty and Nina G. Jablonski, "First Fossil Chimpanzee," *Nature* 437:105–8 (Sept. 1, 2005).

15. Tim D. White, Berhane Asfaw, Yonas Beyene, et al., "*Ardipithecus ramidus* and the Paleobiology of Early Hominids," *Science* 326:64 and 75–86 (Oct. 2, 2009).

16. Hansell H. Stedman, Benjamin W. Kozyak, Anthony Nelson, et al., "Myosin Gene Mutation Correlates with Anatomical Changes in the Human Lineage," *Nature* 428:415–18 (Mar. 25, 2004).

17. The *Nature* paper downplays the evolutionary idea, only mentioning it in passing, but Hansell Stedman, the first author, told me this was an editorial decision by the journal. *Nature* also in the same issue ran an accompanying "News and Views" that explored the notion more fully, as did an article on *ScienceNOW*. Peter Curie, "Muscling in on Hominid Evolution," *Nature* 238: 373–77 (Mar. 25, 2004). See also Elizabeth Pennisi, "Weak Jaw, Big Brain," *ScienceNOW*, Mar. 24, 2005, http://sciencenow.sciencemag.org/cgi/content/full/2004/324/3, last accessed Jan. 12, 2010.

18. Adrienne L. Zihlman, *The Human Evolution Coloring Book* (New York: Harper Resource, 2000), pp. 4–36.

19. Paleoanthropologist John Hawkes of the University of Wisconsin–Madison made this point in "How Strong is a Chimpanzee," *Slate*, Feb. 25, 2009, http://www.slate.com/id/2212232/, last accessed Jan. 12, 2010. Hawkes cites the work of John Bauman, who in the 1920s conducted the studies on chimp strength that put forward the five-to-eight times stronger figure, and then a subsequent—and more carefully done—experiment in the 1940s by Glen Finch of the Yale Laboratories of Primate Biology, which were run by Robert Yerkes, that finds no such thing. John E. Bauman, "The Strength of the Chimpanzee and Orang," *The Scientific Monthly* 16:432–39 (Apr.

1923) and "Observations on the Strength of the Chimpanzee and Its Implications," *Journal of Mammalogy* 7:1–9 (Feb. 1926). See also Glen Finch, "The Bodily Strength of Chimpanzees," *Journal of Mammalogy* 24:224–28 (May 1943).

20. Richard Dawkins, *The Selfish Gene* (New York: Oxford University Press, 1989), endnotes to chap. 9, p. 307.

21. Jeremy M. DeSilva, "Functional Morphology of the Ankle and the Likelihood of Climbing in Early Hominins," *Proceedings of the National Academy of Sciences* 106:6567–72 (Apr. 21, 2009).

22. Catherine Dressler, "Detroit Zoo Plans World-Class Chimpanzee Exhibit," Associated Press, Oct. 28, 1986.

23. Kay Houston, "How the Detroit Zoo's First Day Was Almost Its Last," *Detroit News*, Feb. 24, 1999.

24. Tom Hundley, "New Zoo Display Lets Chimpanzees Be Themselves," *Chicago Tribune*, Dec. 13, 1989.

25. Otto M. J. Adang, Joep A. B. Wensing, and Jan A. R. A. M. van Hooff, "The Arnhem Zoo Colony of Chimpanzees Development and Management Techniques," *International Zoo Yearbook* 26:230–48 (Jan. 1987).

26. Jan A. R. A. M. van Hooff, "The Arnhem Zoo Chimpanzee Consortium: An Attempt to Create an Ecologically and Socially Acceptable Habitat," *International Zoo Yearbook* 13:195–203 (Jan. 1973).

27. Susan McDonald, "The Detroit Zoo Chimpanzees *Pan troglodytes*: Exhibit Design, Group Composition and the Process of Group Formation," *International Zoo Yearbook* 33: 235–41 (Jan. 1994).

28. P. F. M. Ama and S. Ambassa, "Buoyancy of African Black and European White Males," *American Journal of Human Biology* 9:87–92 (1997).

29. Elaine Morgan, *The Scars of Evolution* (London: Souvenir Press, 1990), p. 45.

30. Alister Hardy, "Was Man More Aquatic in the Past?" *New Scientist*, Mar. 17, 1960.

31. Algis Kuliukas, "Bipedal Wading in *Hominoidae* Past and Present," Ph.D. dissertation, University College, London, 2001.

32. The photos were shot by Frans Lanting and the video by Christine Eckstrom. These were in support of a story by Mary Roach, "Almost Human," *National Geographic*, Apr. 2008.

33. J. Gilchrist, K. Gotsch, and G. Ryan, "Nonfatal and Fatal Drownings in Recreational Water Settings—United States, 2001–2002," *Morbidity and Mortality Weekly* 53:447–52 (June 4, 2004).

34. Centers for Disease Control and Prevention, National Center for Injury Prevention and Control. Web-based Injury Statistics Query and Reporting System (WISQARS), www.cdc.gov/ncipc/wisqars, accessed Mar. 23, 2008.

35. R. A. Brenner, G. Saluja Taneja, D. L. Haynie, et al., "Association between Swimming Lessons and Drowning in Childhood: A Case-control Study," *Archives of Pediatric and Adolescent Medicine* 163:203–10 (Mar. 2009).

36. J. R. Moehringer, "Cheeta Speaks," *Los Angeles Times*, Apr. 22, 2007.

37. R. D. Rosen, "Lie of the Jungle," *Washington Post Magazine*, Dec. 7, 2008.

38. Stephen Jay Gould, "Male Nipples and Clitoral Ripples," in *Bully for Brontosaurus: Reflections in Natural History* (New York: Norton, 1991).

39. Sigmund Freud, "The Dissolution of the Oedipus Complex" (1924), in *The Standard Edition of the Complete Psychological Works of Sigmund Freud*, vol. 19 (1923–25); *The Ego and the Id and Other Works*, trans. and ed. James Strachey, (London: Hogarth Press, 1981).

10. Carnal Knowledge

1. World Health Organization, WHO Laboratory Manual: For the Examination of Human Semen and Sperm-cervical Mucus Interaction (New York: Cambridge University Press, 1999), p. 60. See also Aleksander Giwercman, Elisabeth Carlsen, Niels Keiding, et al., "Evidence for Increasing Incidence of Abnormalities of the Human Testis: A Review," Environmental Health Perspectives 101(Supplement 2):65–71 (July 1993).

2. Kenneth G. Gould, Leona G. Young, Eleanor B. Smithwick, and Sarah R. Phythyon, "Semen Characteristics of the Adult Male Chimpanzee (Pan troglodytes)," American Journal of Primatology 29:221–32 (1993).

3. Alexander H. Harcourt, Paul H. Harvey, Susan G. Larson, and Roger V. Short, "Testis Weight, Body Weight and Breeding System in Primates," Nature 293:55–77 (Sept. 3, 1981).

4. Geoffrey A. Parker, "Sperm Competition and Its Evolutionary Consequences in the Insects," Biological Reviews 45:525–67 (1970).

5. Alan F. Dixson and Nicholas I. Mundy, "Sexual Behavior, Sexual Swelling, and Penile Evolution in Chimpanzees (Pan troglodytes)," Archives of Sexual Behavior 23:267–80 (June 1994).

6. Jared Diamond, The Third Chimpanzee: The Evolution and Future of the Human Animal (New York: HarperCollins, 1992), p. 75.

7. Matthew J. Anderson, Shannon J. Chapman, Elaine N. Videan, et al., "Functional Evidence for Differences in Sperm Competition in Humans and Chimpanzees," American Journal of Physical Anthropology 134:274–80 (Oct. 2007). See also Jaclyn M. Nascimento, Linda Z. Shi, Stuart Meyers, et al., "The Use of Optical Tweezers to Study Sperm Competition and Motility in Primates," Journal of the Royal Society Interface 5:297–302 (Mar. 2008).

8. Gagneux later discovered that the Neu5Gc on the human sperm he tested that night actually came from the bovine serum albumin used as a buffer to store the samples. This buffer is routinely used in in vitro fertilization clinics, so the finding of Gc on the sperm may affect fertility success in those settings. When we discussed this in January 2010, he was hoping to obtain more human sperm from men who eat lots of meat and dairy products, and then test them without adding buffer to the samples.

9. Jennifer F. Hughes, Helen Skaletsky, and Tatyana Pyntikova, et al., "Chimpanzee and Human Y Chromosomes Are Remarkably Divergent in Structure and Gene Content," Nature, advance online publication (Jan. 13, 2010), http://www.nature.com/nature/journal/vaop/ncurrent/full/nature08700.html.

10. Lizzie Buchen, "The Fickle Y Chromosome," Nature 463:149 (Jan. 14, 2010).

11. Jocelyne Marson, Sylvain Meuris, F. Moysan, et al., "Cellular and Biochemical Characteristics of Semen Obtained from Pubertal Chimpanzees by Masturbation," Journal of Reproduction and Fertility 82:199–207 (Jan. 1988).

12. Hermann Bader, "Electroejaculation in Chimpanzees and Gorillas and Artificial Insemination in Chimpanzees," Zoo Biology 2:307–14 (1983).

13. Jocelyne Marson, D. Gervais, Sylvain Meuris, et al., "Influence of Ejaculation Frequency on Semen Characteristics in Chimpanzees (Pan troglodytes)," Journal of Reproduction and Fertility 85:43–50 (Jan. 1989).

14. Mel L. Allen and William B. Lemmon, "Orgasm in Female Primates," American Journal of Primatology 1:15–34 (1981).

15. Maurice K. Temerlin, Lucy: Growing Up Human: A Chimpanzee Daughter in a Psychotherapist's Family (Palo Alto, Calif.: Science and Behavior Books, 1976). Lucy

lived with Temerlin and his wife, and he noted that she also tried "to mouth my penis whenever she sees it, whether I am urinating, bathing or having an erection."

16. Elizabeth Hess, *Nim Chimpsky: The Chimp Who Would Be Human* (New York: Bantam Books, 2008), p. 179.

17. Frans B. M. de Waal, "Bonobo Sex and Society," *Scientific American*, Mar. 1995.

18. The word "satyr" was then used to denote that the creature was somewhere between man and beast. W. C. Osman Hill, "The Discovery of the Chimpanzee," in *The Chimpanzee* (Basel: Karger Basel, 1969), pp. 1–21. The "Indian satyr," as the Dutch anatomist Nicolaas Tulp called it, was for many years thought to be a chimpanzee. But the ape was found in what was called Angola—an area that then occupied much more of the western coast of Africa than designated by the Portuguese colony (which has no apes). But as de Waal notes in *Bonobo*, Angola is south of the Congo River, the only place bonobos live today—and Tulp's gravure of the animal looks bonobo-like to some experts. See Frans B. M. de Waal, *Bonobo: The Forgotten Ape* (Berkeley: University of California Press, 1997), p. 7.

19. Amy R. Parish and Frans B. M. de Waal, "The Other 'Closest Living Relative': How Bonobos (*Pan paniscus*) Challenge Traditional Assumptions about Females, Dominance, Intra- and Intersexual Interactions, and Hominid Evolution," *Annals of the New York Academy of Sciences* 907:97–113 (Apr. 2000).

20. Craig B. Stanford, "The Social Behavior of Chimpanzees and Bonobos," *Current Anthropology* 39:399–420 (Aug.–Oct. 1998).

21. Craig B. Stanford, *Significant Others: The Ape-Human Continuum and the Quest for Human Nature* (New York: Basic Books, 2001), p. 27.

22. Ian Parker, "Swingers," *New Yorker*, July 30, 2007. De Waal so disliked the article that he even complained to me about the harsh manner in which Parker depicted Hohmann, his critic. But de Waal laughed when I suggested to him that he cared less about how Hohmann came off than he did about Parker's dismissal of de Waal's idea that the "veneer theory"—which holds that humans are really red in tooth and claw like chimpanzees and all other animals, but disguise it—is challenged by bonobo behavior. "I definitely think there's a difference between chimpanzees and bonobos in the field and in captivity," de Waal told me. "We have not completely sorted out how aggressive bonobos can be, and maybe they're not sexual under all conditions in all circumstances. And there's still something to be discovered about them. But I do think there's a substantial difference."

23. Frans B. M. de Waal, "Bonobos, Left and Right: Primate Politics Heats Up Again as Liberals and Conservatives Spindoctor Science," *eSkeptic*, Aug. 8, 2007, archived at http://www.skeptic.com/eskeptic/07-08-08#note02.

24. Richard Wrangham and Dale Peterson, *Demonic Males* (Boston and New York: Houghton Mifflin, 1996), pp. 26 and 132.

25. Dean Hamer has published several studies, and stirred much controversy, about evidence of links between genes and homosexuality. See Brian S. Mustanski, Michael G. DuPree, et al., "A Genomewide Scan of Male Sexual Orientation," *Human Genetics* 116:272–78 (Mar. 2005).

26. Jon Cohen, *Coming to Term: Uncovering the Truth About Miscarriage* (New York: Houghton Mifflin, 2005).

27. Emil Witschi came up with the overripeness theory after studying frogs. See "Teratogenic Effects from Overripeness of the Egg," in F. Clarke Fraser and Victor A. McKusick, eds., *Congenital Malformations, Proceedings of the Third International Congress, The Hague, The Netherlands, September 7–13, 1969* (New York: Excerpta

Medica International Congress Series no. 204, 1970), pp. 157–69. James German argued that married women had less sex as they aged, which led to more overripe fertilizations and Down syndrome, and while this thesis did not stand the test of time—it is the deterioration of aging oocytes, not the timing of fertilization in older women—his review of the topic is interesting nonetheless. See James German, "Mongolism, Delayed Fertilization and Human Sexual Behavior," *Nature* 217:516–18 (Feb. 10, 1968). For yet another interesting review of the theory see Luc J. Smits, Piet Hein Jongbloet, and Gerhard A. Zielhuis, "Preovulatory Overripeness of the Oocyte as a Cause of Ovarian Dysfunction in the Human Female," *Medical Hypotheses* 45:441–48 (Nov. 1995).

28. For an overview of infanticide in chimps and other great apes, see Wrangham and Peterson, *Demonic Males*, pp. 155–60.

29. There is a vast literature on the evolution of swellings that dates back to Darwin. I like Stanford's synopsis in "The Social Behavior of Chimpanzees and Bonobos," cited above. Also see these reviews: Charles L. Nunn, "The Evolution of Exaggerated Sexual Swellings in Primates and the Graded-signal Hypothesis," *Animal Behaviour* 58:229–46 (Aug. 1999); Robert R. Stallmann and Jeffrey W. Froelich, "Primate Sexual Swellings as Coevolved Signal Systems," *Primates* 41:1–16 (Jan. 2000).

30. In technical jargon, which isn't nearly as much fun, these are respectively called "reliable indicator" and "graded signal."

31. Lynnette Leidy Sievert and Catherine A. Dubois, "Validating Signals of Ovulation: Do Women Who Think They Know, Really Know?" *American Journal of Human Biology*: 17:310–20 (May–June 2005).

32. Many of these ideas are explored by Bogusław Pawłowski, "Loss of Oestrus and Concealed Ovulation in Human Evolution: The Case against the Sexual-Selection Hypothesis," *Current Anthropology* 40:257–76 (June 1999).

11. It's a Chimp's Life

1. BBC Two, "The Demonic Ape" (Jan. 8, 2004).

2. George Owoyesigire, "Survival Strategy or Revenge? Chimpanzee-Human Attacks in Western Uganda," *Canopy* 5:9–11 (Aug. 2006). Tales of chimps abducting human babies often have more anecdote than fact, but this one is told by a worker for the Uganda Wildlife Authority who came onto the scene shortly after it occurred.

3. Jill D. Pruetz and Paco Bertolani, "Savanna Chimpanzees, *Pan troglodytes verus*, Hunt with Tools," *Current Biology* 17:1–6 (Mar. 6, 2007).

4. Kim Hill, Christophe Boesch, Jane Goodall, et al., "Mortality Rates among Wild Chimpanzees," *Journal of Human Evolution* 40:437–50 (May 2001).

5. Jane Goodall, *The Chimpanzees of Gombe: Patterns of Behavior* (Cambridge, Mass.: Harvard University Press, 1986), p. 81.

6. Melissa Emery Thompson, e-mail, Nov. 24, 2009.

7. United Nations, *World Population Prospects: The 2006 Revision* (New York: United Nations, 2007), pp. 80–84, table A.17.

8. Craig R. Whitney, "Jeanne Calment, World's Elder, Dies at 122," *New York Times*, Aug. 5, 1997.

9. Michael Gurven and Hillard Kaplan, "Longevity Among Hunter-Gatherers: A Cross-Cultural Examination," *Population and Development Review* 33:321–65 (June 2007).

10. Bennett Dyke, Timothy B. Gage, Patricia L. Alford, et al., "Model Life Table for Captive Chimpanzees," *American Journal of Primatology* 37:25–37 (1995).

11. Kim-Anh Do, Susan Treloar, Niramala Pandeya, et al., "Predictive Factors of Age at Menopause in a Large Australian Twin Study," *Human Biology* 70:1073–91 (Dec. 1998).

12. Kristen Hawkes and Ken R. Smith, "Do Women Stop Early? Similarities in Fertility Decline in Humans and Chimpanzees," in press.

13. The student, Kim Hill, coauthored the article on the mortality rates of wild chimpanzees cited above and went on to coauthor the definitive book about the Aché. See Kim Hill and A. Magdalena Hurtado, *Aché Life History: The Ecology and Demography of a Foraging People* (New York: Aldine Press, 1996).

14. Kristen Hawkes, Kim Hill, and James F. O'Connell, "Why Hunters Gather: Optimal Foraging and the Aché of Eastern Paraguay," *American Ethnologist* 9:379–98 (May 1982).

15. George C. Williams, "Pleiotropy, Natural Selection, and the Evolution of Senescence," *Evolution* 11:398–411 (Dec. 1957).

16. Kristen Hawkes, James F. O'Connell, and Nicholas G. Blurton Jones, "Hadza Women's Time Allocation, Offspring Provisioning, and the Evolution of Long Postmenopausal Life Spans," *Current Anthropology* 38:551–77 (Aug.–Oct. 1997). See the extensive comments and replies at the end of the article.

17. For a thorough review of the hypothesis, see Kristen Hawkes, "Grandmothers and the Evolution of Human Longevity," *American Journal of Human Biology* 15:380–400 (May–June 2003). For a supporting report from an independent group, see Mirkka Lahdenperä, Virpi Lummaa, Samuli Helle, et al., "Fitness Benefits of Prolonged Post-reproductive Lifespan in Women," *Nature* 428:178–81 (Mar. 11, 2004). See, too, this accompanying editorial, Kristen Hawkes, "The Grandmother Effect," *Nature* 428:128–29 (Mar. 11, 2004).

18. Kristen Hawkes, James F. O'Connell, Nicholas G. Blurton Jones, et al., "Grandmothering, Menopause, and the Evolution of Human Life Histories," *Proceedings of the National Academy of Sciences* 95:1336–39 (Feb. 1998).

19. Katherine A. Roof, William D. Hopkins, M. Kay Izard, et al., "Maternal Age, Parity, and Reproductive Outcome in Captive Chimpanzees (*Pan troglodytes*)," *American Journal of Primatology* 67:199–207 (Oct. 2005).

20. Ibid., 11.66 years. For wild chimpanzees, first birth dropped down to 10.9 years old in Yukimaru Sugiyama's study, "Demographic Parameters and Life History of Chimpanzees at Bossou, Guinea," *American Journal of Physical Anthropology* 124:154–65 (June 2004).

21. Kirtly P. Jones, Lary C. Walker, Daniel Anderson, et al., "Depletion of Ovarian Follicles with Age in Chimpanzees: Similarities to Humans," *Biology of Reproduction* 77:247–51 (Aug. 2007).

22. I discuss many details about human oocyte depletion, with references, in my book *Coming to Term: Uncovering the Truth About Miscarriage* (New York: Houghton Mifflin, 2005), chap. 2.

23. Roof, Hopkins, Izard, et al., "Maternal Age, Parity, and Reproductive Outcome."

24. Melissa Emery Thompson, James H. Jones, Anne E. Pusey, et al., "Aging and Fertility Patterns in Wild Chimpanzees Provide Insights into the Evolution of Menopause," *Current Biology* 17:2150–56 (Dec. 18, 2007).

25. Martin N. Muller, Melissa Emery Thompson, and Richard Wrangham, "Male Chimpanzees Prefer Mating with Old Females," *Current Biology* 16:2234–38 (Nov. 21, 2006).

26. Elaine N. Videan, Christopher B. Heward, Kajal Chowdhury, et al., "Comparison of

Biomarkers of Oxidative Stress and Cardiovascular Disease in Humans and Chimpanzees (*Pan troglodytes*)," *Comparative Medicine* 59:287–96 (June 2009).

27. Martha Tappen and Richard Wrangham, "Recognizing Hominoid-Modified Bones: The Taphonomy of Colobus Bones Partially Digested by Free-Ranging Chimpanzees in the Kibale Forest, Uganda," *American Journal of Physical Anthropology* 113:217–34 (Oct. 2000).

28. Briana L. Pobiner, Jeremy DeSilva, William J. Sanders, and John C. Mitani, "Taphonomic Analysis of Skeletal Remains from Chimpanzee Hunts at Ngogo, Kibale National Park, Uganda," *Journal of Human Evolution* 52:614–36 (June 2007).

29. Richard Wrangham, "Feeding Behavior of Chimpanzees in Gombe National Park Tanzania," in T. H. Clutton-Brock, ed., *Primate Ecology: Studies of Feeding and Ranging Behaviour in Lemurs, Monkeys, and Apes* (London and New York: Academic Press, 1977) pp. 504–38.

30. Richard Wrangham, *Catching Fire: How Cooking Made Us Human* (New York: Basic Books, 2009), pp. 140–42.

31. Charles Darwin, *The Descent of Man, and Selection in Relation to Sex* (London: John Murray, 1871), chap. 4, p. 137. Note that in subsequent editions, chapter 4, "Manner of Development," became chapter 2.

32. Richard Wrangham, James Holland Jones, Greg Laden, David Pilbeam, and NancyLou Conklin-Brittain, "The Raw and the Stolen: Cooking and the Ecology of Human Origins," *Current Anthropology* 40:567–94 (Dec. 1999). In a delicious back-and-forth, several colleagues write extended comments that follow the paper, and the authors then reply.

33. Corinna Koebnick, Carola Strassner, Ingrid Hoffmann, and Claus Leitzmann, "Consequences of a Long-term Raw Food Diet on Body Weight and Menstruation: Results of a Questionnaire Survey," *Annals of Nutrition and Metabolism* 43:69–79 (1999). Some raw foodists, as Wrangham wryly noted in *Catching Fire*, believe that by reducing the toxins in their bodies, they need fewer ejaculations and menstruation to cleanse the system. "Perhaps it is unnecessary to note that medical science finds no support for the idea that toxins are removed by seminal emissions or menstruation," he wrote.

34. Richard Wrangham and NancyLou Conklin-Brittain, "Cooking as a Biological Trait," *Comparative Biochemistry and Physiology Part A* 136:35–46 (Sept. 2003).

35. Scott M. Bobacka, Christian L. Coxa, Brian D. Otta, et al., "Comparative Cooking and Grinding Reduces the Cost of Meat Digestion," *Biochemistry and Physiology—Part A: Molecular and Integrative Physiology* 148:651–56 (Nov. 2007).

36. Rachel N. Carmody and Richard Wrangham, "The Energetic Significance of Cooking," *Journal of Human Evolution* 57:379–91 (Oct. 2009).

37. For a sharply critical review of *Catching Fire*, see Pat Shipman, "Cooking Debate Goes Off the Boil," *Nature* 459:1059–60 (June 25, 2009). The notion that raw meat alone could have expanded the brain is eloquently spelled out in what is known as the Expensive-Tissue Hypothesis, which says that the energy-gobbling brain expanded after the energy-gobbling gut shrank in response to a higher-quality diet (including raw meat). See Leslie C. Aiello and Peter Wheeler, "The Expensive-Tissue Hypothesis: The Brain and the Digestive System in Human and Primate Evolution," *Current Anthropology* 36:199–221 (Apr. 1995).

38. A. John Iafrate, Lars Feuk, Miguel N. Rivera, et al., "Detection of Large-scale Variation in the Human Genome," *Nature Genetics* 36:949–51 (Sept. 2004).

39. George H. Perry, Nathaniel J. Dominy, Katrina G. Claw, et al., "Diet and the Evolution

of Human Amylase Gene Copy Number Variation," *Nature Genetics* 39:1256–60 (Oct. 2007).

40. Caleb E. Finch and Robert M. Sapolsky, "The Evolution of Alzheimer Disease, the Reproductive Schedule, and *apoE* Isoforms," *Neurobiology of Aging* 20:407–28 (July–Aug. 1999).

41. Caleb E. Finch and Craig B. Stanford, "Meat-Adaptive Genes and the Evolution of Slower Aging in Humans," *Quarterly Review of Biology* 79:3–50 (Mar. 2004).

42. The *apoE* story is filled with weird twists and turns. The function of the chimp *apoE* allele most closely resembles *E3*, but the DNA sequence has more in common with *E4*, suggesting that it is the ancestral gene in both humans and chimps. But this ancestral version of *E4* did not lead to cardiovascular or Alzheimer's disease. That only occurred later in humans, when the human *E4* evolved away from the chimp version, possibly because of improved resistance to infection. As we started eating more meat, the *E3* allele evolved. This is a dizzying chain of events, and, in the end, I like the idea more than the evidence.

12. Born to Be Wild

1. Percival Bailey, Gerhardt Von Bonin, and Warren S. McCulloch, *The Isocortex of the Chimpanzee* (Urbana-Champaign: University of Illinois Press, 1950).

2. Paola Cavalieri and Peter Singer, eds., *The Great Ape Project: Equality Beyond Humanity* (New York: St. Martin's Press, 1993). See "A Declaration on Great Apes," pp. 4–7.

3. Paul Recer, "New AIDS Vaccine Candidate Approved for Early Clinical Trials," Associated Press, Nov. 20, 1990. The press conference was held at the Madison Hotel in Washington, D.C.

4. Sam Howe Verhovek, "Court Rules Letters to the Editor Deserve Protection from Libel Suits," *New York Times*, Jan. 16, 1991. See also Anthony Lewis, "Abroad at Home; Abusing the Law," *New York Times*, May 10, 1991. The case ended on June 3, 1991, when the Supreme Court denied a petition from Immuno to review the New York State of Appeals decision, on remand, against the company. For the history of Teleki and the animal trade, see passages about Franz Sitter in Dale Peterson and Jane Goodall, *Visions of Caliban: On Chimpanzees and People* (Athens: University of Georgia Press, 2000).

5. Geza Teleki, "Committee for Conservation and Care of Chimpanzees—An International Association of Professionals Devoted to Improving the Survival Prospects and Living Conditions of All Chimpanzees," *Primates* 29: 579–81 (Oct. 1988).

6. It was published in 1989 as part of a book from the 1986 Chicago chimpanzee conference. Geza Teleki, "Population Status of Wild Chimpanzees (*Pan troglodytes*) and Threats to Survival," in Paul G. Heltne and Linda A. Marquardt, eds., *Understanding Chimpanzees* (Cambridge, Mass.: Harvard University Press, 1989), pp. 312–53.

7. Thomas M. Butynski, "The Robust Chimpanzee *Pan troglodytes*: Taxonomy, Distribution, Abundance and Conservation Status," in Rebecca Kormos, Christophe Boesch, Mohamed I. Bakkar, and Thomas M. Butynski, eds., *West African Chimpanzee Status Survey and Conservation Action Plan* (Gland, Switzerland: IUCN/SSC Primate Specialist Group, 2003), pp. 5–12.

8. U.S. Department of the Interior, Fish and Wildlife Service, "Endangered and Threatened Wildlife and Plants; Endangered Status for Chimpanzee and Pygmy Chimpanzee," Final Rule, *Federal Register* 55:9129–36 (Mar. 12, 1990).

9. William Booth, "Chimps and Research: Endangered?" *Science* 241:777–78 (Aug. 12, 1988).

10. Ibid.

11. See Teleki, "Population Status of Wild Chimpanzees," in *Understanding Chimpanzees*, p. 331, in which he claimed that 406 chimpanzees left Sierra Leone and went, for the most part, to the United States between 1975 and 1979. Teleki backed this up with a "not for distribution" memo that he shared with me, "Notes on the Use and Abuse of Chimpanzees by the U.S. National Institutes of Health" (May 26, 1988), which documents specific importations in U.S. Fish and Wildlife Service records, as well as misleading claims by NIH officials that it had not imported chimps since CITES was signed in 1974 [*sic*]. The NIH's "Report of the Task Force on the Use of and Need for Chimpanzees" (May 22 and July 24, 1978) further claimed, with skimpy evidence, that the country needed to add 300 to 350 animals each year. "If the chimpanzees now in the United States are not sufficient for a self-sustaining breeding program, additional chimpanzees should be imported for this purpose as soon as possible," the report stated. It noted the CITES regulations but said "it is reasonable to expect that a limited number could be imported if they were to be used in the establishment of breeding colonies."

12. Dale Peterson, *Jane Goodall: The Woman Who Redefined Man* (New York: Mariner Books, 2008), pp. 446–48 and 476.

13. Jon Cohen, "Born to Be Wild," *Washington City Paper*, Nov. 15–21, 1991.

14. Jim, the fourth chimpanzee returned to Uganda, had died.

15. Doug Cress, Pan African Sanctuary Alliance secretariat, e-mail, Dec. 9, 2009.

16. Letter from John G. Gold, colonel, USAF, BSC, chief Veterinary Science Division, to Avrett S. Thombs, vice president of research at New Mexico State University, which housed the chimpanzees. "The U.S. Air Force has initiated the process of finding an institution by the grant process to assume ownership of these chimpanzees," wrote Gold in his letter of Mar. 15, 1993.

17. National Research Council, *Chimpanzees in Research: Strategies for Their Ethical Care, Management, and Use* (Washington, D.C.: National Academy Press, 1997).

18. "Chimpanzee Health Improvement, Maintenance, and Protection Act," Public Law 106-551, 106th Congress.

19. The nonprofit secured the chimps' purchase with support from the Arcus Foundation. For more details, see the "Save the Chimps' Mission" Web page, http://www.savethechimps.org/about-us.aspx#begining, accessed Dec. 9, 2009.

20. "Chimp Haven Is Home Act," Public Law 110-170, was signed by President George W. Bush on Dec. 26, 2007.

21. UK Home Office Research, Development and Statistics, "Frequently Asked Questions about the Use of Animals in Scientific Procedures," http://scienceandresearch.homeoffice.gov.uk/animal-research/animal-testing-faqs/, accessed Dec. 9, 2009.

22. UK Home Office Research, Development and Statistics, "Statistics of Scientific Procedures on Living Animals," Dec. 2005, p. 14. Also see David Pearson, "Science: An End to Genocide for Our Ape Cousins," *Independent*, Dec. 2, 1997.

23. The move to the safari park kicked off still more controversies, as Four Paws and other animal advocates did not believe this was a suitable home. See Great Ape Protection, "Uncertain Destiny for 41 Chimpanzees in Austria," June 4, 2009, http://www.projetogap.com.br/en-US/noticias/Show/2076,uncertain-destiny-for-41-chimpanzees-in-austria; and Four Paws, "Animal Welfare Success: Transfer of 'Baxter Chimpanzees' from Ganserdorf to Hungary Stopped!" http://www.fourpaws.org.uk/

website/output.php?id=1224&idcontent=2036&somany=30&keywords=chim panzees, accessed Dec. 9, 2009. Ironically, an attempt to donate the animals to Chimp Haven was blocked in part by CITES regulations. A Baxter employee who worked on the transfer published a detailed account of the absurd events. See Janet C. Gonder, "Resocialization and Retirement of Laboratory Chimpanzees," *Journal of Medial Primatology* 33:60–62 (Feb. 2004).

24. Biomedical Primate Research Centre, "Animal Welfare/Relocation of Chimpanzees," http://www.bprc.nl/BPRCE/L3/RelocChimps.html, accessed Dec. 9, 2009.

25. AAP Sanctuary for Exotic Animals, "Chimpanzee Complex," http://www.aap.nl/ english/chimpanseecomplex.html, accessed Dec. 9, 2009.

26. The Uto Chimpanzee Sanctuary Web site, http://cs-uto.org, was only available in Japanese as of Dec. 2009.

27. Project R&R, "International Bans," http://www.releasechimps.org/mission/end -chimpanzee-research/country-bans/, accessed Dec. 9, 2009.

28. Pascal Gagneux, James J. Moore, and Ajit Varki, "The Ethics of Research on Great Apes," *Nature* 437:27–29 (Sept. 1, 2005).

29. VandeBerg, for his part, rejected the idea that researchers should only do to chimpanzees what they would do to humans. "I don't agree with that argument—I don't agree at all," he told me. He argued that chimpanzees remain the best model to test drugs for hepatitis B and vaccines for hepatitis C. "We're talking about billions of people who are going to benefit," he said. And he allowed that some monoclonal antibodies destined for human studies never went forward because they were tested in chimpanzees and killed the animals. "It's unethical from a human standpoint not to do the research," said VandeBerg. "I consider humans to be very different from chimps." Jon Cohen, "The Endangered Lab Chimp," *Science* 315:450–52 (Jan. 26, 2007). Also see John L. VandeBerg and Stuart M. Zola, "A Unique Biomedical Resource at Risk," *Nature* 437:30–32 (Sept. 1, 2005).

30. Jon Cohen, "NIH to End Breeding for Research," *Science* 316:1265 (June 1, 2007).

31. Humane Society of the United States, "Breeding of Government Owned Chimpanzees for Research Comes to an End," press release, May 24, 2007.

32. Lisa Fletcher and Arash Ghadishah, "EXCLUSIVE: Ex-Employees Claim 'Horrific' Treatment of Primates at Lab; Hidden-Camera Investigation Goes Behind Closed Doors at New Iberia Research Center," ABCNews.com, Mar. 4, 2009, http://abcnews .go.com/Nightline/story?id=6997869&page=1.

33. The U.S. National Institute of Allergy and Infectious Diseases had a $6.2-million contract (N01-AO-22754) with New Iberia to supply four to twelve infant chimpanzees each year between Sept. 2002 and 2009 for research on several viral diseases. "The moratorium was not intended for privately owned chimpanzees, or to apply to the other NIH Institutes," an NIH statement read.

34. Amanda McElfresh, "UL Fights Back: Primate Center Staff Disputes Images in Video," *Daily Advertiser*, Mar. 6, 2009.

35. Jon Cohen, "Humane Society Launches Offensive to Ban Invasive Chimp Research," *Science* 323:1414–15 (Mar. 13, 2009).

36. Many press accounts document Hiasl's fate. For a thorough review of the case through Sept. 2007, see Martin Balluch and Eberhard Theuer, "Trial on Personhood for Chimp 'Hiasl,'" *Altex* 24:335–42 (2007). The lawyers Lee Hall and Anthony Jon Waters also published a "model brief" of a fictional chimpanzee case for personhood, "From Property to Person: The Case of Evelyn Hart," 11 *Seton Hall Constitutional Law Journal* 1 (Fall 2000).

37. Jeffrey Stinson, "Activists Pursue Basic Legal Rights for Great Apes," *USA Today*, July 15, 2008, http://www.usatoday.com/news/offbeat/2008-07-15-chimp_N.htm.

38. Thomas Catan, "Apes Get Legal Rights in Spain, to Surprise of Bullfight Critics," *TimesOnline*, June 27, 2008, http://www.timesonline.co.uk/tol/news/world/europe/article4220884.ece.

39. Gary L. Francione, "The Great Ape Project: Not So Great," Animal Rights: The Abolitionist Approach blog, Dec. 12, 2006, http://www.abolitionistapproach.com/the-great-ape-project-not-so-great/.

40. John C. Waller and Vernon Reynolds, "Limb Injuries Resulting from Snares and Traps in Chimpanzees (*Pan troglodytes schweinfurthii*) of the Budongo Forest, Uganda," *Primates* 42:135–39 (Apr. 2001). See also "Proposal for Bugoma Snare Removal Project (JGI Uganda)," http://www.janegoodall.nl/projecten/Bugoma SnareRemovalProject.pdf, accessed Dec. 11, 2009.

41. Tatyana Humle, Charles T. Snowdon, and Tetsuro Matsuzawa, "Social Influences on Ant-dipping Acquisition in the Wild Chimpanzees (*Pan troglodytes verus*) of Bossou, Guinea, West Africa," *Animal Cognition* Suppl. 1:S37–48 (Oct. 2009).

42. John C. Mitani, William McGrew, and Richard Wrangham, "Toshisada Nishida's Contributions to Primatology," *Primates* 47:2–5 (Jan. 2006).

43. The IUCN Red List of Threatened Species, Version 2009.2, http://www.iucnredlist.org/apps/redlist/details/15933/0.

44. Ibid.

45. John F. Oates, "Is the Chimpanzee, *Pan troglodytes*, an Endangered Species? It Depends on What 'Endangered' Means," *Primates* 47:102–12 (Jan. 2006).

46. Frands Carlsen and Tom de Jongh, "Getting It Right for Chimpanzees: EAZA Has Developed a Clear Strategy for the Future Management of Chimpanzees, and It's One that Needs Full Attention," *EAZA News* 67:22–23 (Autumn 2009).

47. Pascal Gagneux, "The Genus Pan: Population Genetics of an Endangered Outgroup," *TRENDS in Genetics* 18:327–30 (July 2002). Steve Ross, a zoologist at the Lincoln Park Zoo who chairs the American Zoo and Aquarium Association's Chimpanzee Species Survival Plan, told me that in 2008 he worked with his European counterpart and they calculated there were about 815 chimps in accredited European zoos.

48. Habitat Ecologie et Liberté des Primates (HELP), "What Do We Do?," http://www.help-primates.org/new/en/actions.html, accessed Dec. 11, 2009.

49. Benoît Goossens, Joanna M. Setchell, E. Tchidongo, et al., "Survival, Interactions with Conspecifics and Reproduction in 37 Chimpanzees Released into the Wild," *Biological Conservation* 123:461–75 (June 2005). See also Benoît Goossens, Joanna M. Setchell, et al., "Successful Reproduction in Wild-released Orphan Chimpanzees (*Pan troglodytes troglodytes*)," *Primates* 44: 67–69 (Jan. 2003).

ACKNOWLEDGMENTS

A great benefit to being a journalist is access. I routinely read journal articles and books and then speak with the authors. I have passed through guarded gates to enter nature reserves, sanctuaries, and primate centers that are allergic to outsiders. Archivists have helped me comb through century-old letters, newspaper articles, and charts, and I have received other hard-to-find documents through the Freedom of Information Act. I have held chimp skulls and femurs and even young, living chimps, bonobos, and orangutans. Along the way, I have received an education in chimpanzees and our differences from the best teachers in the world. I am abundantly grateful to my hundreds of professors and, rather than list all of these fine people in what would resemble the dizzying wine list at a fancy restaurant, I offer a collective thank-you to everyone who has, at no clear benefit to themselves, taken time out of their days to speak, e-mail, or Internet chat or video with me. Any and all errors are of course my own.

I owe special thanks to several people who have gone further still and repeatedly spoken with me over several years, allowed me to watch them work in the field or the lab, and reviewed early drafts. Foremost among those who have gone out of their way to endure my odd questions and encourage this project is Pascal Gagneux, who has crossed the line from source into friend. On top of allowing me to watch him work in the lab and sharing with me many papers and books from his spectacular library, Pascal has been my main sounding board for advice and has shared with me many stories, meals, and waves.

I have benefited in many ways from the fact that Pascal lives in San

Diego, my own home base, as does his frequent collaborator Ajit Varki, who also deserves a most special thanks. Ajit invited me to the off-the-record gatherings that he organized several times each year under the auspices of the Project for Explaining the Origin of Humans. He and his wife (thank you, Nissi) allowed me to accompany them to the Primate Foundation of Arizona. He also frequently discussed new research with me, tipping me off to interesting findings from his own lab and others' labs.

The origins of this book date back to 1989, when I started to cover the AIDS epidemic closely, but the spark of the actual idea was an e-mail from the news editor at *Science*, Colin Norman, one of the most thoughtful, discerning, and judicious journalists in the business. In April 2006 Colin flagged an upcoming paper by Varki and Gagneux that focused on different immune responses in humans and chimpanzees. While writing a *Science* story about that paper, I realized that the recent publication of the chimpanzee genome promised to bring many other differences to light, several of which I subsequently wrote about for the magazine. I similarly stand on the shoulders of Leslie Roberts, my main editor at *Science* for the past decade, who day in, day out wisely helps me sort through my ideas and move the better ones forward. Several of the other writers and editors at *Science* are expert observers of several fields I cover in the book, and I often have turned to them for advice and editing. Thanks to Ann Gibbons, Elizabeth Culotta, Michael Balter, Elizabeth Pennisi, David Grimm, Erik Stokstad, and Martin Enserink.

I wrote a few stories for other magazines that I used as a springboard for this book, and the editors at those publications have my heartfelt gratitude. At *Outside*, Alex Heard provided superb editing for a story on hybrids that pushed the envelope for a publication that's best known for young people falling off mountains. At *Technology Review*, Erika Jonietz helped me put into transparent English the complicated workings of *FOXP2* and the genetics of language. At the online video magazine SlateV, Bill Smee expertly edited three separate pieces on chimpanzee research.

Malcolm Linton, an exceptionally talented photographer, shot the material for SlateV and *Science* and traveled with me to observe wild chimps in Uganda and captive animals involved with language research at the Great Ape Trust in Des Moines, Iowa. Malcolm, who provided several photos for the book, has become one of my brothers by choice (as opposed to by blood), and I have profound respect and gratitude for all he has done to complement my words with images since we first worked together in the Democratic Republic of Congo in 1997.

Jack Shafer, *Slate*'s press critic, also is one of my brothers by choice. My former editor and comrade in arms at the *Washington City Paper*, Shafer spars with me constantly about ideas, arguing about everything and anything for both sport and something approaching illumination. Jack is my rabbi in chief, the ultimate arbiter and visionary I look to when I am weighing life options, and at the inception of this project he gave me an enthusiastic shove forward that meant the world to me. He also regularly peppered me with related news stories and books, providing me with an extra set of ears and eyes. I can never thank him enough for his enduring friendship and all that he has done to sharpen my thinking and my attempts to communicate those thoughts.

On the brothers-by-choice front, I also have had the great fortune to have met Paul Shapiro more than thirty years ago, and he is my on-retainer scientist (check is in the mail) for questions large and small. Paul thinks critically at every turn, and his reading of my drafts was invaluable.

I traveled extensively doing research for the book, and I am deeply grateful to everyone who helped arrange my trips and welcomed me when I was far from home. Richard Wrangham, in addition to answering many questions about his own studies and opinions about the work of other groups, gave the green light for my trip to Uganda's Kibale National Park. At Kibale, I much appreciate my warm reception from Emily Otali, and many thanks to Paco Bertolani and the field assistants there for introducing me to chimping. My visit to Uganda's Budongo Forest Reserve would not have taken place without help from Klaus Zuberbühler and Fred Babweteera, and, when I arrived, Zinta Zommers, Cathy Crockford, Roman Wittig, and their field assistants generously allowed me to go chimping with them, taking time to explain what I was seeing and their own studies. Vernon Reynolds kindly followed up afterward.

In Japan, Tetsuro Matsuzawa rolled out the red carpet at the

Primate Research Institute. In addition to being a most gracious host and allowing me to observe research in action, Matsuzawa gave up many hours to explain his group's findings to me. Chris Martin, Tatyana Humle, Masayuki Tanaka, Masaki Tomonaga, and Laura Martinez also made me feel at home when I was in Inuyama. Toshisada Nishida and Yukimaru Sugiyama opened their doors for me, too, and I much appreciated their historical perspectives.

I spent a week meeting with researchers at the Max Planck Institute for Evolutionary Anthropology in Leipzig, Germany, and it would not have been possible without Sandra Jacob's assistance. At the institute and the affiliated zoo, everyone I met was kind and offered food for thought, which I dined on repeatedly when it came time to write. I especially want to thank Svante Pääbo, Christophe Boesch, Michael Tomasello, Jean-Jacques Hublin, Gottfried Hohmann, Wolfgang Enard, Josep Call, Tobias Deschner, Crickette Sanz, David Morgan, Katerina Havarti, and Tanya Smith.

For the FOXP2 story, I met with Simon Fisher, Anthony Monaco, and Jane Hurst in Oxford, England, and each of them provided valuable insights and details to the discovery of the gene. Similarly, Faraneh Vargha-Khadem in London kindly offered her perspectives about the KE family and the genetics of speech and language. Stewart Young, a member of the KE family, candidly shared his personal perspective on having the rare mutation. Across the pond and closer to my home, Daniel Geschwind at the University of California, Los Angeles, devoted many hours to bring me up to speed with his work. I also thank his collaborator, Stephanie White, who allowed me to observe zebra finch experiments run by Michael Condro and Elizabeth Wong.

I made a special trip to Moscow to meet with Kirill Rossiianov, the historian of science who fleshed out the remarkable story of Il'ya Ivanov. Kirill took a day off to show me around Moscow and share his views about the would-be humanzee maker.

At the Yerkes National Primate Research Center in Atlanta, Georgia, Todd Preuss and Jim Rilling went the extra yard and not only invited me into their labs but allowed me to watch them perform a brain scan on a living chimpanzee. Todd and Jim also had the patience to help me unravel the complicated intricacies of the brain. At the field station, Victoria Horner and Devyn Carter introduced me to their culture studies with the chimpanzees there. Thanks to Lisa Parr and Frank Novembre, too, and to Frans de Waal at Emory University. The public relations team, led by Lisa Newbern, greased many wheels to make sure my visits went smoothly.

The Great Ape Trust in Des Moines allowed Malcolm Linton and me to visit their apes and researchers for two days. Ben Beck, Rob Shumaker, William Fields, Karyl Swartz, and Serge Wich each opened my eyes to different aspects of ape language research. Thanks as well to Al Setka for arranging the visit and to animal handlers Liz Pugh and Susannah Maisel.

Jo Fritz welcomed me at the Primate Foundation of Arizona (and indulged many questions following my visit), as did Elizabeth Videan and James Murphy. Bhagavan "Doc" Antle opened the gates for a private tour of his menagerie, including ligers, at his Myrtle Beach, South Carolina, TIGERS, and at his home outside Miami. He then introduced me to Jungle Island, where he stages a cage-free show of chimps, orangutans, and several other species. Many thanks to Doc and his staff.

Evan Eichler at the University of Washington in Seattle entertained a visit to his lab and also clarified for me the history of the chimpanzee genome project. In Cambridge, Massachusetts, I had the good fortune to tour Daniel Lieberman's Harvard University lab and watch him teach an undergraduate class. Harvard's David Reich and the Broad Institute's Nick Patterson suffered through interminable queries about their hybridization theory. Elizabeth Lonsdorf and Steve Ross of the Lincoln Park Zoo in Chicago allowed me to attend the mind-expanding Mind of the Chimpanzee conference they organized, and Steve also gave me a copy of his invaluable chimpanzee studbook.

Closer to home, I am much obliged to Tuck Finch and Joe Hacia, who met with me at their University of Southern California labs. In my backyard, at the University of California at San Diego, Katerina Semenderi showed me her wonderful collection of ape brains, and Margaret Schoenginger allowed Alyssa Crittenden and Andy Froehle to walk me through their collection of chimpanzee bones. Kurt Benirschke once again served as a wise mentor, a job he never sought nor wanted.

UCSD's Jim Moore deserves thanks, too, for fielding many inquiries over the years.

Danny Povinelli and Kristen Hawkes deserve special thanks, and we established long-term relationships that I am confident will outlive this project. My many years covering chimpanzee research and HIV/AIDS meant that I tapped some of my previous professors for help here, including Beatrice Hahn, Preston Marx, Larry Arthur, and Bette Korber. Steven Wolinsky especially helped me with arcane genetics and meticulously combed through a near-final draft. Michael Allin, writer extraordinaire and my next-door neighbor, also critiqued early drafts and indulged in many related over-the-fence gabfests.

Although my Africa trips did not include visits to Karl Ammann in Kenya, Crickette Sanz and David Morgan in the Republic of Congo, or Lilly Ajarova and Andrew Seguya at Ngamba Island in Uganda, it was not for lack of trying, and they went to some effort to help me make plans. Mary Lewis, Jane Goodall's assistant, provided important historical information, as did Rick Swope, Linda Culpepper at the *Detroit News*, the staff at the Smithsonian Institution's National Anthropological Archives, Jeremy Rich, and Gregory Radick.

I also am grateful to the scientists who allowed me to reprint their graphics, many of which they altered to accommodate my needs.

To Geza Teleki, I owe more than a thank you. Had I never met Geza in 1990, I suspect I never would have written this book. His passion for all things chimp infected me.

At Henry Holt, Robin Dennis grasped the ambitious scope of this book from the outset and provided superb guidance from beginning to end. She is smart, tactful, and knows the difference between simply identifying a problem and suggesting a workable solution. I also cannot thank her enough for making the special effort to take the book across the finish line even though as I was completing the manuscript she started a new chapter in her own life in a far-off land. I similarly am grateful to Paul Golob, the editorial director, for his stalwart enthusiasm for the project.

My agent Gail Ross, and her assistant Howard Yoon, provided excellent criticisms of the initial drafts of my book proposal. This is my third book with Gail, which in itself speaks to my admiration for her skills, integrity, and friendship.

Finally, many thanks to my extraordinary family. I have dedicated this book to my parents, Esther and Avshalom, and my older brother Ronnie (who would be a brother by choice if he was not one by blood). Then there is the second family I have had the great fortune to be part of, the one created with Shannon Bradley, my wife and best friend. As a journalist herself, Shannon intimately understands the process, and has lent me a hand more times than I had any right to request, listening to my leads, fetching books for me at the library, rereading and rereading drafts. This last Thanksgiving, she gave me the great gift of taking our children away to *my* relatives so that I could stay at home and write in peace for an entire week. That is love. Our children, Aidan, Ryan, and Erin, also have my everlasting gratitude for putting up with my long work hours and absences, which I hope fade into the background and are replaced by memories of the very good times we have had and will continue to enjoy together.

INDEX

Page numbers in *italics* refer to illustrations.

ABOUT THE AUTHOR

JON COHEN is the author of *Shots in the Dark* and *Coming to Term*. He is a correspondent at the internationally renowned *Science* magazine and has also written for *The Atlantic*, *The New Yorker*, *The New York Times Magazine*, *The Washington Post*, *Discover*, *Smithsonian*, and *Slate*. He lives in Cardiff-by-the-Sea, California.

FEB 2 4 2020

9 780312 611767